Lecture Notes in Mathematics 2011

Ben Andrews · Christopher Hopper

The Ricci Flow in Riemannian Geometry

A Complete Proof of the Differentiable 1/4-Pinching Sphere Theorem

Ben Andrews
Australian National University
Mathematical Sciences Institute
ACT 0200 Australia
Ben.Andrews@anu.edu.au

Christopher Hopper
University of Oxford
Mathematical Institute
St Giles' 24-29
OX1 3LB Oxford
United Kingdom
hopper@maths.ox.ac.uk

ISBN: 978-3-642-16285-5 e-ISBN: 978-3-642-16286-2
DOI: 10.1007/978-3-642-16286-2
Springer Heidelberg Dordrecht London New York

Lecture Notes in Mathematics ISSN print edition: 0075-8434
ISSN electronic edition: 1617-9692

Mathematics Subject Classification (2010): 35-XX, 53-XX, 58-XX

Cover design: SPi Publisher Services

Printed on acid-free paper

Springer is part of Springer Science+Business Media (www.springer.com)

For in the very torrent, tempest, and as I may say, whirlwind of your passion, you must acquire and beget a temperance that may give it smoothness.

— Shakespeare, *Hamlet.*

Preface

There is a famous theorem by Rauch, Klingenberg and Berger which states that a complete simply connected n-dimensional Riemannian manifold, for which the sectional curvatures are strictly between 1 and 4, is homeomorphic to a n-sphere. It has been a longstanding open conjecture as to whether or not the 'homeomorphism' conclusion could be strengthened to a 'diffeomorphism'.

Since the introduction of the Ricci flow by Hamilton [Ham82b] some two decades ago, there have been several inroads into this problem – particularly in dimensions three and four – which have thrown light upon a possible proof of this result. Only recently has this conjecture (and a considerably stronger generalisation) been proved by Simon Brendle and Richard Schoen. The aim of the present book is to provide a unified expository account of the differentiable 1/4-pinching sphere theorem together with the necessary background material and recent convergence theory for the Ricci flow in n-dimensions. This account should be accessible to anyone familiar with enough differential geometry to feel comfortable with tensors, covariant derivatives, and normal coordinates; and enough analysis to follow standard PDE arguments. The proof we present is self-contained (except for the quoted Cheeger–Gromov compactness theorem for Riemannian metrics), and incorporates several improvements on what is currently available in the literature.

Broadly speaking, the structure of this book falls into three main topics. The first centres around the introduction and analysis the Ricci flow as a geometric heat-type partial differential equation. The second concerns Perel'man's monotonicity formulæ and the 'blow up' analysis of singularities associated with the Ricci flow. The final topic focuses on the recent contributions made – particularly by Böhm and Wilking [BW08], and by Brendle and Schoen [BS09a] – in developing the necessary convergence theory for the Ricci flow in n-dimensions. These topics are developed over several chapters, the final of which aims to prove the differentiable version of the sphere theorem.

The book begins with an introduction chapter which motivates the pinching problem. A survey of the sphere theorem's long historical development is discussed as well as possible future applications of the Ricci flow.

As with any discussion in differential geometry, there is always a labyrinth of machinery needed before any non-trivial analysis can take place. We present some of the standard and non-standard aspects of this in Chap. 1. The chapter's focus is to set the notational conventions used throughout, as well as provide supplementary material needed for future computations – particular for those in Chaps. 2 and 3. Careful attention is paid to the construction of the connection and curvature on various bundles together with some non-standard aspects of the pullback bundle structure. We refer the reader to [Lee02, Lee97, Pet06, dC92, Jos08] as additional references with respect to this background material.

In Chap. 2 we look at some classical results related to Harmonic map heat-flow between Riemannian manifolds. The inclusion of this chapter serves as a gently introduction to the techniques of geometric analysis as well as provides good motivation for the Ricci flow. Within, we present the convergence result of Eells and Sampson [ES64] with improvements made by Hartman [Har67].

After establishing this, Chap. 3 introduces the Ricci flow as a geometric parabolic equation. Some basic properties of the flow are discussed followed by a detailed derivation of the associated evolution equations for the curvature tensor and its various traces. Thereafter we give a brief survey of the sphere theorem of Huisken [Hui85], Nishikawa [Nis86] and Margerin [Mar86] together with the algebraic decomposition of the curvature tensor.

Short-time existence for the Ricci flow is discussed in Chap. 4. We follow the approach first outlined by DeTurck [DeT83] which relates Ricci flow to Ricci–DeTurck flow via a Lie derivative. A discussion on the ellipticity failure of the Ricci tensor due to the diffeomorphism invariance of the curvature is also included.

Chapter 5 discusses the so-called Uhlenbeck trick, which simplifies the evolution equation of the curvature so that it can be written as a reaction-diffusion type equation. This will motivate the development of the vector bundle maximum principle of the next chapter. We present the original method first discussed in [Ham86], and improved in [Ham93], which uses an abstract bundle and constructs an identification with the tangent bundle at each time. Thereafter we introduce a new method that looks to place a natural connection on a 'spatial' vector bundle over the space-time manifold $M \times \mathbb{R}$. We will build upon this space-time construction in subsequent chapters.

In Chap. 6 we discuss the maximum principle for parabolic PDE as a very powerful tool central to our understanding of the Ricci flow. A new general vector bundle version, for heat-type PDE of section $u \in \Gamma(E \to M \times \mathbb{R})$ over the space-time manifold, is discussed here. The stated vector bundle maximum principle, Theorem 7.15 and the related Corollary 7.17, will provide the main tools for the convergence theory of the Ricci flow discussed in later chapters. Emphasis is placed on the 'vector field points into the set' condition as it correctly generalises the null-eigenvector condition of Hamilton [Ham82b]. The related convex analysis necessary for the vector bundle maximum principle is discussed in Appendix B – where we use the same conventions as that of the classic text [Roc70]. The maximum principle for symmetric 2-tensors is also discussed as well as applications of the Ricci flow for 3-manifolds.

The parabolic nature of the Ricci flow is further developed in Chap. 7 where regularity and long-time existence is discussed. We see that the Ricci flow enjoys excellent regularity properties by deriving global Shi estimates [Shi89]. They are used to prove long-time existence soon thereafter.

Chapter 8 look at a compactness theorem for sequences of solutions to the Ricci flow. The result originates in the convergence theory developed by Cheeger and Gromov. We use the regularity of previous chapter to give a proof of the compactness theorem for the Ricci flow, given the compactness theorem for metrics; however, we will not give a proof of the general Cheeger–Gromov compactness result. It has natural applications in the analysis of singularities of the Ricci flow by 'blow-up' – which we will employ in the proof of the differentiable sphere theorem.

Chapters 9 and 10 aim to establish Perel'man's local noncollapsing result for the Ricci flow [Per02]. This will provide a positive lower bound on the injectivity radius for the Ricci flow under blow-up analysis. We also discuss the gradient flow formalism of the Ricci flow and Perel'man's motivation from physics [OSW06, Car10].

The work of Böhm and Wilking [BW08], in which whole families of preserved convex sets for the Ricci flow are derived from an initial one, is presented in Chaps. 11 and 12. Using this we will be able to argue, in conjunction with the vector bundle maximum principle, that solutions of the Ricci flow which have their curvature in a given initial cone will evolve to have constant curvature as they approach their limiting time. Chapter 11 focuses on the algebraic decomposition of the curvature and the required family of scaled transforms. In particular we use the inherent Lie algebra structure (discussed in Appendix C) related to the curvature to elucidate the nature of the reaction term in the evolution equation for the curvature. The key result (Theorem 12.33) is that these transforms induce a change in the reaction terms which does not depend on the Weyl curvature, and so can be computed entirely from the Ricci tensor and expressed in terms of its eigenvalues. Chapter 12 uses these explicit eigenvalues to generate a one-parameter family of preserved convex cones that are parameterised piecewise into two parts; one to accommodate the initial behaviour of the cone, the other to accommodate the required limiting behaviour. Thereafter we discuss the formulation of generalised pinching sets. The main result of this section, Theorem 13.8, provides the existence of a pinching set simply from the existence of a suitable family of cones.

In Chap. 13 we discuss the positive curvature condition on totally isotropic 2-planes, first introduced by Micallef and Moore [MM88], as a possible initial convex cone. We show, using ideas from [BS09a, Ngu08, Ngu10, AN07], that the positive isotropic curvature (PIC) condition is preserved by the Ricci flow; as is the positive complex sectional curvature (PCSC) condition. We also give a simplified proof that PIC is preserved by the Ricci flow by working directly with the complexification of the tangent bundle. In order to relate the PIC condition to the 1/4-pinching sphere theorem, we present the argument of Brendle and Schoen [BS09a] that relates the 1/4-pinching condition with the PIC condition on $M \times \mathbb{R}^2$. The result, i.e. Corollary 14.13, shows that M is a compact manifold with pointwise 1/4-pinched sectional curvature, then $M \times \mathbb{R}^2$ has positive isotropic curvature.

Chapter 14 brings the discussion to a climax. Here we finally give a proof of the differentiable 1/4-pinching sphere theorem from the material presented in earlier chapters. In the final section we outline a general convergence result due to Brendle [Bre08] which looks at the weaker condition of PIC on $M \times \mathbb{R}$.

A synopsis of the chapter progressions and inter-relationships is summarised by the following diagram:

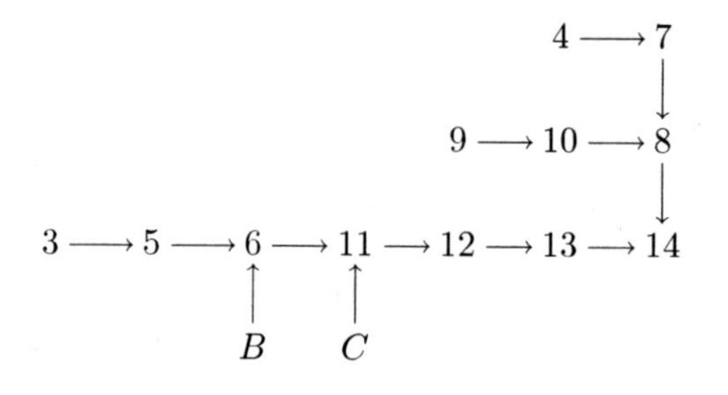

Here the main argument is represented along the horizontal together with the supplementary appendices. The regularity, existence theory and blow-up analysis are shown above this.

This book grew from my honours thesis completed in 2008 at the Australian National University. I would like to express my deepest gratitudes to my supervisor, Dr Ben Andrews, for without his supervision, assistance and immeasurable input this book would not be possible. Finally, I would like to thank my parents for their continuous support and encouragement over the years.

July, 2009 *Christopher Hopper*

Contents

Notation and List of Symbols

Γ_{ij}^k	*Christoffel symbol* of a connection ∇ w.r.t. a local frame (∂_i).
(x^i)	A *local coordinate chart* for a neighbourhood U with a local chart $x = (x^i) : U \to \mathbb{R}^n$.
Curv	*Algebraic curvature operators.*
I, id	The identities on $S^2(\bigwedge^2 V^*)$ and $S^2(V)$. N.B. $I = \mathrm{id} \wedge \mathrm{id}$.
inj	*Injectivity radius.*
$\Delta_{g,h}$	*Harmonic map Laplacian* w.r.t. the domain metric g and codomain metric h.
$\bigwedge^2 V$	*Second exterior power* of the vector space V.
$\mathcal{L}_X$	*Lie derivative* w.r.t. the vector field X.
$\mathfrak{Met}$	*Space of metrics.*
$\mathcal{N}_x A$	*Normal cone* to A at x.
ν	*Unit outward normal.*
$\owedge$	*Kulkarni-Nomizu product.*
${}^f\nabla$	*Pullback connection* on f^*E.
f^*E	*Pullback bundle* of E by f.
$\xi_f, (\xi_i)_f$	*Restriction* of $\xi, \xi_i \in \Gamma(E)$ to f.
R, $R_{ijk\ell}$	*Riemannian curvature tensor.*
Ric, R_{ij}	*Ricci curvature tensor.*
Scal	*Scalar curvature tensor.*
$\Gamma(E)$	The space of *smooth sections* of a vector bundle $\pi : E \to M$.
$\mathfrak{S}$	*Spatial tangent bundle.*
$\mathrm{Sym}^2 T^*M$	*Symmetric* $(2,0)$*-bundle* over M.
$S^2(U)$	*Symmetric tensor space* of U.
$T_\ell^k(V)$	The set of all multilinear maps $(V^*)^\ell \times V^k \to \mathbb{R}$ over V.
$T_\ell^k M$	(k,ℓ)*-tensor bundle* over a manifold M.
$\mathcal{T}_\ell^k(M)$	The space of (k,ℓ)*-tensor fields* over M, i.e. $\Gamma(\otimes^k T^*M \otimes^\ell TM)$.
$\mathcal{T}_x A$	*Tangent cone* to A at x.
$d\mu$, $d\mu(g)$	*Volume form* with respect to a metric g.
$d\sigma$	*Volume form on a hypersurface or boundary* of a manifold.
$\mathscr{X}(M)$	The space of *vector fields*, i.e. $\Gamma(TM)$.

Chapter 1
Introduction

The relationship between curvature and topology has traditionally been one of the most popular and highly developed topics in Riemannian geometry. In this area, a central issue of concern is that of determining global topological structures from local metric properties. Of particular interest to us the so-called pinching problem and related sphere theorems in geometry. We begin with a brief overview of this problem, from Hopf's inspiration to the latest developments in Hamilton's Ricci flow.

1.1 Manifolds with Constant Sectional Curvature

One of the earliest insights into the relationship between curvature and topology is the problem of classifying complete Riemannian manifolds with constant sectional curvature, referred to as *space forms*. In the late 1920s Heinz Hopf studied the global properties of such space forms and proved, in his PhD dissertation [Hop25] (see also [Hop26]), the following:

Theorem 1.1 (Uniqueness of Constant Curvature Metrics). *Let M be a complete, simply-connected, n-dimensional Riemannian manifold with constant sectional curvature. Then M is isometric to either $\mathbb{R}^n$, S^n or $\mathbb{H}^n$.*

Furthermore, if the manifold is compact then the space forms are compact quotients of the either $\mathbb{R}^n$, S^n or $\mathbb{H}^n$. Placing this result on solid ground was one of Hopf's tasks during the 1930s, however the classification is still incomplete (the categorisation of hyperbolic space quotients has been extremely problematic).

Given these developments, curiosity permits one to ask if a similar result would hold under a relaxation of the curvature hypothesis. In other words, assuming a compact manifold has a sectional curvature 'varying not too much' (we will later say the manifold is 'pinched'), can one deduce that the underlying manifold is topologically (one would even hope differentiably) identical to one of the above space forms? After rescaling the metric there are three

B. Andrews and C. Hopper, *The Ricci Flow in Riemannian Geometry*, Lecture Notes in Mathematics 2011, DOI 10.1007/978-3-642-16286-2_1,

cases: the pinching problem around $\kappa_0 = +1, 0, -1$. Therefore if the sectional curvature K satisfies $|K - \kappa_0| \leq \varepsilon$, the question becomes one of finding an optimal $\varepsilon > 0$ in which the manifold is identical (in some sense) to a particular space form.

For our purposes, the question of interest is that of positive pinching around $\kappa_0 = 1$; our subsequent discussion will focus entirely on this case. The problem has enjoyed a great deal interest over the years due to its historical importance both in triggering new results and as motivation for creating new mathematical tools.

1.2 The Topological Sphere Theorem

The only simply connected manifold of constant positive sectional curvature is, by the above theorem, the sphere. A heuristic sense of continuity leads one to hope that if the sectional curvature of a manifold is close to a positive constant, then the underlying manifold will still be a sphere. Hopf himself repeatedly put forward this problem, in particular when Harry Rauch (an analyst and expert in Riemannian surfaces) visited him in Zürich throughout 1948–1949. Rauch was so enthusiastic about Hopf's pinching that, back at the Institute for Advanced Study in Princeton, he finally managed to prove Hopf's conjecture with a pinching constant of roughly $\delta \approx 3/4$. Specifically Rauch [Rau51] proved:

Theorem 1.2 (Rauch, 1951). *Let (M^n, g) be a complete Riemannian manifold with $n \geq 2$. If the sectional curvature $K(p, \Pi)$ (where $p \in M$ and Π is a 2-dimensional plane through p) satisfies*

$$\delta k \leq K(p, \Pi) \leq k - \varepsilon$$

for some constant $k > 0$, some $\varepsilon > 0$, all $p \in M$; and all planes Π, where $\delta \approx 3/4$ is the root of the equation $\sin \pi\sqrt{\delta} = \sqrt{\delta}/2$. Then the simply connected covering space of M is homeomorphic to the n-dimensional sphere S^n. In particular if M is simply connected, then M is homeomorphic to S^n.

Rauch's paper is seminal in two respects. First of all, he was able to control the metric on both sides. Secondly, to get a global result, he made a subtle geometric study in which he proved that (under the pinching assumption) one can build a covering of the manifold from the sphere.

Thereafter Klingenberg [Kli59] sharpened the constant – in the even dimensional case – to the solution of $\sin \pi\sqrt{\delta} = \sqrt{\delta}$ (i.e. $\delta \approx 0.54$). Crucially, the precise notion of an injectivity radius was introduced to the pinching problem. Using this, Berger [Ber60a] improved Klingenberg's result, under the even dimension assumption, with $\delta = 1/4$. Finally, Klingenberg [Kli61] extended the injectivity radius lower estimate to odd dimensions. This proved the sphere theorem in its entirely. The resulting *quarter pinched sphere theorem* is stated as follows:

Theorem 1.3 (Rauch–Klingenberg–Berger Sphere Theorem). *If a simply connected, complete Riemannian manifold has sectional curvature K satisfying*

$$\frac{1}{4} < K \leq 1,$$

then it is homeomorphic to a sphere.[1]

The theorem's pinching constant is optimal since the conclusion is false if the inequality is no longer strict. The standard counterexample is complex projective space with the Fubini-Study metric (sectional curvatures of this metric take on values between $1/4$ and 1 inclusive). Other counterexamples may be found among the rank one symmetric spaces (see e.g. [Hel62]).

As a matter of convention, we say a manifold is *strictly* δ-pinched in the *global sense* if $0 < \delta < K \leq 1$ and *weakly* δ-pinched in the *global sense* if $0 < \delta \leq K \leq 1$. In which case the sphere theorem can be reformulated as: 'Any complete, simply-connected, strictly $1/4$-pinched Riemannian manifold is homeomorphic to the sphere.'

1.2.1 Remarks on the Classical Proof

The proof from the 1960s consisted in arguing that the manifold can be covered with only two topological balls (e.g. see [AM97, Sect. 1]). The proof relies heavily on classical comparison techniques.

Idea of Proof. Choose points p and q in such a way that the distance $d(p, q)$ is maximal, that is $d(p,q) = \mathrm{diam}(M)$. The key property of such a pair of points is that for any unit tangent vector $v \in T_qM$, there exists a minimising geodesic γ from q to p making an acute angle with v. From this one can apply Toponogov's triangle comparison theorem [Top58] (see also [CE08, Chap. 2]) and Klingenberg's injectivity radius estimates to show that

$$M = B(p, r) \cup B(q, r)$$

---

[1] In the literature some quote the sectional curvature as taking values in the interval $(1, 4]$. This is only a matter of scaling. For instance, if the metric is scaled by a factor λ then the sectional curvatures are scaled by $1/\lambda$. Thus one can adjust the maximum principal curvature to 1, say. What *is* important about the condition is that the ratio of minimum to maximum curvature remains greater than $1/4$.

for $\pi/2\sqrt{\delta} < r < \pi$ and $1/4 < \delta \leq K \leq 1$. In other words, under the hypothesis of the sphere theorem, we can write M as a union of two balls. One then concludes, by classical topological arguments, that M is *homeomorphic* to the sphere. □

1.2.2 Manifolds with Positive Isotropic Curvature

One cannot overlook the contributions made by Micallef and Moore [MM88] in generalising the classical Rauch–Klingenberg–Berger sphere theorem. By introducing the so-called positive isotropic curvature condition to the problem together with harmonic map theory, they managed to prove the following:

Theorem 1.4 (Micallef and Moore, 1988). *Let M be a compact simply connected n-dimensional Riemannian manifold which has positive curvature on totally isotropic 2-planes, where $n \geq 4$. Then M is homeomorphic to a sphere.*

This is achieved as follows: Firstly, for a n-dimensional Riemannian manifold M with positive curvature on totally isotropic two-planes one can show that any nonconstant conformal harmonic map $f\colon S^2 \to M$ has index at least $\frac{n-3}{2}$. The proof uses classical results of Grothendieck on the decomposition of holomorphic bundles over S^2. Secondly, one can show: If M is a compact Riemannian manifold such that $\pi_k(M) \neq 0$, where $k \geq 2$, then there exists a nonconstant harmonic 2-sphere in M of index $\leq k-2$. This fact is a modification of the Sacks–Uhlenbeck theory of minimal 2-spheres in Riemannian manifolds. From this, the above theorem easily follows by using duality with these two results, and using the higher-dimensional Poincaré conjecture, which was proved for $n \geq 5$ by Smale and for $n = 4$ by Freedman.

A refined version of the classical sphere theorem now follows by replacing the global pinching hypotheses on the curvature by a pointwise one. In particular we say that a manifold M is *strictly* δ-pinched in the *pointwise sense* if $0 < \delta K(\Pi_1) < K(\Pi_2)$ for all points $p \in M$ and all 2-planes $\Pi_1, \Pi_2 \subset T_pM$. Furthermore, we say M is *weakly* δ-pinched in the *pointwise sense* if $0 \leq \delta K(\Pi_1) \leq K(\Pi_2)$ for all point $p \in M$ and all 2-planes $\Pi_1, \Pi_2 \subset T_pM$. In which case it follows from Berger's Lemma (see Sect. 2.7.7) that any manifold which is strictly 1/4-pinched in the pointwise sense has positive isotropic curvature (and so is homeomorphic to S^n).

1.2.3 A Question of Optimality

Given the above stated Rauch–Klingenberg–Berger sphere theorem, a natural question to ask is: *Can the homeomorphic condition can be replaced by a diffeomorphic one?* In answering this, there are some dramatic provisos.

The biggest concerns the method used in the above classical proof – as the argument presents M as a union of two balls glued along their common boundary so that M is homeomorphic to S^n. However, Milnor [Mil56] famously showed that there are 'exotic' structures on S^7, namely:

Theorem 1.5 (Milnor, 1956). *There are 7-manifolds that are homeomorphic to, but not diffeomorphic to, the 7-sphere.*[2]

Moreover, some exotic spheres are precisely obtained by gluing two half spheres along their equator with a weird identification. Hence the above classical proof cannot do better since it does not allow one to obtain a diffeomorphism between M and S^n in general. As a result, it remained an open question as to whether or not the sphere theorem's conclusion was optimal in this sense.

To add to matters, it has also been a long-standing open problem as to whether there actually are any concrete examples of exotic spheres with positive curvature. This has now apparently been resolved by Petersen and Wilhelm [PW08] who show there is a metric on the Gromoll–Meyer Sphere with positive sectional curvature.

1.3 The Differentiable Sphere Theorem

There have been many attempts at proving the differentiable version, most of which have been under sub-optimal pinching assumptions. The first attempt was by Gromoll [Gro66] with a pinching constant $\delta = \delta(n)$ that depended on the dimension n and converged to 1 as n goes to infinity. The result for δ independent of n was obtained by Sugimoto, Shiohama, and Karcher [SSK71] with $\delta = 0.87$. The pinching constant was subsequently improved by Ruh [Ruh71, Ruh73] with $\delta = 0.80$ and by Grove, Karcher and Ruh [GKR74] with $\delta = 0.76$. Furthermore, Ruh [Ruh82] proved the differentiable sphere theorem under a pointwise pinching condition with a pinching constant converging to 1 as $n \to \infty$.

1.3.1 The Ricci Flow

In 1982 Hamilton introduced fundamental new ideas to the differentiable pinching problem. His seminal work [Ham82b] studied the evolution of a heat-type geometric evolution equation:

$$\frac{\partial}{\partial t} g(t) = -2\,\mathrm{Ric}(g(t)) \qquad g(0) = g_0,$$

[2] In fact Milnor and Kervaire [KM63] showed that S^7 has exactly 28 non-diffeomorphic smooth structures.

herein referred to as the Ricci flow. This intrinsic geometric flow has, over the years, served to be an invaluable tool in obtaining global results within the differentiable category. The first of which is following result due to Hamilton [Ham82b].

Theorem 1.6 (Hamilton, 1982). *Suppose M is a simply-connected compact Riemannian 3-manifold with strictly positive Ricci curvature. Then M is diffeomorphic to S^3.*

Following this, Hamilton [Ham86] developed powerful techniques to analyse the global behaviour of the Ricci flow. This enabled him to extend his previous result to 4-manifolds by showing:

Theorem 1.7 (Hamilton, 1986). *A compact 4-manifold M with a positive curvature operator is diffeomorphic to the sphere S^4 or the real projective space $\mathbb{R}P^4$.*

Note that we say a manifold has a positive curvature operator if the eigenvalues if R are all positive, or alternatively $R(\phi, \phi) > 0$ for all 2-forms $\phi \neq 0$ (when considering the curvature operator as a self-adjoint operator on two-forms).

Following this, Chen [Che91] show that the conclusion of Hamilton's holds under the weaker assumption that the manifold has a 2-positive curvature operator. Specifically, he proved that: *If M is a compact 4-manifold with a 2-positive curvature operator, then M is diffeomorphic to either S^4 or $\mathbb{R}P^4$.* Moreover, Chen showed that the 2-positive curvature condition is implied by pointwise 1/4-pinching. In which case he managed to show:

Theorem 1.8 (Chen, 1991). *If M^4 is a compact pointwise 1/4-pinched 4-manifold, then M is either diffeomorphic to the sphere S^4 or the real projective space $\mathbb{R}P^4$, or isometric to the complex projective space $\mathbb{C}P^2$.*

Note that we say a manifold has a 2-positive curvature operator if the sum of the first two eigenvalues of R are positive.[3] It is easy to see that when $\dim M = 3$, the condition of positive Ricci and 2-positive curvature operator are equivalent (in fact, 2-positive implies positive Ricci in n-dimensions).

However despite this progress it still remained an open problem for over a decade as to whether or not these results could be extended for manifolds M with $\dim M \geq 4$.

[3] Equivalently one could also say that $R(\phi, \phi) + R(\psi, \psi) > 0$ for all 2-forms ϕ and ψ satisfying $|\phi|^2 = |\psi|^2$ and $\langle \phi, \psi \rangle = 0$.

1.3.2 Ricci Flow in Higher Dimensions

Huisken [Hui85] was one of the first to study the Ricci flow on manifolds of dimension $n \geq 4$. His analysis focuses on decomposing the curvature tensor into $R_{ijk\ell} = U_{ijk\ell} + V_{ijk\ell} + W_{ijk\ell}$, where $U_{ijk\ell}$ denotes the part of the curvature tensor associated with the scalar curvature, $V_{ijk\ell}$ is the part of the curvature associated with the trace-free Ricci curvature, and $W_{ijk\ell}$ denotes the Weyl tensor. By following the techniques outlined in [Ham82b], Huisken showed that if the scalar-curvature-free part of the sectional curvature tensor is small compared to the scalar curvature, then the same evolution equation for the curvature yields the same result – a deformation to a constant positive sectional curvature metric.

Theorem 1.9 (Huisken, 1985). *Let $n \geq 4$. If the curvature tensor of a smooth compact n-dimensional Riemannian manifold of positive scalar curvature satisfies*

$$|W|^2 + |V|^2 < \delta_n |U|^2$$

with $\delta_4 = \frac{1}{5}$, $\delta_5 = \frac{1}{10}$, and $\delta_n = \frac{2}{(n-2)(n+1)}$ for $n \geq 6$, then the Ricci flow has a solution $g(t)$ for all times $0 \leq t < \infty$ and $g(t)$ converges to a smooth metric of constant positive curvature in the C^∞-topology as $t \to \infty$.

There are also similar results by Nishikawa [Nis86] and Margerin [Mar86] with more recent sharp results by Margerin [Mar94a, Mar94b] as well.

In 2006, Böhm and Wilking [BW08] managed to generalise Chen's work in four dimensions by constructed a new family of cones (in the space of algebraic curvature operators) which is invariant under Ricci flow. This allowed them to prove:

Theorem 1.10 (Böhm and Wilking, 2006). *On a compact manifold M^n the Ricci flow evolves a Riemannian metric with 2-positive curvature operator to a limit metric with constant sectional curvature.*

Their paper [BW08] overcomes some major technical problems, particular those related to controlling the Weyl part of the curvature.

Soon thereafter, Brendle and Schoen [BS09a] finally managed to prove:

Theorem 1.11 (Brendle and Schoen, 2007). *Let M be a compact Riemannian manifold of dimension $n \geq 4$ with sectional curvature strictly 1/4-pinched in the pointwise sense. Then M admits a metric of constant curvature and therefore is diffeomorphic to a spherical space form.*

The key novel step in the proof is to show that nonnegative isotropic curvature is invariant under the Ricci flow. This was also independently proved by Nguyen [Ngu08, Ngu10]. Brendle and Schoen obtain the abovementioned result by working with the nonnegative isotropic curvature condition on

Chapter 2
Background Material

This chapter establishes the notational conventions used throughout while also providing results and computations needed for later analyses. Readers familiar with differential geometry may wish to skip this chapter and refer back when necessary.

2.1 Smooth Manifolds

Suppose M is a topological space. We say M is a *n-dimensional topological manifold* if it is Hausdorff, second countable and is 'locally Euclidean of dimension n' (i.e. every point $p \in M$ has a neighbourhood U homeomorphic to an open subset of $\mathbb{R}^n$). A *coordinate chart* is a pair (U, ϕ) where $U \subset M$ is open and $\phi : U \to \phi(U) \subset \mathbb{R}^n$ is a homeomorphism. If (U, ϕ) and (V, ψ) are two charts, the composition $\psi \circ \phi^{-1} : \phi(U \cap V) \to \psi(U \cap V)$ is called the *transition map* from ϕ to ψ. It is a homeomorphism since both ϕ and ψ are.

In order for calculus ideas to pass to the setting of manifolds we need to impose an extra smoothness condition on the chart structure. We say two charts (U, ϕ) and (V, ψ) are *smoothly compatible* if the transition map $\psi \circ \phi^{-1}$, as a map between open sets of $\mathbb{R}^n$, is a diffeomorphism.

We define a *smooth atlas* for M to be a collection of smoothly compatible charts whose domains cover M. We say two smooth atlases are *compatible* if their union is also a smooth atlas. As compatibility is an equivalence relation, we define a *differentiable structure* for M to be an equivalence class of smooth atlases. Thus a *smooth manifold* is a pair $(M, \mathscr{A})$ where M is a topological manifold and $\mathscr{A}$ is a smooth differentiable structure for M. When there is no ambiguity, we usually abuse notation and simply refer to a 'differentiable manifold M' without reference to the atlas. From here on, manifolds will always be of the differentiable kind.

B. Andrews and C. Hopper, *The Ricci Flow in Riemannian Geometry*, Lecture Notes in Mathematics 2011, DOI 10.1007/978-3-642-16286-2_2,

2.1.1 Tangent Space

There are various equivalent ways of defining the tangent space of a manifold. For our purposes, we emphasis the construction of the tangent space as derivations on the algebra $C^\infty(M)$.

Definition 2.1. Let M be a smooth manifold with a point p. A $\mathbb{R}$-linear map $X : C^\infty(M) \to \mathbb{R}$ is called a *derivation at* p if it satisfies the Leibniz rule: $X(fg) = f(p)Xg + g(p)Xf$. The *tangent space* at p, denoted by T_pM, is the set of all derivations at p.

The tangent space T_pM is clearly a vector space under the canonical operations $(X + Y)f = Xf + Yf$ and $(\lambda X)f = \lambda(Xf)$ where $\lambda \in \mathbb{R}$. In fact T_pM is of finite dimension and isomorphic to $\mathbb{R}^n$. By removing the pointwise dependence in the above definition, we define:

Definition 2.2. A *derivation* is an $\mathbb{R}$-linear map $Y : C^\infty(M) \to C^\infty(M)$ which satisfies the Leibniz rule: $Y(fg) = fYg + gYf$.

We identify such derivations with vector fields on M (see Remark 2.10 below).

Remark 2.3. In the setting of abstract algebra, a derivation is a function on an algebra which generalises certain features of the derivative operator. Specifically, given an associative algebra $\mathscr{A}$ over a ring or field R, a R-derivation is a R-linear map $D : \mathscr{A} \to \mathscr{A}$ that satisfies the product rule: $D(ab) = (Da)b + a(Db)$ for $a, b \in \mathscr{A}$. In our case, the algebra $\mathscr{A} = C^\infty(M)$ and the field is $\mathbb{R}$.

2.2 Vector Bundles

Definition 2.4. Let F and M be smooth manifolds. A *fibre bundle* over M with fibre F is a smooth manifold E, together with a surjective submersion $\pi : E \to M$ satisfying a local triviality condition: For any $p \in M$ there exists an open set U in M containing p, and a diffeomorphism $\phi : \pi^{-1}(U) \to U \times F$ (called a local trivialization) such that $\pi = \pi_1 \circ \phi$ on $\pi^{-1}(U)$, where $\pi_1(x, y) = x$ is the projection onto the first factor. The fibre at p, denoted E_p, is the set $\pi^{-1}(p)$, which is diffeomorphic to F for each p.

Although a fibre bundle E is *locally* a product $U \times F$, this may not be true globally. The space E is called the total space, M the base space and π the projection. Occasionally we refer to the bundle by saying: 'let $\pi : E \to M$ be a (smooth) fibre bundle'. In most cases the fibre bundles we consider will be vector bundles in which the fibre F is a vector space and the local trivializations induce a well-defined linear structure on E_p for each p:

Definition 2.5. Let M be a differentiable manifold. A *smooth vector bundle* of rank k over M is a fibre bundle $\pi : E \to M$ with fibre $\mathbb{R}^k$, such that

1. The fibres $E_p = \pi^{-1}(p)$ have a k-dimensional vector space structure.
2. The local trivializations $\phi : \pi^{-1}(U) \to U \times \mathbb{R}^k$ are such that $\pi_2 \circ \phi|_{E_p}$ is a linear isomorphism for each $p \in U$, where $\pi_2(x, y) = y$.

One of the most fundamental vector bundles over a manifold M is the tangent bundle $TM = \bigcup_{p \in M} T_pM$. It is a vector bundle of rank equal to $\dim M$. Other examples include the tensor bundles constructed from TM (see Sect. 2.3.3).

Definition 2.6. A *section* of a fibre bundle $\pi : E \to M$ is a smooth map $X : M \to E$, written $p \mapsto X_p$, such that $\pi \circ X = \mathrm{id}_M$. If E is a vector bundle then the collection of all *smooth sections* over M, denoted by $\Gamma(E)$, is a real vector space under pointwise addition and scalar multiplication.

Definition 2.7. A *local frame* for a vector bundle E of rank k is a k-tuple (ξ_i) of pointwise linearly independent sections of E over open $U \subset M$, that is linear independent $\xi_1, \ldots, \xi_k \in \Gamma(E\big|_U)$.

Given such a local frame, any section α of E over U can be written in the form $\sum_{i=1}^k \alpha^i \xi_i$, where $\alpha^i \in C^\infty(U)$. Local frames correspond naturally to local trivializations, since the map $(p, \sum_{i=1}^k \alpha^i \xi_i(p)) \mapsto (p, \alpha^1(p), \ldots, \alpha^k(p))$ is a local trivialization, while the inverse images of a standard basis in a local trivialization defines a local frame. Moreover we recall the local frame criterion for smoothness of sections.

Proposition 2.8. *A section $\alpha \in \Gamma(E|_U)$ is a smooth if and only if its component functions α^i, with respect to (ξ_i), on U are smooth.*

Remark 2.9. In fact $\Gamma(E)$ is a module over the ring $C^\infty(M)$ since for each $X \in \Gamma(E)$ we define $fX \in \Gamma(E)$ by $(fX)(p) = f(p)X(p)$. For instance the space of sections of any tensor bundle is a module over $C^\infty(M)$.[1]

Remark 2.10. There is an important identification between derivations (as in Definition 2.2) and smooth sections of the tangent bundle:

Proposition 2.11. *Smooth sections of $TM \to M$ are in one-to-one correspondence with derivations of $C^\infty(M)$.*

---

[1] Recall that a module over a ring generalises the notion of a vector space. Instead of requiring the scalars to lie in a field, the 'scalars' may lie in an arbitrary ring. Formally a left R-module over a ring R is an Abelian group $(G, +)$ with scalar multiplication $\cdot : R \times G \to G$ that is associative and distributive.

Proof. If $X \in \Gamma(TM)$ is a smooth section, define a derivation $\mathcal{X} : f \to Xf$ by $(\mathcal{X}f)(p) = X_p(f)$. Conversely, given a derivation $\mathcal{Y} : C^\infty(M) \to C^\infty(M)$ define a section Y of TM by $Y_p f = (\mathcal{Y}f)(p)$. An easy exercise shows that this section is smooth. □

As a result, we can either think of a smooth section $X \in \Gamma(TM)$ as a smooth map $X : M \to TM$ with $X \circ \pi = \mathrm{id}_M$ or as a derivation – that is, a $\mathbb{R}$-linear map $X : C^\infty(M) \to C^\infty(M)$ that satisfies the Leibniz rule. We call such an X a *vector field* and let the set of vector fields be denoted by $\mathscr{X}(M)$.

2.2.1 Subbundles

Definition 2.12. For a vector bundle $\pi : E \to M$, a *subbundle* of E is a vector bundle E' over M with an injective vector bundle homomorphism $i : E' \to E$ covering the identity map on M (so that $\pi_E \circ i = \pi_{E'}$, where π_E and $\pi_{E'}$ are the projections on E and E' respectively).

The essential idea of a subbundle of a vector bundle $E \to M$ is that it should be a smoothly varying family of linear subspaces E'_p of the fibres E_p that constitutes a vector bundle in their own right. However it is convenient to distinguish sections of the subbundle from sections of the larger bundle, and for this reason we use the definition above. One can think of the map i as an inclusion of E' into E.

Example 2.13. Let $f : M \hookrightarrow N$ be a smooth immersion between manifolds. The pushforward $f_* : TM \to TN$ (Sect. 2.8.2) over $f : M \to N$ induces a vector bundle mapping $i : TM \to f^*(TN)$ over M. On fibres over $p \in M$ this is the map $f_*|_p : T_pM \to T_{f(p)}N = (f^*TN)_p$ which is injective since f is an immersion. Hence, i exhibits TM as a subbundle of f^*TN over M.

It is also useful to note, using a rank type theorem, that:

Proposition 2.14. *If $f : E \to E'$ is a smooth bundle surjection over M. Then there exists a subbundle $j : E_0 \to E$ such that $j(E_0(p)) = \ker(f|_p)$ for each $p \in M$.*

In which case we have a well-defined subbundle $E_0 = \ker f$ inside E.

2.2.2 Frame Bundles

For a vector bundle $\pi : E \to M$ of rank k, there is an associated fibre bundle over M with fibre $\mathrm{GL}(k)$ called the *general linear frame bundle* $F(E)$. The fibre $F(E)_x$ over $x \in M$ consists of all linear isomorphisms $Y : \mathbb{R}^k \to E_x$, or equivalently the set of all ordered bases for E_x (by identifying the map Y with the basis (Y_a), where $Y_a = Y(e_a)$ for $a = 1, \ldots, k$). The group $\mathrm{GL}(k)$ acts on each fibre by composition, so that $\mathbb{A} \in \mathrm{GL}(k)$ acts on a frame $Y: \mathbb{R}^k \to E_x$ to give $Y^{\mathbb{A}} = Y \circ \mathbb{A}: \mathbb{R}^k \to E_x$ (alternatively, the basis $(Y_a) \in F(E)_x$ maps to the basis $(\mathbb{A}_b^a Y_a)$). From standard linear algebra, this action

$$\mathrm{GL}(k) \times F(E) \to F(E); \quad (\mathbb{A}, Y) \mapsto Y^{\mathbb{A}}$$

is simply transitive on each fibre (that is, for any $Y, Z \in F(E)_x$ there exists a unique $\mathbb{A} \in \mathrm{GL}(k)$ such that $Y^{\mathbb{A}} = Z$).

Note that a local trivialization $\phi: \pi^{-1}(U) \to U \times \mathbb{R}^k$ for E, together with a local chart $\eta: U \to \mathbb{R}^n$ for M, produces a chart for E compatible with the bundle structure: We take $(x, v) \mapsto \phi(x, v) = (x, \pi_2\phi(x, v)) \mapsto (\eta(x), \pi_2\phi(x, v)) \in \mathbb{R}^n \times \mathbb{R}^k$. Any such chart also produces a chart for $F(E)$ giving it the structure of a manifold of dimension $n + k^2$: We take $(x, Y) \mapsto (\eta(x), \pi_2 \circ \phi_x \circ Y) \in \mathbb{R}^n \times \mathrm{GL}(k) \subset \mathbb{R}^{n+k^2}$, where $\phi_x(\cdot) = \phi(x, \cdot)$. Similarly a local trivialization of $F(E)$ is defined by $(x, Y) \mapsto (x, \pi_2 \circ \phi_x \circ Y)$, giving $F(E)$ the structure of a fibre bundle with fibre $\mathrm{GL}(k)$ as claimed.

If the bundle E is equipped with a metric g (Sect. 2.4.4) – so that E_x is an inner product space – then one can introduce the *orthonormal frame bundle* $O(E)$. Specifically $O(E)$ is the subset of $F(E)$ defined by

$$O(E) = \{Y \in F(E) : g(Y_a, Y_b) = \delta_{ab}\}.$$

The orthogonal group $O(k)$ acts on $O(E)$ by

$$O(k) \times O(E) \to O(E); \quad (\mathbb{O}, Y) \mapsto Y^{\mathbb{O}}$$

where $Y^{\mathbb{O}}(u) = Y(\mathbb{O}u)$ for each $u \in \mathbb{R}^k$ and $O(E)$ is a fibre bundle over M with fibre $O(k)$.

Remark 2.15. A local frame for E consists of k pointwise linearly independent smooth sections of E over an open set U, say $p \mapsto \xi_i(p) \in E_p$ for $i = 1, \ldots, k$. This corresponds to a section Y of $F(E)$ over U, defined by $Y_p(u^i e_i) = u^i \xi_i(p)$ for each $p \in U$ and $u \in \mathbb{R}^k$.

2.3 Tensors

Let V be a finite dimensional vector space. A *covariant k-tensor* on V is a multilinear map $F: V^k \to \mathbb{R}$. Similarly, a *contravariant ℓ-tensor* is a multilinear map $F: (V^*)^\ell \to \mathbb{R}$. A mixed tensor of type $\binom{k}{\ell}$ or (k, ℓ) is a multilinear map

$$F: \underbrace{V^* \times \cdots \times V^*}_{\ell \text{ times}} \times \underbrace{V \times \cdots \times V}_{k \text{ times}} \to \mathbb{R}.$$

We denote the space of all k-tensors on V by $T^k(V)$, the space of contravariant ℓ-tensors by $T_\ell(V)$, and the space of all mixed (k,ℓ) tensors by $T^k_\ell(V)$.

The following canonical isomorphism is frequently useful:

Lemma 2.16. *The tensor space $T^1_1(V)$ is canonically isomorphic to* End(V), *where the (bases independent) isomorphism $\Phi : End(V) \to T^1_1(V)$ is given by*

$$\Phi(A) : (\omega, X) \mapsto \omega(A(X))$$

for all $A \in \mathrm{End}(V)$, $\omega \in V^*$, and $X \in V$.

Remark 2.17. Alternatively one could state this lemma by saying: If V and W are vector spaces then $V \otimes W^* \simeq \mathrm{End}(W,V)$ where the isomorphism $\Psi : V \otimes W^* \to \mathrm{End}(W,V)$ is given by $\Psi(v \otimes \xi) : w \to \xi(w)v$.

A general version of this identification is expressed as follows.

Lemma 2.18. *The tensor space $T^k_{\ell+1}(V)$ is canonically isomorphic to the space* Mult$((V^*)^\ell \times V^k, V)$, *where the isomorphism $\Phi : \mathrm{Mult}((V^*)^\ell \times V^k, V) \to T^k_{\ell+1}(V)$ is given by*

$$\Phi(A) : (\omega^0, \omega^1, \ldots, \omega^\ell, X_1, \ldots, X_k) \mapsto \omega^0(A(\omega^1, \ldots, \omega^\ell, X_1, \ldots, X_k))$$

for all $A \in \mathrm{Mult}((V^*)^\ell \times V^k, V)$, $\omega_i \in V^*$, and $X_j \in V$.[2]

2.3.1 Tensor Products

There is a natural product that links the various tensor spaces over V. If $F \in T^k_\ell(V)$ and $G \in T^p_q(V)$, then the *tensor product* $F \otimes G \in T^{k+p}_{\ell+q}(V)$ is defined to be

$$\begin{aligned}(F \otimes G)&(\omega^1, \ldots, \omega^{\ell+q}, X_1, \ldots, X_{k+p}) \\ &= F(\omega^1, \ldots, \omega^\ell, X_1, \ldots, X_k)\, G(\omega^{\ell+1}, \ldots, \omega^{\ell+q}, X_{k+1}, \ldots, X_{k+p}).\end{aligned}$$

Moreover, if $(e_1, \ldots, e_n)$ is a basis for V and $(\varphi^1, \ldots, \varphi^n)$ is the corresponding dual basis, defined by $\varphi^i(e_j) = \delta^i_j$, then it can be shown that a *basis* for $T^k_\ell(V)$ takes the form

$$e_{j_1} \otimes \cdots \otimes e_{j_\ell} \otimes \varphi^{i_1} \otimes \cdots \otimes \varphi^{i_k},$$

where

$$e_{j_1} \otimes \cdots \otimes e_{j_\ell} \otimes \varphi^{i_1} \otimes \cdots \otimes \varphi^{i_k}(\varphi^{s_1}, \ldots, \varphi^{s_\ell}, e_{r_1}, \ldots, e_{r_k}) = \delta^{s_1}_{j_1} \cdots \delta^{s_\ell}_{j_\ell}\, \delta^{i_1}_{r_1} \cdots \delta^{i_k}_{r_k}.$$

---

[2] Here Mult$((V^*)^\ell \times V^k, V)$ is the set of multilinear maps from $(V^*)^\ell \times V^k$ to V.

Therefore any tensor $F \in T^k_\ell(V)$ can be written, with respect to this basis, as

$$F = F^{j_1,\dots,j_\ell}{}_{i_1,\dots,i_k} e_{j_1} \otimes \cdots \otimes e_{j_\ell} \otimes \varphi^{i_1} \otimes \cdots \otimes \varphi^{i_k}$$

where $F^{j_1,\dots,j_\ell}{}_{i_1,\dots,i_k} = F(\varphi^{j_1},\dots,\varphi^{j_\ell}, e_{i_1},\dots,e_{i_k})$.

2.3.2 Tensor Contractions

A tensor contraction is an operation on one or more tensors that arises from the natural pairing of a (finite-dimensional) vector space with its dual.

Intuitively there is a natural notion of 'the trace of a matrix' $A = (A^i_j) \in \mathrm{Mat}_{n\times n}(\mathbb{R})$ given by $\operatorname{tr} A = \sum_i A^i_i$. It is $\mathbb{R}$-linear and commutative in the sense that $\operatorname{tr} AB = \operatorname{tr} BA$. From the latter property, the trace is also cyclic (in the sense that $\operatorname{tr} ABC = \operatorname{tr} BCA = \operatorname{tr} CAB$). Therefore the trace is similarity-invariant, which means for any $P \in \mathrm{GL}(n)$ the trace $\operatorname{tr} P^{-1}AP = \operatorname{tr} PP^{-1}A = \operatorname{tr} A$. Whence we can extend tr over $\mathrm{End}(V)$ by taking the trace of a matrix representation – this definition is basis independent since different bases give rise to similar matrices – and so by Lemma 2.16 tr can act on tensors as well.

Naturally, we define the *contraction* of any $F \in T^1_1(V)$ by taking the trace of F as a linear map in $\mathrm{End}(V)$. In which case $\operatorname{tr} : T^1_1(V) \to \mathbb{R}$ is given by $\operatorname{tr} F = F(\varphi^i, e_i) = \sum_i F^i_i$, since $\Phi^{-1}(F) = (F(\varphi^i, e_j))^n_{i,j=1} \in \mathrm{End}(V)$. In general we define

$$\operatorname{tr} : T^{k+1}_{\ell+1}(V) \to T^k_\ell(V)$$

by

$$(\operatorname{tr} F)(\omega^1,\dots,\omega^\ell, X_1,\dots,X_k) = \operatorname{tr}\big(F(\omega^1,\dots,\omega^\ell,\ \cdot\ ,X_1,\dots,X_k,\cdot\ )\big).$$

That is, we define $(\operatorname{tr} F)(\omega^1,\dots,\omega^\ell, X_1,\dots,X_k)$ to be the trace of the endomorphism $F(\omega^1,\dots,\omega^\ell,\ \cdot\ ,X_1,\dots,X_k,\cdot\ ) \in T^1_1(V) \simeq \mathrm{End}(V)$. In components this is equivalent to

$$(\operatorname{tr} F)^{j_1\dots j_\ell}{}_{i_1\dots i_k} = F^{j_1\dots j_\ell m}{}_{i_1\dots i_k m}.$$

It is clear that the contraction is linear and lowers the rank of a tensor by 2. Unfortunately there is no general notation for this operation! So it is best to explicitly describe the contraction in words each time it arises. We give some simple examples of how this might occur.

There are several variations of tr: Firstly, there is nothing special about which particular component pairs the contraction is taken over. For instance if $F = F_i{}^j{}_k \varphi^i \otimes e_j \otimes \varphi^k \in T_1^2(V)$ then one could take the contraction of F over the first two components: $(\operatorname{tr}_{12} F)_k = \operatorname{tr} F(\cdot, \cdot, e_k) = F_i{}^i{}_k$ or the last two components: $(\operatorname{tr}_{23} F)_k = \operatorname{tr} F(e_k, \cdot, \cdot) = F_k{}^i{}_i$.

Another variation occurs if one wants to take the contraction over multiple component pairs. For example if $F = F_{ij}{}^{k\ell} \varphi^i \otimes \varphi^j \otimes e_k \otimes e_\ell \in T_2^2(V)$, one could take the trace over the 1st and 3rd with the 2nd and 4th so that $\operatorname{tr} F = \operatorname{tr}_{13} \operatorname{tr}_{24} F = \operatorname{tr} F(\star, \cdot, \star, \cdot) = F_{pq}{}^{pq}$.

Furthermore, if one has a metric g then it is possible to take contractions over two indices that are either both vectors or covectors. This is done by taking a tensor product with the metric tensor (or its inverse) and contracting each of the two indices with one of the indices of the metric. This operation is known as *metric contraction* (see Sect. 2.4.3 for further details).

Finally, one of the most important applications arises when $F = \omega \otimes X \in T_1^1(V)$ for some (fixed) vector X and covector ω. In this case

$$\operatorname{tr} F = F(\partial_i, dx^i) = \omega(\partial_i) X(dx^i) = \omega_i X^i = \omega(X).$$

The idea can be extended as follows: If $F \in T_\ell^k(V)$ with $\omega^1, \ldots, \omega^\ell \in V^*$ and $X_1, \ldots, X_k \in V$ (fixed) then

$$F \otimes \omega^1 \otimes \cdots \otimes \omega^\ell \otimes X_1 \otimes \cdots \otimes X_k \in T_{2\ell}^{2k}(V).$$

So the contraction of this tensor over all indexes becomes

$$\begin{aligned} &\operatorname{tr}(F \otimes \omega^1 \otimes \cdots \otimes \omega^\ell \otimes X_1 \otimes \cdots \otimes X_k) \\ &= \omega_{j_1} \cdots \omega_{j_\ell} F^{j_1 \cdots j_\ell}{}_{i_1 \cdots i_k} X^{i_1} \cdots X^{i_k} \\ &= F(\omega^1, \ldots, \omega^\ell, X_1, \ldots, X_k). \end{aligned} \tag{2.1}$$

2.3.3 Tensor Bundles and Tensor Fields

For a manifold M we can apply the above tensor construction pointwise on each tangent space T_pM. In which case a (k, ℓ)-tensor at $p \in M$ is an element $T_\ell^k(T_pM)$. We define the *bundle of* (k, ℓ)*-tensors* on M by

$$T_\ell^k M = \bigcup_{p \in M} T_\ell^k(T_pM) = \bigcup_{p \in M} \otimes^k T_p^* M \otimes^\ell T_pM.$$

In particular, $T_1 M = TM$ and $T^1 M = T^* M$. An important subbundle of $T^2 M$ is $\operatorname{Sym}^2 T^* M$, the space of all symmetric $(2, 0)$-tensors on M. A (k, ℓ)-*tensor field* is an element of $\Gamma(T_\ell^k M) = \Gamma(\otimes^k T^* M \otimes^\ell TM)$ – we sometimes use the notation $\mathscr{T}_\ell^k(M)$ as a synonym for $\Gamma(T_\ell^k M)$.

To check that $T^k_\ell M$ is a vector bundle, let $\pi : T^k_\ell M \to M$ send $F \in T^k_\ell(T_pM)$ to the base point p. If (x^i) is a local chart on open $U \subset M$ around point p, then any tensor $F \in T^k_\ell(T_pM)$ can be expressed as

$$F = F^{j_1,\ldots,j_\ell}{}_{i_1,\ldots,i_k} \partial_{j_1} \otimes \cdots \otimes \partial_{j_\ell} \otimes dx^{i_1} \otimes \cdots \otimes dx^{i_k}.$$

The local trivialisation $\phi : \pi^{-1}(U) \to U \times \mathbb{R}^{n^{k+\ell}}$ is given by

$$\phi : T^k_\ell(T_pM) \ni F \longmapsto (p, F^{j_1,\ldots,j_\ell}{}_{i_1,\ldots,i_k}).$$

2.3.4 Dual Bundles

If E is a vector bundle over M, the *dual bundle* E^* is the bundle whose fibres are the dual spaces of the fibres of E:

$$E^* = \{(p, \omega) : \ \omega \in E^*_p\}.$$

If (ξ_i) is a local frame for E over an open set $U \subset M$, then the map $\phi : \pi^{-1}_{E^*}(U) \to U \times \mathbb{R}^k$ defined by $(p, \omega) \mapsto (p, \omega(\xi_1(p)), \ldots, \omega(\xi_k(p)))$ is a local trivialisation of E^* over U. The corresponding local frame for E^* is given by the sections θ^i defined by $\theta^i(\xi_j) = \delta^i_j$.

2.3.5 Tensor Products of Bundles

If $E_1, \ldots, E_k$ are vector bundles over M, the tensor product $E_1 \otimes \cdots \otimes E_k$ is the vector bundle whose fibres are the tensor products $(E_1)_p \otimes \cdots \otimes (E_k)_p$. If U is an open set in M and $\{\xi^j_i : 1 \le i \le n_j\}$ is a local frame for E_j over U for $j = 1, \ldots, k$, then $\{\xi^1_{i_1} \otimes \cdots \otimes \xi^k_{i_k} : 1 \le i_j \le n_j, \ 1 \le j \le k\}$ forms a local frame for $E_1 \otimes \cdots \otimes E_k$. Taking tensor products commutes with taking duals (in the sense of Sect. 2.3.4) and is associative. That is, $E^*_1 \otimes E^*_2 \simeq (E_1 \otimes E_2)^*$ and $(E_1 \otimes E_2) \otimes E_3 \simeq E_1 \otimes (E_2 \otimes E_3)$.

2.3.6 A Test for Tensorality

Let $E_1, \ldots, E_k$ be vector bundles over M. Given a tensor field $F \in \Gamma(E^*_1 \otimes \cdots \otimes E^*_k)$ and sections $X_i \in \Gamma(E_i)$, Proposition 2.8 implies that the function on U defined by

$$F(X_1, \ldots, X_k) : p \mapsto F_p(X_1\big|_p, \ldots, X_k\big|_p),$$

is smooth, so that F induces a mapping $F : \Gamma(E_1) \times \cdots \times \Gamma(E_k) \to C^\infty(M)$. It can easily be seen that this map is multilinear over $C^\infty(M)$ in the sense that

$$F(f_1X_1, \ldots, f_kX_k) = f_1 \cdots f_k F(X_1, \ldots, X_k)$$

for any $f_i \in C^\infty(M)$ and $X_i \in \Gamma(E_i)$. In fact the converse holds as well.

Proposition 2.19 (Tensor Test). *For vector bundles $E_1, \ldots, E_k$ over M, the mapping $F : \Gamma(E_1) \times \cdots \times \Gamma(E_k) \to C^\infty(M)$ is a tensor field, i.e. $F \in \Gamma(E_1^* \otimes \cdots \otimes E_k^*)$, if and only if F is multilinear over $C^\infty(M)$.*

By Lemma 2.18 we also have:

Proposition 2.20 (Bundle Valued Tensor Test). *For vector bundles E_0, $E_1, \ldots, E_k$ over M, the mapping $F : \Gamma(E_1) \times \cdots \times \Gamma(E_k) \to \Gamma(E_0)$ is a tensor field, i.e. $F \in \Gamma(E_1^* \otimes \cdots \otimes E_k^* \otimes E_0)$, if and only if F is multilinear over $C^\infty(M)$.*

Remark 2.21. This proposition leaves one to interpret

$$F \in \Gamma(E_1^* \otimes \cdots \otimes E_k^* \otimes E_0)$$

as an E_0-valued tensor acting on $E_1 \otimes \cdots \otimes E_k$.

The importance of Propositions 2.19 and 2.20 is that it allows one to work with tensors *without* referring to their pointwise attributes. For example, the metric g on M (as we shall see) can be consider as a pointwise inner product $g_p : T_pM \times T_pM \to \mathbb{R}$ that smoothly depends on its base point. By our identification we can also think of this tensor as a map

$$g : \mathscr{X}(M) \times \mathscr{X}(M) \to C^\infty(M).$$

Therefore if X, Y and Z are vector fields, $g(X, Y) \in C^\infty(M)$ and so by Remark 2.10 we also have $Zg(X, Y) \in C^\infty(M)$.

2.4 Metric Tensors

An inner product on a vector space allows one to define lengths of vectors and angles between them. Riemannian metrics bring this structure onto the tangent space of a manifold.

2.4.1 Riemannian Metrics

A *Riemannian metric* g on a manifold M is a symmetric positive definite $(2,0)$-tensor field (i.e. $g \in \Gamma(\mathrm{Sym}^2 T^*M)$ and g_p is an inner product for each $p \in M$). A manifold M together with a given Riemannian metric g is called a *Riemannian manifold* (M, g).

In local coordinates (x^i), $g = g_{ij} dx^i \otimes dx^j$. The archetypical Riemannian manifold is $(\mathbb{R}^n, \delta_{ij})$. As the name suggests, the concept of the metric was first introduced by Bernhard Riemann in his 1854 habilitation dissertation.

2.4.1.1 Geodesics

We want to think of geodesics as length minimising curves. From this point of view, we seek to minimise the length functional

$$L(\gamma) = \int_0^1 \|\dot{\gamma}(t)\|_g dt$$

amongst all curves $\gamma : [0,1] \to M$. There is also a natural 'energy' functional:

$$E(\gamma) = \frac{1}{2} \int_0^1 \|\dot{\gamma}(t)\|_g^2 dt.$$

As $L(\gamma)^2 \leq 2E(\gamma)$, for any smooth curve $\gamma : [0,1] \to M$, the problem of minimising $L(\gamma)$ amongst all smooth curves γ is equivalent to minimising $E(\gamma)$. By doing so we find:

Theorem 2.22. *The Euler–Lagrange equations for the energy functional are*

$$\ddot{\gamma}^i(t) + \Gamma^i_{jk}(\gamma(t))\dot{\gamma}^j(t)\dot{\gamma}^k(t) = 0, \tag{2.2}$$

where the connection coefficients Γ^i_{jk} *are given by* (2.9).

Hence any smooth curve $\gamma : [0,1] \to M$ satisfying (2.2) is called a geodesic. By definition they are critical points of the energy functional. Moreover, by the Picard–Lindelöf theorem we recall:

Lemma 2.23 (Short-Time Existence of Geodesics). *Suppose* (M, g) *is a Riemannian manifold. Let* $p \in M$ *and* $v \in T_pM$ *be given. Then there exists* $\varepsilon > 0$ *and precisely one geodesic* $\gamma : [0, \varepsilon] \to M$ *with* $\gamma(0) = p$, $\dot{\gamma}(0) = v$ *and* γ *depends smoothly on* p *and* v.

2.4.2 The Product Metric

If $(M_1, g^{(1)})$ and $(M_2, g^{(2)})$ are two Riemannian manifolds then, by the natural identification $T_{(p_1,p_2)}M_1 \times M_2 \simeq T_{p_1}M_1 \oplus T_{p_2}M_2$, there is a canonical Riemannian metric $g = g^{(1)} \oplus g^{(2)}$ on $M_1 \times M_2$ defined by

$$g_{(p_1,p_2)}(u_1 + u_2, v_1 + v_2) = g^{(1)}_{p_1}(u_1, u_2) + g^{(2)}_{p_2}(v_1, v_2),$$

where $u_1, u_2 \in T_{p_1}M_1$ and $v_1, v_2 \in T_{p_2}M_2$. If $\dim M_1 = n$ and $\dim M_2 = m$, the product metric, in local coordinates $(x^1, \ldots, x^{n+m})$ about (p_1, p_2), is the block diagonal matrix:

$$(g_{ij}) = \begin{pmatrix} \boxed{g^{(1)}_{ij}} & \\ & \boxed{g^{(2)}_{ij}} \end{pmatrix}$$

where $(g^{(1)}_{ij})$ is an $n \times n$ block and $(g^{(2)}_{ij})$ is an $m \times m$ block.

2.4.3 Metric Contractions

As the Riemannian metric g is non-degenerate, there is a canonical g-dependent isomorphism between TM and T^*M.[3] By using this it is possible to take tensor contractions over two indices that are either both vectors or covectors.

For example, if h is a symmetric $(2,0)$-tensor on a Riemannian manifold then $h^\sharp$ is a $(1,1)$-tensor. In which case the trace of h with respect to g, denoted by $\mathrm{tr}_g h$, is

$$\mathrm{tr}_g\, h = \mathrm{tr}\, h^\sharp = h_i^{\ i} = g^{ij} h_{ij}.$$

Equivalently one could also write

$$\mathrm{tr}_g\, h = \mathrm{tr}_{13}\, \mathrm{tr}_{24}\, g^{-1} \otimes h = (g^{-1} \otimes h)(dx^i, dx^j, \partial_i, \partial_j) = g^{ij} h_{ij}.$$

2.4.4 Metrics on Bundles

A metric g on a vector bundle $\pi : E \to M$ is a section of $E^* \otimes E^*$ such that at each point p of M, g_p is an inner product on E_p (that is, g_p is symmetric and

---

[3] Specifically, the isomorphism $\sharp : T^*M \to TM$ sends a covector ω to $\omega^\sharp = \omega^i \partial_i = g^{ij}\omega_j \partial_i$, and $\flat : TM \to T^*M$ sends a vector X to $X^\flat = X_i dx^i = g_{ij} X^j dx^i$.

positive definite for each p: $g_p(\xi,\eta) = g_p(\eta,\xi)$ for all $\xi,\eta \in E_p$; $g_p(\xi,\xi) \geq 0$ for all $\xi \in E_p$, and $g_p(\xi,\xi) = 0 \Rightarrow \xi = 0$).

A metric on E defines a bundle isomorphism $\iota_g : E \to E^*$ given by $\iota_g(\xi) : \eta \mapsto g_p(\xi,\eta)$ for all $\xi,\eta \in E_p$.

2.4.5 Metric on Dual Bundles

If g is a metric on E, there is a unique metric on E^* (also denoted g) such that ι_g is a bundle isometry:

$$g(\iota_g(\xi), \iota_g(v)) = g(\xi,\eta)$$

for all $\xi,\eta \in E_p$; or equivalently $g(\omega,\sigma) = g(\iota_g^{-1}\omega, \iota_g^{-1}\sigma)$ for all $\omega,\sigma \in E_p^* = (E_p)^*$.

2.4.6 Metric on Tensor Product Bundles

If g_1 is a metric on E_1 and g_2 is a metric on E_2, then

$$g = g_1 \otimes g_2 \in \Gamma((E_1^* \otimes E_1^*) \otimes (E_2^* \otimes E_2^*)) \simeq \Gamma((E_1 \otimes E_2)^* \otimes (E_1 \otimes E_2)^*)$$

is the unique metric on $E_1 \otimes E_2$ such that

$$g(\xi_1 \otimes \eta_1, \xi_2 \otimes \eta_2) = g_1(\xi_1,\xi_2) g_2(\eta_1,\eta_2).$$

The construction of metrics on tensor bundles now follows. It is well-defined since the metric constructed on a tensor product of dual bundles agrees with that constructed on the dual bundle of a tensor product.

Example 2.24. Given tensors $S, T \in \mathscr{T}_\ell^k(M)$, the inner product, denoted by $\langle\cdot,\cdot\rangle$, at p is

$$\langle S, T\rangle = g^{a_1b_1} \cdots g^{a_kb_k} g_{i_1j_1} \cdots g_{i_\ell j_\ell} S_{a_1\ldots a_k}{}^{i_1\ldots i_\ell} T_{b_1\ldots b_k}{}^{j_1\ldots j_\ell} . \tag{2.3}$$

2.5 Connections

Connections provide a coordinate invariant way of taking directional derivatives of vector fields. In $\mathbb{R}^n$, the derivative of a vector field $X = X^i e_i$ in direction v is given by $D_v X = v(X^i)e_i$. Simply put, D_v differentiates the coefficient functions X^i and thinks of the basis vectors e_i as being held constant. However there is no canonical way to compare vectors from different vector spaces, hence there is no natural coordinate invariant analogy applicable to abstract manifolds.[4] To circumnavigate this, we impose an additional structure – in the form of a connection operator – that provides a way to 'connect' these tangent spaces.

Our method here is to directly specify how a connection acts on elements of $\Gamma(E)$ as a module over $C^\infty(M)$. They are of central importance in modern geometry largely because they allow a comparison between the local geometry at one point and the local geometry at another point.

Definition 2.25. A *connection* ∇ on a vector bundle E over M is a map

$$\nabla : \mathscr{X}(M) \times \Gamma(E) \to \Gamma(E),$$

written as $(X, \sigma) \mapsto \nabla_X \sigma$, that satisfies the following properties:

1. ∇ is $C^\infty(M)$-linear in X:

$$\nabla_{f_1 X_1 + f_2 X_2} \sigma = f_1 \nabla_{X_1} \sigma + f_1 \nabla_{X_1} \sigma$$

2. ∇ is $\mathbb{R}$-linear in σ:

$$\nabla_X (\lambda_1 \sigma_1 + \lambda_2 \sigma_2) = \lambda_1 \nabla_X \sigma_1 + \lambda_2 \nabla_X \sigma_2$$

and ∇ satisfies the product rule:

$$\nabla_X (f\sigma) = (Xf)\,\sigma + f \nabla_X \sigma$$

We say $\nabla_X \sigma$ is the *covariant derivative* of σ in the direction X.

Remark 2.26. Equivalently, we could take the connection to be

$$\nabla : \Gamma(E) \to \Gamma(T^*M \otimes E)$$

which is linear and satisfies the product rule. For if $\sigma \in \Gamma(E)$ then $\nabla\sigma \in \Gamma(T^*M \otimes E)$, where $(\nabla\sigma)(X) = \nabla_X \sigma$ is $C^\infty(M)$-linear in X by Property 1 of Definition 2.25. Thus Proposition 2.20 implies that $\nabla\sigma : \Gamma(TM) \to \Gamma(E)$ is an E-valued tensor acting on TM.

---

[4] There is however another generalisation of directional derivatives which is canonical: the Lie derivative. The Lie derivative evaluates the change of one vector field along the flow of another vector field. Thus, one must know both vector fields in an open neighbourhood. The covariant derivative on the other hand only depends on the vector direction at a single point, rather than a vector field in an open neighbourhood of a point. In other words, the covariant derivative is linear over $C^\infty(M)$ in the direction argument, while the Lie derivative is $C^\infty(M)$-linear in neither argument.

For a connection ∇ on the tangent bundle, we define the *connection coefficients* or *Christoffel symbols* of ∇ in a given set of local coordinates (x^i) by defining

$$\Gamma_{ij}{}^k = dx^k\left(\nabla_{\partial_i}\partial_j\right)$$

or equivalently,

$$\nabla_{\partial_i}\partial_j = \Gamma_{ij}{}^k\partial_k.$$

More generally, the connection coefficients of a connection ∇ on a bundle E can be defined with respect to a given local frame $\{\xi_\alpha\}$ for E by the equation

$$\nabla_{\partial_i}\xi_\alpha = \Gamma_{i\alpha}{}^\beta\xi_\beta.$$

2.5.1 Covariant Derivative of Tensor Fields

In applications one is often interested in computing the covariant derivative on the tensor bundles $T^k_\ell M$. This is a special case of a more general construction (see Sect. 2.5.3).

Proposition 2.27. *Given a connection ∇ on TM, there is a unique connection on the tensor bundle, also denoted by ∇, that satisfies the following properties:*

1. *On TM, ∇ agrees with the given connection.*
2. *On $C^\infty(M) = T^0 M$, ∇ is the action of a vector as a derivation:*

$$\nabla_X f = Xf,$$

for any smooth function f.

3. *∇ obeys the product rule with respect to tensor products:*

$$\nabla_X(F \otimes G) = (\nabla_X F) \otimes G + F \otimes (\nabla_X G),$$

for any tensors F and G.

4. *∇ commutes with all contractions:*

$$\nabla_X(\operatorname{tr} F) = \operatorname{tr}(\nabla_X F),$$

for any tensor F.

Example 2.28. We compute the covariant derivative of a 1-form ω with respect to the vector field X. Since $\operatorname{tr} dx^j \otimes \partial_i = \delta_i^j$, $\nabla_X(\operatorname{tr} dx^j \otimes \partial_i) = 0$. Thus $(\nabla_X dx^j)(\partial_i) = -dx^j(\nabla_X \partial_i)$ and so

$$\begin{aligned}(\nabla_X \omega)(\partial_k) &= \nabla_X \omega_k + \omega_j\,(\nabla_X dx^j)(\partial_k)\\ &= \nabla_X \omega_k - \omega_j dx^j(\nabla_X \partial_k)\\ &= X^i \partial_i \omega_k - \omega_j X^i\, \Gamma_{ik}^j.\end{aligned}$$

Therefore $\nabla_X \omega = (X^i \partial_i \omega_k - \omega_j X^i\, \Gamma_{ik}^j) dx^k$.

In general we have the following useful formulas.

Proposition 2.29. *For any tensor field $F \in \mathscr{T}_\ell^k(M)$, vector fields Y_i and 1-forms ω^j we have*

$$\begin{aligned}(\nabla_X F)(\omega^1, \dots, \omega^\ell, Y_1, \dots, Y_k) &= X(F(\omega^1, \dots, \omega^\ell, Y_1, \dots, Y_k))\\ &\quad - \sum_{j=1}^{\ell} F(\omega^1, \dots, \nabla_X \omega^j, \dots, \omega^\ell, Y_1, \dots, Y_k)\\ &\quad - \sum_{i=1}^{k} F(\omega^1, \dots, \omega^\ell, Y_1, \dots, \nabla_X Y_i, \dots, Y_k).\end{aligned}$$

Proof. By (2.1) we have

$$\begin{aligned}\operatorname{tr}(F \otimes \omega^1 \otimes \cdots \otimes \omega^\ell \otimes Y_1 \otimes \cdots \otimes Y_k) &= \omega_{i_1} \cdots \omega_{i_k} F^{i_1 \cdots i_k}{}_{j_1 \cdots j_k} Y^{j_1} \cdots Y^{j_k}\\ &= F(\omega^1, \dots, \omega^\ell, Y_1, \dots, Y_k).\end{aligned}$$

Thus by Proposition 2.27,

$$\begin{aligned}&\nabla_X(F(\omega^1, \dots, \omega^\ell, Y_1, \dots, Y_k))\\ &\quad = \operatorname{tr}\Big[(\nabla_X F) \otimes \omega^1 \otimes \cdots \otimes Y_k + F \otimes (\nabla_X \omega^1) \otimes \cdots \otimes Y_k\\ &\qquad + \cdots + F \otimes \omega^1 \otimes \cdots \otimes (\nabla_X Y_k)\Big]\\ &\quad = (\nabla_X F)(\omega^1, \dots, \omega^\ell, Y_1, \dots, Y_k) + F(\nabla_X \omega^1, \dots, \omega^\ell, Y_1, \dots, Y_k)\\ &\qquad + \cdots + F(\omega^1, \dots, \omega^\ell, Y_1, \dots, \nabla_X Y_k).\end{aligned}$$

□

As the covariant derivative is $C^\infty(M)$-linear over X, we define $\nabla F \in \Gamma(\otimes^{k+1} T^*M \otimes^\ell TM)$ by

$$(\nabla F)(X, Y_1, \dots, Y_k, \omega^1, \dots, \omega^\ell) = \nabla_X F(Y_1, \dots, Y_k, \omega^1, \dots, \omega^\ell),$$

for any $F \in \mathscr{T}_\ell^k(M)$. Hence (in this case) ∇ is an $\mathbb{R}$-linear map $\nabla : \mathscr{T}_\ell^k(M) \to \mathscr{T}_\ell^{k+1}(M)$ that takes a (k, ℓ)-tensor field and gives a $(k+1, \ell)$-tensor field.

2.5.2 The Second Covariant Derivative of Tensor Fields

By utilising the results of the previous section, we can make sense of the second covariant derivative ∇^2.

To do this, suppose the vector bundle $E = T^k_\ell M$ with associated connection ∇. By Remark 2.26, if $F \in \Gamma(E)$ then $\nabla F \in \Gamma(T^*M \otimes E)$ and so $\nabla^2 F \in \Gamma(T^*M \otimes T^*M \otimes E)$. Therefore, for any vector fields X, Y we find that

$$\begin{aligned}(\nabla^2 F)(X,Y) &= \big(\nabla_X(\nabla F)\big)(Y) \\ &= \nabla_X\big((\nabla F)(Y)\big) - (\nabla F)(\nabla_X Y) \\ &= \nabla_X(\nabla_Y F)) - (\nabla_{\nabla_X Y} F). \end{aligned} \tag{2.4}$$

Example 2.30. If $f \in C^\infty(M)$ is a $(0,0)$-tensor, then $\nabla^2 f$ is a $(2,0)$-tensor. In local coordinates:

$$\begin{aligned}\nabla^2_{\partial_i,\partial_j} f &= (\nabla_{\partial_i}(\nabla f))(\partial_j) \\ &= \partial_i((\nabla f)(\partial_j)) - (\nabla f)(\nabla_{\partial_i}\partial_j) \\ &= \partial_i(\partial_j f) - (\nabla f)(\Gamma^k_{ij}\partial_k) \\ &= \partial_i\,\partial_j f - \Gamma^k_{ij}\nabla_{\partial_k} f \\ &= \frac{\partial^2 f}{\partial x^i \partial x^j} - \Gamma^k_{ij}\frac{\partial f}{\partial x^k}. \end{aligned} \tag{2.5}$$

In general equation (2.4) amounts to the following useful formula:

Proposition 2.31. *If ∇ is a connection on TM, then*

$$\nabla^2_{Y,X} = \nabla_Y \circ \nabla_X - \nabla_{\nabla_Y X} : \mathscr{T}^k_\ell(M) \to \mathscr{T}^k_\ell(M) \tag{2.6}$$

where $X, Y \in \mathscr{X}(M)$ are given vector fields.

2.5.2.1 Notational Convention

It is important to note that we interpret

$$\nabla_X \nabla_Y F = \nabla^2_{X,Y} F = (\nabla\nabla F)(X, Y, \cdots) = (\nabla_X(\nabla F))(Y, \ldots),$$

whenever *no brackets* are specified; this is differs from $\nabla_X(\nabla_Y F)$ with brackets [KN96, pp. 124–125]. Furthermore, for notational simplicity we often write ∇_{∂_p} as just ∇_p and $(\nabla_{\partial_p} F)(dx^{i_1}, \ldots, dx^{i_\ell}, \partial_{j_1}, \ldots, \partial_{j_k})$ simply as $\nabla_p F^{i_1\cdots i_\ell}{}_{j_1\cdots j_k}$.

2.5.2.2 The Hessian

We define the Hessian of $f \in C^\infty(M)$ to be

$$\mathrm{Hess}(f) = \nabla df.$$

When applied to vector fields $X, Y \in \mathscr{X}(M)$, we find that $\mathrm{Hess}(f)(X,Y) = (\nabla df)(X,Y) = \nabla^2_{X,Y} f$. Also note that the Hessian is symmetric precisely when the connection is symmetric (see Example 2.30).

2.5.3 *Connections on Dual and Tensor Product Bundles*

So far we have looked at the covariant derivative on the tensor bundle $\bigotimes^k T^*M \otimes \bigotimes^\ell TM$. In fact much of the same structure works on a general vector bundle as well.

Proposition 2.32. *If ∇ is a connection on E, then there is a unique connection on E^*, also denoted by ∇, such that*

$$X(\omega(\xi)) = (\nabla_X \omega)(\xi) + \omega(\nabla_X \xi)$$

for any $\xi \in \Gamma(E)$, $\omega \in \Gamma(E^*)$ and $X \in \mathscr{X}(M)$.

Proposition 2.33. *If $\nabla^{(i)}$ is a connection on E_i for $i = 1, 2$, then there is a unique connection ∇ on $E_1 \otimes E_2$ such that*

$$\nabla_X(\xi_1 \otimes \xi_2) = (\nabla^{(1)}_X \xi_1) \otimes \xi_2 + \xi_1 \otimes (\nabla^{(2)}_X \xi_2)$$

for all $X \in \mathscr{X}(M)$ and $\xi_i \in \Gamma(E_i)$.

Propositions 2.32 and 2.33 define a canonical connection on any tensor bundle constructed from E by taking duals and tensor products. In particular, if $S \in \Gamma(E_1^* \otimes E_2)$ is an E_2-valued tensor acting on E_1, then $\nabla S \in \Gamma(T^*M \otimes E_1^* \otimes E_2)$ is given by

$$(\nabla_X S)(\xi) = {}^{E_2}\nabla_X \left(S(\xi)\right) - S\left({}^{E_1}\nabla_X \xi\right) \tag{2.7}$$

where $\xi \in \Gamma(E_1)$ and $X \in \mathscr{X}(M)$.

Moreover if we also have a connection $\widehat{\nabla}$ on TM, then $\nabla^2 S \in \Gamma(T^*M \otimes T^*M \otimes E_1^* \otimes E_2)$ – since we can construct this connection from the connections on TM, E_1 and E_2 by taking duals and tensor products. Explicitly,

$$\begin{aligned}(\nabla^2 S)(X,Y,\xi) &= {}^{E_2}\nabla_X((\nabla_Y S)(\xi)) - (\nabla_{\widehat{\nabla}_X Y} S)(\xi) - (\nabla_Y S)({}^{E_1}\nabla_X \xi) \\ &= \left(\nabla_X(\nabla_Y S)\right)(\xi) - (\nabla_{\widehat{\nabla}_X Y} S)(\xi)\end{aligned} \tag{2.8}$$

where $X, Y \in \mathscr{X}(M)$ and $\xi \in \Gamma(E_1)$.

2.5.4 The Levi–Civita Connection

When working on a Riemannian manifold, it is desirable to work with a particular connection that reflects the geometric properties of the metric. To do so, one needs the notions of compatibility and symmetry.

Definition 2.34. A connection ∇ on a vector bundle E is said to be *compatible* with a metric g on E if for any $\xi, \eta \in \Gamma(E)$ and $X \in \mathscr{X}(M)$,

$$X(g(\xi,\eta)) = g(\nabla_X \xi, \eta) + g(\xi, \nabla_X \eta).$$

Moreover, if ∇ is compatible with a metric g on E, then the induced connection on E^* is compatible with the induced metric on E^*. Also, if the connections on two vector bundles are compatible with given metrics, then the connection on the tensor product is compatible with the tensor product metric.

Unfortunately compatibility by itself is not enough to determine a unique connection. To get uniqueness we also need the connection to be symmetric.

Definition 2.35. A connection ∇ on TM is *symmetric* if its torsion vanishes.[5] That is, if $\nabla_X Y - \nabla_Y X = [X, Y]$ or equivalently $\Gamma_{ij}^k = \Gamma_{ji}^k$.

We can now state the fundamental theorem of Riemannian geometry.

Theorem 2.36. *Let (M, g) be a Riemannian manifold. There exists a unique connection ∇ on TM which is symmetric and compatible with g. This connection is referred to as* the Levi–Civita connection *of g.*

The reason why this connection has been anointed *the* Riemannian connection is that the symmetry and compatibility conditions are invariantly defined natural properties that force the connection to coincide with the tangential connection, whenever M is realised as a submanifold of $\mathbb{R}^n$ with the induced metric (which is always possible by the Nash embedding).

Proposition 2.37. *In local coordinate (x^i), the Christoffel symbols of the Levi–Civita connection are given by*

$$\Gamma_{ij}^k = \frac{1}{2} g^{k\ell} (\partial_j g_{i\ell} + \partial_i g_{j\ell} - \partial_\ell g_{ij}). \tag{2.9}$$

[5] The torsion τ of ∇ is defined by $\tau(X,Y) = \nabla_X Y - \nabla_Y X - [X,Y]$. τ is a $(2,1)$-tensor field since $\nabla_{fX}(gY) - \nabla_{gY} fX - [fX, gY] = fg(\nabla_X Y - \nabla_Y X - [X,Y])$.

2.6 Connection Laplacian

In its simplest form, the Laplacian Δ of $f \in C^\infty(M)$ is defined by $\Delta f = \operatorname{div} \operatorname{grad} f$. In fact, the Laplacian can be extended to act on tensor bundles over a Riemannian manifold (M, g). The resulting differential operator is referred to as the *connection Laplacian.* Note that there are a number of other second-order, linear, elliptic differential operators bearing the name Laplacian which have alternative definitions.

Definition 2.38. For any tensor field $F \in \mathscr{T}_\ell^k(M)$, the connection Laplacian

$$\Delta F = \operatorname{tr}_g \nabla^2 F \tag{2.10}$$

is the trace of the second covariant derivative with the metric g.

Explicitly,

$$\begin{aligned}
(\Delta F)^{j_1 \ldots j_\ell}{}_{i_1 \ldots i_k} &= (\operatorname{tr}_g \nabla^2 F)^{j_1 \ldots j_\ell}{}_{i_1 \ldots i_k} \\
&= \left(\operatorname{tr}_{13} \operatorname{tr}_{24} g^{-1} \otimes \nabla^2 F\right)^{j_1 \ldots j_\ell}{}_{i_1 \ldots i_k} \\
&= g^{pq} (\nabla_{\partial_p} \nabla_{\partial_q} F)(\partial_{j_1}, \ldots, \partial_{j_\ell}, dx^{i_1}, \ldots, dx^{i_k}).
\end{aligned}$$

Example 2.39. If the tensor bundle is $T^0 M = C^\infty(M)$, then (2.5) implies that

$$\begin{aligned}
\Delta f &= g^{ij} \nabla_{\partial_i} \nabla_{\partial_j} f \\
&= g^{ij} \left(\frac{\partial^2 f}{\partial x^i \partial x^j} - \Gamma_{ij}^k \frac{\partial f}{\partial x^k} \right).
\end{aligned}$$

2.7 Curvature

We introduce the curvature tensor as a purely algebraic object that arises from a connection on a vector bundle. From this we will look at the curvature on specific bundle structures.

2.7.1 Curvature on Vector Bundles

Definition 2.40. Let E be a vector bundle over M. If ∇ is a connection on E, then the curvature of the connection ∇ on the bundle E is the section $R_\nabla \in \Gamma(T^*M \otimes T^*M \otimes E^* \otimes E)$ defined by

$$R_\nabla(X, Y)\xi = \nabla_Y (\nabla_X \xi) - \nabla_X (\nabla_Y \xi) + \nabla_{[X,Y]} \xi. \tag{2.11}$$

In the literature, there is much variation in the sign convention; some define the curvature to be of opposite sign to ours.

2.7.2 Curvature on Dual and Tensor Product Bundles

The curvature on a dual bundle E^*, with respect to the dual connection, is characterised by the formula

$$0 = (R(X,Y)\omega)(\xi) + \omega(R(X,Y)\xi)$$

for all $X, Y \in \mathscr{X}(M)$, $\omega \in \Gamma(E^*)$ and $\xi \in \Gamma(E)$.

The curvature on a tensor product bundle $E_1 \otimes E_2$, with connection ∇ given by Proposition 2.33, can be computed in terms of the curvatures on each of the factors by the formula

$$R_\nabla(X,Y)(\xi_1 \otimes \xi_2) = (R_{\nabla^{(1)}}(X,Y)\xi_1) \otimes \xi_2 + \xi_1 \otimes (R_{\nabla^{(2)}}(X,Y)\xi_2)\,,$$

where $X, Y \in \mathscr{X}(M)$ and $\xi_i \in \Gamma(E_i)$, $i = 1, 2$.

Example 2.41. Of particular interest, the curvature on $E_1^* \otimes E_2$ (E_2-valued tensors acting on E_1) is given by

$$(R(X,Y)S)(\xi) = R_{\nabla^{(2)}}(X,Y)\big(S(\xi)\big) - S\big(R_{\nabla^{(1)}}(X,Y)\xi\big), \tag{2.12}$$

where $S \in \Gamma(E_1^* \otimes E_2)$, $\xi \in \Gamma(E_1)$ and $X, Y \in \mathscr{X}(M)$.

2.7.3 Curvature on the Tensor Bundle

One of the most important applications is the curvature of the tensor bundle. By (2.11) and Sect. 2.5.1, we have the following:

Proposition 2.42. *Let R be the curvature on the (k,ℓ)-tensor bundle. If $F, G \in \mathscr{T}_\ell^k(M)$ are tensors, then*

$$R(X,Y)(\operatorname{tr} F) = \operatorname{tr}(R(X,Y)F)$$
$$R(X,Y)(F \otimes G) = (R(X,Y)F) \otimes G + F \otimes (R(X,Y)G)$$

for any vector fields X and Y.

Moreover we also have the following important formulas.

Proposition 2.43. *Let R be the curvature on the (k,ℓ)-tensor bundle. If $F \in \mathscr{T}_\ell^k(M)$, then*

$$\begin{aligned} R(X,Y)\big(F(\omega^1,\dots,\omega^\ell,Z_1,\dots,Z_k)\big) &= (R(X,Y)F)(\omega^1,\dots,\omega^\ell,Z_1,\dots,Z_k) \\ &\quad+\sum_{j=1}^{\ell} F(\omega^1,\dots,R(X,Y)\omega^j,\dots,\omega^\ell,Z_1,\dots,Z_k) \\ &\quad+\sum_{i=1}^{k} F(\omega^1,\dots,\omega^\ell,Z_1,\dots,R(X,Y)Z_i,\dots,Z_k) \end{aligned}$$

for any vector fields X, Y, Z_i and 1-forms ω^j.

Proof. Let the vector bundle $E = T_\ell^k M$, so for any $\xi \in \Gamma(E)$ we find that

$$\begin{aligned} R(X,Y)\big(F(\xi)\big) &= R(X,Y)\big(\operatorname{tr} F \otimes \xi\big) \\ &= \operatorname{tr}\big[(R(X,Y)F)\otimes\xi + F\otimes(R(X,Y)\xi)\big] \\ &= (R(X,Y)F)(\xi) + F(R(X,Y)\xi). \end{aligned}$$

As ξ takes the form

$$\xi = \omega^1 \otimes \cdots \otimes \omega^\ell \otimes Z_1 \otimes \cdots \otimes Z_k,$$

a similar argument shows that $F(R(X,Y)\xi) = F(R(X,Y)\omega^1,\dots,Z_k) + \cdots + F(\omega^1,\dots,R(X,Y)Z_k)$ from which the result follows. □

Proposition 2.44. *Let R be the curvature on the (k,ℓ)-tensor bundle. If the connection ∇ on TM is symmetric, then*

$$R(X,Y) = \nabla^2_{Y,X} - \nabla^2_{X,Y} \tag{2.13}$$

and so $R(X,Y) : \mathscr{T}_\ell^k(M) \to \mathscr{T}_\ell^k(M)$.

Proof. For any (k,ℓ)-tensor F, we see by (2.6) that

$$\begin{aligned} \nabla^2_{Y,X}F - \nabla^2_{X,Y}F &= \nabla_Y(\nabla_X F) - \nabla_{\nabla_Y X}F - \nabla_X(\nabla_Y F) + \nabla_{\nabla_X Y}F \\ &= \nabla_Y(\nabla_X F) - \nabla_X(\nabla_Y F) + \nabla_{[X,Y]}F. \end{aligned}$$

□

Example 2.45. The curvature of $C^\infty(M) = T^0 M$ vanishes since

$$R(\partial_i,\partial_j)f = \nabla_{\partial_i}(\nabla_{\partial_j}f) - \nabla_{\partial_j}(\nabla_{\partial_i}f) + \cancel{\nabla_{[\partial_i,\partial_j]}f} = \partial_i\partial_j f - \partial_j\partial_i f = 0$$

for any $f \in C^\infty(M)$.

2.7.4 Riemannian Curvature

If (M, g) is a Riemannian manifold, the curvature $R \in \Gamma(\otimes^3 T^*M \otimes TM)$ of the Levi–Civita connection ∇ on TM is a $(3,1)$-tensor field that, in local coordinates (x^i), takes the form

$$R = R_{ijk}{}^{\ell}\, dx^i \otimes dx^j \otimes dx^k \otimes \partial_\ell,$$

where $R(\partial_i, \partial_j)\partial_k = R_{ijk}{}^{\ell}\partial_\ell$. Accompanying this is the *Riemann curvature tensor*, also denoted by R. It is a covariant $(4,0)$-tensor field defined by

$$R(X, Y, Z, W) = g(R(X, Y)Z, W)$$

for all $W, X, Y, Z \in \mathscr{X}(M)$. In local coordinates (x^i) it can be expressed as

$$R = R_{ijk\ell}\, dx^i \otimes dx^j \otimes dx^k \otimes dx^\ell,$$

where $R_{ijk\ell} = g_{\ell p} R_{ijk}{}^{p}$.

Lemma 2.46. *In local coordinates (x^i), the curvature of the Levi–Civita connection can be expressed as follows:*

$$R_{ijk}{}^{\ell} = \partial_j \Gamma^\ell_{ik} - \partial_i \Gamma^\ell_{jk} + \Gamma^m_{ik}\Gamma^\ell_{jm} - \Gamma^m_{jk}\Gamma^\ell_{im}$$

$$R_{ijk\ell} = \frac{1}{2}\big(\partial_j\partial_k g_{i\ell} + \partial_i\partial_\ell g_{jk} - \partial_i\partial_k g_{j\ell} - \partial_j\partial_\ell g_{ik}\big) + g_{\ell p}(\Gamma^m_{ik}\Gamma^p_{jm} - \Gamma^m_{jk}\Gamma^p_{im})$$

2.7.4.1 Symmetries of the Curvature Tensor

The curvature tensor possesses a number of important symmetry properties. They are:

(a) Antisymmetric in first two arguments: $R_{ijk\ell} + R_{jik\ell} = 0$
(b) Antisymmetric in last two arguments: $R_{ijk\ell} + R_{ji\ell k} = 0$
(c) Symmetry between the first and last pair of arguments: $R_{ijk\ell} = R_{k\ell ij}$

In addition to this, there are also the 'cyclic' Bianchi identities:

(d) First Bianchi identity: $R_{ijk\ell} + R_{jki\ell} + R_{kij\ell} = 0$
(e) Second Bianchi identity: $\nabla_m R_{ijk\ell} + \nabla_k R_{ij\ell m} + \nabla_\ell R_{ijmk} = 0$

2.7.5 Ricci and Scalar Curvature

As the curvature tensor can be quite complicated, it is useful to consider various contractions that summarise some of the information contained in the curvature tensor. The first of these contractions is the *Ricci tensor*, denoted by Ric. It is defined as

$$\operatorname{Ric}(X,Y) = \operatorname{tr}_g R(X,\cdot,Y,\cdot) = (\operatorname{tr}_{14}\operatorname{tr}_{26}\, g^{-1} \otimes R)(X,Y).$$

In component form, $\operatorname{Ric}(\partial_i, \partial_j) = R_{ij} = R_{ikj}{}^{k} = g^{pq}R_{ipjq}$. From the symmetry properties of R it is clear that Ric is symmetric.

A further trace of the Ricci tensor gives a scalar quantity called the *scalar curvature*, denoted by Scal:

$$\operatorname{Scal} = \operatorname{tr}_g \operatorname{Ric} = \operatorname{Ric}_i{}^i = g^{ij}R_{ij}.$$

It is important to note that if the curvature tensor is defined with opposite sign, the contraction is defined so that the Ricci tensor matches the one given here. Hence the Ricci tensor has the same meaning for everyone. By Lemma 2.46, the Ricci tensor can be expressed locally as follows.

Lemma 2.47. *In local coordinates (x^i), the Ricci tensor takes the form*

$$R_{ik} = \frac{1}{2} g^{j\ell}\left(\frac{\partial^2 g_{i\ell}}{\partial x^j \partial x^k} + \frac{\partial^2 g_{jk}}{\partial x^i \partial x^\ell} - \frac{\partial^2 g_{j\ell}}{\partial x^i \partial x^k} - \frac{\partial^2 g_{ik}}{\partial x^j \partial x^\ell}\right) + \Gamma^m_{ik}\Gamma^j_{jm} - \Gamma^m_{jk}\Gamma^j_{im}.$$

2.7.5.1 Contraction Commuting with Covariant Derivative

As we are working with a compatible connection, $\nabla g \equiv 0$. Thus one can commute covariant derivatives with metric contractions.

Proposition 2.48. *If ∇ is the Levi–Civita connection, then*

$$\nabla_k R_{ij} = g^{pq}\,\nabla_k R_{ipjq} \tag{2.14}$$

$$\nabla^2_{k,\ell} R_{ij} = g^{pq}\,\nabla^2_{k,\ell} R_{ipjq}. \tag{2.15}$$

Proof. To show (2.14), let $X, Y, Z \in \mathscr{X}(M)$ so that

$$\begin{aligned}
(\nabla_Z \operatorname{Ric})(X,Y) &= (\nabla_Z\,(\operatorname{tr} g^{-1} \otimes R))(X,Y) \\
&= \big(\operatorname{tr} \nabla_Z (g^{-1} \otimes R)\big)(X,Y) \\
&= \big(\operatorname{tr} \cancel{\nabla_Z g^{-1}} \otimes R + \operatorname{tr} g^{-1} \otimes \nabla_Z R\big)(X,Y) \\
&= (\operatorname{tr}_{14}\operatorname{tr}_{26}\, g^{-1} \otimes \nabla_Z R)(X,Y) \\
&= (\operatorname{tr}_g \nabla_Z R)(X,\cdot,Y,\cdot).
\end{aligned}$$

Similarly, to show (2.15) note that

$$\nabla^2 \mathrm{Ric} = \nabla^2(\mathrm{tr}\, g^{-1} \otimes R) = \mathrm{tr}\left(\nabla^2 g^{-1} \otimes R + 2\,\nabla g^{-1} \otimes \nabla R + g^{-1} \otimes \nabla^2 R\right)$$
$$= \mathrm{tr}\, g^{-1} \otimes \nabla^2 R. \qquad \square$$

In later applications we will need the *contracted second Bianchi identity*:

$$g^{jk}\nabla_k \mathrm{Ric}_{ij} = \frac{1}{2}\nabla_i \mathrm{Scal}. \tag{2.16}$$

This follows easily from (2.14) and the second Bianchi identity, since

$$\begin{aligned} 0 &= g^{am}g^{bn}(\nabla_\ell R_{abmn} + \nabla_m R_{abn\ell} + \nabla_n R_{ab\ell m}) \\ &= g^{am}(\nabla_\ell \mathrm{Ric}_{am} - \nabla_m \mathrm{Ric}_{a\ell}) + g^{am}g^{bn}\nabla_n R_{ab\ell m} \\ &= \nabla_\ell \mathrm{Ric}_a{}^a - g^{am}\nabla_m \mathrm{Ric}_{a\ell} - g^{bn}\nabla_n R_{\ell mb}{}^m \\ &= \nabla_\ell \mathrm{Scal} - g^{am}\nabla_m \mathrm{Ric}_{a\ell} - g^{bn}\nabla_n \mathrm{Ric}_{\ell b} \end{aligned}$$

from which (2.16) now follows.

2.7.6 Sectional Curvature

Suppose (M, g) is a Riemannian manifold. If Π is a two-dimensional subspace of T_pM, we define the *sectional curvature* K of Π to be

$$K(\Pi) = R(e_1, e_2, e_1, e_2),$$

where $\{e_1, e_2\}$ is an orthonormal basis for Π. By a rotation or reflection in the plane, one can show K is independent of the choice of basis. We refer to the oriented plane generated from e_i and e_j by the notation $e_i \wedge e_j$ (cf. Sect. C.3). Furthermore if $\{u, v\}$ is any basis for the 2-plane Π, one has

$$K(u \wedge v) = \frac{R(u, v, u, v)}{|u|^2|v|^2 - g(u, v)^2}.$$

If $U \subset T_pM$ is a neighbourhood of zero on which $\exp_p$ is a diffeomorphism, then $S_\Pi := \exp_p(\Pi \cap U)$ is a 2-dimensional submanifold of M containing p, called the *plane of section* determined by Π. That is, it is the surface swept out by geodesics whose initial tangent vectors lie in Π. One can geometrically interpret the sectional curvature of M associated to Π to be the Gaussian curvature of the surface S_Π at p with the induced metric.

By computing the sectional curvature of the plane $\frac{1}{2}(e_i + e_k) \wedge (e_j + e_\ell)$ one can show:

Proposition 2.49. *The curvature tensor R is completely determined by the sectional curvature. In particular,*

$$
\begin{aligned}
R_{ijk\ell} = &\frac{1}{3}K\Big(\frac{(e_i+e_k)\wedge(e_j+e_\ell)}{2}\Big) + \frac{1}{3}K\Big(\frac{(e_i-e_k)\wedge(e_j-e_\ell)}{2}\Big) \\
&-\frac{1}{3}K\Big(\frac{(e_j+e_k)\wedge(e_i+e_\ell)}{2}\Big) - \frac{1}{3}K\Big(\frac{(e_j-e_k)\wedge(e_i-e_\ell)}{2}\Big) \\
&-\frac{1}{6}K(e_j\wedge e_\ell) - \frac{1}{6}K(e_i\wedge e_k) + \frac{1}{6}K(e_i\wedge e_\ell) + \frac{1}{6}K(e_j\wedge e_k).
\end{aligned}
$$

One can also show that the scalar curvature $\mathrm{Scal} = \mathrm{Ric}_j{}^j = \sum_{j\neq k} K(e_j \wedge e_k)$, where (e_i) is orthonormal basis for T_pM.

Each of the model spaces $\mathbb{R}^n$, S^n and $\mathbb{H}^n$ has an isometry group that acts transitively on orthonormal frames, and so acts transitively on 2-planes in the tangent bundle. Therefore each has a *constant sectional curvature* – in the sense that the sectional curvatures are the same for all planes at all points.

It is well known that the Euclidean space $\mathbb{R}^n$ has constant zero sectional curvature (this is geometrically intuitive as each 2-plane section has zero Gaussian curvature). The sphere S^n of radius 1 has constant sectional curvature equal to 1 and the hyperbolic space $\mathbb{H}^n$ has constant sectional curvature equal to -1.

2.7.7 Berger's Lemma

A simple but important result is the so-called lemma of Berger [Ber60b, Sect. 6]. Following [Kar70], we show the curvature can be bounded whenever the sectional curvature is bounded from above and below. That is, if a Riemannian manifold $(M, \langle\cdot,\cdot\rangle)$ has sectional curvature bounds $\delta = \min_{u,v\in T_pM} K(u\wedge v)$ and $\Delta = \max_{u,v\in T_pM} K(u\wedge v)$ with the assumption $\delta \geq 0$, we prove the following bounds on the curvature tensor:

Lemma 2.50 (Berger). *For orthonormal $u, v, w, x \in T_pM$, one can bound the curvature tensor by*

$$|R(u,v,w,v)| \leq \frac{1}{2}(\Delta - \delta) \tag{2.17}$$

$$|R(u,v,w,x)| \leq \frac{2}{3}(\Delta - \delta). \tag{2.18}$$

Proof. From the symmetries of the curvature tensor we find that:

$$\begin{aligned} 4R(u,v,w,v) &= R(u+w,v,u+w,v) - R(u-w,v,u-w,v) \\ 6R(u,v,w,x) &= R(u,v+x,w,v+x) - R(u,v-x,w,v-x) \\ &\quad - R(v,u+x,w,u+x) + R(v,u-x,w,u-x). \end{aligned}$$

Using the definition of the sectional curvature together with the first identity gives

$$\begin{aligned} 2|R(u,v,w,v)| &= \frac{1}{2}\Big|K\big((u+w)\wedge v\big)\big(|u+w|^2|v|^2 - \cancel{\langle u+w,v\rangle}\big) \\ &\quad - K\big((u-w)\wedge v\big)\big(|u-w|^2|v|^2 - \cancel{\langle u-w,v\rangle}\big)\Big| \\ &= \big|K\big((u+w)\wedge v\big) - K\big((u-w)\wedge v\big)\big| \\ &\leq (\Delta - \delta) \end{aligned}$$

which is identity (2.17). To prove (2.18), use the second identity and apply (2.17) to the four terms – whilst taking into consideration $|v \pm x|^2 = |u \pm x|^2 = 2$, for orthonormal u, v, w and x. □

2.8 Pullback Bundle Structure

Let M and N be smooth manifolds, let E be a vector bundle over N and f be a smooth map from M to N.

Definition 2.51. The pullback bundle of E by f, denoted f^*E, is the smooth vector bundle over M defined by $f^*E = \{(p,\xi) : p \in M, \xi \in E, \pi(\xi) = f(p)\}$. If $\xi_1, \ldots, \xi_k$ are a local frame for E near $f(p) \in N$, then $\Xi_i(p) = \xi_i(f(p))$ are a local frame for $f^*(E)$ near p.

Lemma 2.52. *Pullbacks commute with taking duals and tensor products:*

$$(f^*E)^* = f^*(E^*) \qquad \text{and} \qquad (f^*E_1) \otimes (f^*E_2) = f^*(E_1 \otimes E_2).$$

2.8.1 Restrictions

The *restriction* $\xi_f \in \Gamma(f^*E)$ of $\xi \in \Gamma(E)$ to f is defined by

$$\xi_f(p) = \xi(f(p)) \in E_{f(p)} = (f^*E)_p,$$

for all points $p \in M$.

Example 2.53. Suppose g is a metric on E. Then $g \in \Gamma(E^* \otimes E^*)$, and by restriction we obtain $g_f \in \Gamma((f^*E)^* \otimes (f^*E)^*)$, which is a metric on f^*E (the 'restriction of g to f'): If $\xi, \eta \in (f^*E)_p = E_{f(p)}$, then $(g_f(p))(\xi, \eta) = (g(f(p))(\xi, \eta)$.

Remark 2.54. In using this terminology one wants to distinguish the restriction of a tensor field on E (which is a section of a tensor bundle over f^*E) with the pullback of a tensor on the tangent bundle, which is discussed below. Thus the metric in the above example should not be called the 'pullback metric'. Notice that we can restrict both covariant and contravariant tensors, in contrast to the situation with pullbacks.

2.8.2 Pushforwards

If $f : M \to N$ is smooth, then for each $p \in M$, we have the linear map $f_*(p) : T_pM \to T_{f(p)}N = (f^*TN)_p$. That is, $f_*(p) \in T_p^*M \otimes (f^*TN)_p$, so f_* is a smooth section of $T^*M \otimes f^*TN$. Given a section $X \in \Gamma(TM) = \mathscr{X}(M)$, the *pushforward* of X is the section $f_*X \in \Gamma(f^*TN)$ given by applying f_* to X.

2.8.3 Pullbacks of Tensors

By duality (combined with restriction) we can define an operation taking $(k, 0)$-tensors on N to $(k, 0)$-tensors on M, which we call the *pullback* operation: If S is a $(k, 0)$-tensor on N (i.e. $S \in \Gamma(\otimes^k T^*N)$), then by restriction we have $S_f \in \Gamma(\otimes^k(f^*T^*N))$, and we define $f^*S \in \Gamma(\otimes^k T^*M)$ by

$$f^*S(X_1, \ldots, X_k) = S_f(f_*X_1, \ldots, f_*X_k) \tag{2.19}$$

where X_i are vector fields on M.

Example 2.55. If f is an embedding and g is a Riemannian metric on N, then f^*g is the pullback metric on M (often called the 'induced metric').

This definition can be extended a little to include bundle-valued tensors: Suppose $S \in \Gamma(\otimes^k T^*N \otimes E)$ is an E-valued $(k, 0)$-tensor field on N. Then restriction gives $S_f \in \Gamma(\otimes^k(f^*T^*N) \otimes f^*E)$, and we will denote by f^*S the f^*E-valued k-tensor on M defined by

$$f^*S(X_1, \ldots, X_k) = S_f(f^*X_1, \ldots, f^*X_k).$$

That is, the same formula as before except now both sides are f^*E-valued.

2.8.4 The Pullback Connection

Let ∇ be a connection on E over N, and $f : M \to N$ a smooth map.

Theorem 2.56. *There is a unique connection* ${}^f\nabla$ *on* f^*E*, referred to as the pullback connection, such that*

$$^f\nabla_v(\xi_f) = \nabla_{f_*v}\xi$$

for any $v \in TM$ *and* $\xi \in \Gamma(E)$.

Remark 2.57. To justify the term 'pullback connection': If $\xi \in \Gamma(E)$, then $\nabla\xi \in \Gamma(T^*N \otimes E)$, so the pullback gives

$$^f\nabla\xi_f := f^*(\nabla\xi) \in \Gamma(T^*M \otimes f^*E).$$

To define ${}^f\nabla_v\xi$ for arbitrary $\xi \in \Gamma(f^*E)$, we fix $p \in M$ and choose a local frame $\sigma_1, \ldots, \sigma_k$ about $f(p)$ for E. Then we can write $\xi = \sum_{i=1}^k \xi^i\,(\sigma_i)_f$ with each ξ^i a smooth function defined near p, so the rules for a connection together with the pullback connection condition give

$$\begin{aligned} {}^f\nabla_v\xi &= {}^f\nabla_v(\xi^i\,(\sigma_i)_f) \\ &= \xi^i\,{}^f\nabla_v(\sigma_i)_f + v(\xi^i)\,(\sigma_i)_f \\ &= \xi^i\nabla_{f_*v}\sigma_i + v(\xi^i)\,(\sigma_i)_f. \end{aligned} \tag{2.20}$$

Note that pullback connections on duals and tensor products of pullback bundles agree with those obtained by applying the previous constructions to the pullback bundles on the factors.

There are two other important properties of the pullback connection:

Proposition 2.58. *If* g *is a metric on* E *and* ∇ *is a connection on* E *compatible with* g*, then* ${}^f\nabla$ *is compatible with the restriction metric* g_f.

Proof. As ∇ is compatible with g if and only if $\nabla g = 0$, we therefore must show that ${}^f\nabla g_f = 0$ if $\nabla g = 0$. However this is immediate, since ${}^f\nabla_v(g_f) = \nabla_{f_*v}g = 0$. □

Proposition 2.59. *The curvature of the pullback connection is the pullback of the curvature of the original connection. That is,*

$$R_{^f\nabla}(X, Y)\xi_f = (f^*R_\nabla)(X, Y)\xi$$

where $X, Y \in \mathscr{X}(M)$ *and* $\xi \in \Gamma(E)$. *Note that* $R_\nabla \in \Gamma(T^*N \otimes T^*N \otimes E^* \otimes E)$ *here, so that*

$$\begin{aligned} f^*(R_\nabla) &\in \Gamma(T^*M \otimes T^*M \otimes f^*(E^* \otimes E)) \\ &= \Gamma(T^*M \otimes T^*M \otimes (f^*E)^* \otimes f^*E). \end{aligned}$$

Proof. Since curvature is tensorial, it is enough to check the formula for a basis. Choose a local frame $\{\sigma_p\}_{p=1}^k$ for E, so that $\{(\sigma_p)_f\}$ is a local frame for f^*E. Also choose local coordinates $\{y^\alpha\}$ for N near $f(p)$ and $\{x^i\}$ for M near p, and write $f^\alpha = y^\alpha \circ f$. Then

$$\begin{aligned} R_{{}^f\nabla}(\partial_i, \partial_j)(\sigma_p)_f &= {}^f\nabla_{\partial_j}\left({}^f\nabla_{\partial_i}(\sigma_p)_f\right) - (i \leftrightarrow j) \\ &= {}^f\nabla_j\left(\nabla_{f_*\partial_i}\sigma_p\right) - (i \leftrightarrow j) \\ &= {}^f\nabla_j\left(\partial_i f^\alpha \nabla_\alpha \sigma_p\right) - (i \leftrightarrow j) \\ &= (\partial_j \partial_i f^\alpha)\nabla_\alpha \sigma_p + \partial_i f^\alpha\, {}^f\nabla_j\left((\nabla_\alpha \sigma_p)_f\right) - (i \leftrightarrow j) \\ &= \partial_i f^\alpha \nabla_{f_*\partial_j}\left(\nabla_\alpha \sigma_p\right) - (i \leftrightarrow j) \\ &= \partial_i f^\alpha \partial_j f^\beta \left(\nabla_\beta\left(\nabla_\alpha \sigma_p\right) - (\alpha \leftrightarrow \beta)\right) \\ &= \partial_i f^\alpha \partial_j f^\beta R_\nabla(\partial_\alpha, \partial_\beta)\sigma_p \\ &= R_\nabla(f_*\partial_i, f_*\partial_j)\sigma_p. \end{aligned}$$

□

When pulling back a tangent bundle, there is another important property:

Proposition 2.60. *If ∇ is a symmetric connection on TN, then the pull-back connection ${}^f\nabla$ on f^*TN is symmetric in the sense that*

$${}^f\nabla_U(f_*V) - {}^f\nabla_V(f_*U) = f_*([U,V])$$

for any $U, V \in \Gamma(TM)$.

Proof. As before choose local coordinates x^i for M near p, and y^α for N near $f(p)$. By writing $U = U^i\partial_i$ and $V = V^j\partial_j$ we find that

$$\begin{aligned} &{}^f\nabla_U(f_*V) - {}^f\nabla_V(f_*U) \\ &\quad = {}^f\nabla_U\left(V^j\partial_j f^\alpha \partial_\alpha\right) - (U \leftrightarrow V) \\ &\quad = U^i\partial_i\left(V^j\partial_j f^\alpha\right)\partial_\alpha + V^j\partial_j f^\alpha\, {}^f\nabla_U \partial_\alpha - (U \leftrightarrow V) \\ &\quad = (U^i\partial_i V^j - V^i\partial_i U^j)\partial_j f^\alpha \partial_\alpha + U^i V^j(\partial_i\partial_j f^\alpha - \partial_j\partial_i f^\alpha)\partial_\alpha \\ &\qquad + V^j U^i \partial_j f^\alpha \partial_i f^\beta\left(\nabla_\beta \partial_\alpha - \nabla_\alpha \partial_\beta\right) \\ &\quad = f_*\left([U,V]\right). \end{aligned}$$

□

2.8.5 Parallel Transport

Parallel transport is a way of using a connection to compare geometrical data at different points along smooth curves:

Let ∇ be a connection on the bundle $\pi : E \to M$. If $\gamma : I \to M$ is a smooth curve, then a *smooth section along* γ is a section of γ^*E. Associated to this is the pullback connection ${}^{\gamma}\nabla$. A section V along a curve γ is *parallel along* γ if ${}^{\gamma}\nabla_{\partial_t} V \equiv 0$. In a local frame (e_j) over a neighbourhood of $\gamma(t_0)$:

$$ {}^{\gamma}\nabla_{\partial_t} V(t_0) = \big(\dot{V}^k(t_0) + \Gamma_{ij}^k(\gamma(t_0))V^j(t_0)\dot{\gamma}^i(t_0)\big)e_k, $$

where $V(t) = V^j(t)e_j$ and Γ_{ij}^k is the Christoffel symbol of ∇ in this frame.

Theorem 2.61. *Given a curve $\gamma : I \to M$ and a vector $V_0 \in E_{\gamma(0)}$, there exists a unique parallel section V along γ such that $V(0) = V_0$. Such a V is called the* parallel translate *of V_0 along γ.*

Moreover, for such a curve γ there exists a unique family of linear isomorphisms $P_t : E_{\gamma(0)} \to E_{\gamma(t)}$ such that a vector field V along γ is parallel if and only if $V(t) = P_t(V_0)$ for all t.

2.8.6 Product Manifolds' Tangent Space Decomposition

Given manifolds M_1 and M_2, let $\pi_j : M_1 \times M_2 \to M_j$ be the standard smooth projection maps. The pushforward $(\pi_j)_* : T(M_1 \times M_2) \to TM_j$ over $\pi_j : M_1 \times M_2 \to M_j$ induces, via the pullback bundle, a smooth bundle morphism

$$ \Pi_j : T(M_1 \times M_2) \to \pi_j^*(TM_j) $$

over $M_1 \times M_2$. In which case one has the bundle morphism

$$ \Pi_1 \oplus \Pi_2 : T(M_1 \times M_2) \to \pi_1^*(TM_1) \oplus \pi_2^*(TM_2) $$

over $M_1 \times M_2$. On fibres over (x_1, x_2) this is simply the pointwise isomorphism $T_{(x_1,x_2)}(M_1 \times M_2) \simeq T_{x_1}M_1 \oplus T_{x_2}M_2$, so $\Pi_1 \oplus \Pi_2$ is in fact a smooth bundle isomorphism. Furthermore, $\Pi_j : T(M_1 \times M_2) \to \pi_j^*(TM_j)$ is bundle surjection as π_j is a projection. Thus by Proposition 2.14 there exists a well-defined subbundle E_1 inside $T(M_1 \times M_2)$ given by

$$ E_1 = \ker \Pi_1 = \{v \in T(M_1 \times M_2) : \Pi_1(v) = 0\}. $$

We observe that this is in fact equal to $\pi_2^*(TM_2)$, since E_1 consists of all vectors such that Π_1 projection vanishes. Therefore one has isomorphic vector bundles

$$ T(M_1 \times M_2) \simeq \ker \Pi_1 \oplus \ker \Pi_2. $$

Example 2.62. If we let $M_1 = M$ and $M_2 = \mathbb{R}$ with $\pi_1 = \pi$ projection from $M \times \mathbb{R}$ onto M and $\pi_2 = t$ be the projection from $M \times \mathbb{R}$ onto $\mathbb{R}$, then on fibres over $(x_1, x_2) = (x_0, t_0)$ we have that

$$\Pi_2(v) = t_*(v) = dt(v)$$

since t is a $\mathbb{R}$-valued function on $M \times \mathbb{R}$. So by letting $\mathfrak{S} = \ker dt = \{v \in T(M \times \mathbb{R}) : dt(v) = 0\}$ we have that

$$T(M \times \mathbb{R}) \simeq \mathfrak{S} \oplus \mathbb{R}\, \partial_t, \tag{2.21}$$

since $(\pi_2^* T\mathbb{R})_{(x_0,t_0)} = (t^* T\mathbb{R})_{(x_0,t_0)} = T_{t_0}\mathbb{R} = \mathbb{R}\, \partial_t|_{t_0}$ where $\partial_t|_{t_0}$ is the standard coordinate basis for $T_{t_0}\mathbb{R}$.

2.8.7 Connections and Metrics on Subbundles

Suppose F is a subbundle of a vector bundle E over a manifold M, as defined in Definition 2.12. If E is equipped with a metric g, then there is a natural metric induced on F by the inclusion: If $\iota : F \to E$ is the inclusion of F in E, then the induced metric on F is defined by $g_F(\xi, \eta) = g(\iota(\xi), \iota(\eta))$.

There is not in general any natural way to induce a connection on F from a connection ∇ on E. We will consider only the following special case:[6]

Definition 2.63. A subbundle F of a vector bundle E is called *parallel* if F is invariant under parallel transport, i.e. for any smooth curve $\sigma : [0,1] \to M$, and any parallel section ξ of $\sigma^* E$ over $[0,1]$ with $\xi(0) \in F_{\sigma(0)}$, we have $\xi(t) \in F_{\sigma(t)}$ for all $t \in [0,1]$.

One can check that a subbundle F is parallel if and only if the connection on E maps sections of F to F, i.e. $\nabla_u(\iota\xi) \in \iota(F_p)$ for any $u \in T_pM$ and $\xi \in \Gamma(F)$. If F is parallel there is a unique connection ∇^F on F such that

$$\iota \nabla^F_u \xi = \nabla_u (\iota\xi)$$

for every $u \in TM$ and $\xi \in \Gamma(F)$. Note also that if ∇ is compatible with a metric g on E, and F is a parallel subbundle of E, then ∇^F is compatible with the induced metric g_F.

Example 2.64. An important example which will reappear later (cf. Sect. 7.5) is the following: Let E be a vector bundle with connection ∇. Then the bundle of symmetric 2-tensors on E is a parallel subbundle of the bundle of 2-tensors

[6] Although natural constructions can be done much more generally, for example when a pair of complementary subbundles is supplied.

on E. To prove this we need to check that $\nabla_U T$ is symmetric whenever T is a symmetric 2-tensor. But this is immediate: We have for any $X, Y \in \Gamma(E)$ and $U \in \mathscr{X}(M)$,

$$\begin{aligned}(\nabla_U T)(X,Y) &= U(T(X,Y)) - T(\nabla_U X, Y) - T(X, \nabla_U Y) \\ &= U(T(Y,X)) - T(\nabla_U Y, X) - T(Y, \nabla_U X) \\ &= (\nabla_U T)(Y,X),\end{aligned}$$

so $\nabla_U T$ is symmetric as required.

2.8.8 The Taylor Expansion of a Riemannian Metric

As an application of the pullback structure seen in this section, we compute the Taylor expansion of the Riemannian metric in exponential normal coordinates. In particular the curvature tensor is an obstruction to the existence of local coordinates in which the second derivatives of the metric tensor vanish.

Theorem 2.65. *Let (M, g) be a Riemannian manifold. With respect to a geodesic normal coordinates system about $p \in M$, the metric g_{ij} may be expressed as:*

$$g_{ij}(u^1, \ldots, u^n) = \delta_{ij} - \frac{1}{3} R_{ikj\ell} u^k u^\ell + O(\|u\|^3).$$

Remark 2.66. The proof produces a complete Taylor expansion about $u = 0$ (see also [LP87, pp. 60-1]).

Proof. Consider $\varphi : \mathbb{R}^2 \to M$ defined by $\varphi(s,t) = \exp_p(tV(s))$ where $V(s) \in S^{n-1} \subset T_pM$. Let $V(0) = u$, $V'(0) = v$.

As $\varphi_* : T_{(s,t)}\mathbb{R}^2 \to (\varphi^*TM)_{(s,t)}$, $\varphi_*\partial_s = (\exp_p)_*(tV'(s)) = t\,\partial_s V^i(s)(\partial_i)_\varphi$. Thus we find that the pullback metric of g via φ is

$$\begin{aligned}(\varphi^* g)(\partial_s, \partial_s)\big|_{(s,t)} &= g_\varphi(\varphi_*\partial_s, \varphi_*\partial_s)\big|_{(s,t)} \\ &= t^2\, \partial_s V^i \partial_s V^j g_{ij}(\varphi(s,t)).\end{aligned} \tag{2.22}$$

Note that ${}^\varphi\nabla g_\varphi \in \Gamma(T\mathbb{R}^2 \otimes (\varphi^*TM)^* \otimes (\varphi^*TM)^*)$, so by Proposition 2.58:

$$\begin{aligned}0 &= ({}^\varphi\nabla g_\varphi)(\partial_t, \varphi_*\partial_s, \varphi_*\partial_s) \\ &= \partial_t g_\varphi(\varphi_*\partial_s, \varphi_*\partial_s) - g_\varphi({}^\varphi\nabla_{\partial_t}\varphi_*\partial_s, \varphi_*\partial_s) - g_\varphi(\varphi_*\partial_s, {}^\varphi\nabla_{\partial_t}\varphi_*\partial_s).\end{aligned}$$

So by taking $(\partial_t)^k$ derivatives of (2.22), we find that

$$
\begin{aligned}
&(\partial_t)^k g_\varphi(\varphi_*\partial_s, \varphi_*\partial_s) \\
&= \sum_{\ell=0}^{k} \binom{k}{\ell} g_\varphi({}^\varphi\nabla_{\partial_t}^{(k-\ell)}\varphi_*\partial_s, {}^\varphi\nabla_{\partial_t}^{(\ell)}\varphi_*\partial_s) \\
&= \Big(k(k-1)g_{ij}^{(k-2)}(t) + 2k\,t g_{ij}^{(k-1)}(t) + t^2 g_{ij}^{(k)}(t)\Big)\partial_s V^i \partial_s V^j.
\end{aligned}
$$

Evaluating this expression at $(s,t) = (0,0)$ gives

$$
k(k-1)g_{ij}^{(k-2)}(0)v^i v^j = \sum_{\ell=0}^{k} \binom{k}{\ell} g_\varphi({}^\varphi\nabla_{\partial_t}^{(k-\ell)}\varphi_*\partial_s, {}^\varphi\nabla_{\partial_t}^{(\ell)}\varphi_*\partial_s). \tag{2.23}
$$

Note that $\varphi_*\partial_s\big|_{(0,0)} = 0$, $\varphi_*\partial_t\big|_{(0,0)} = u$, ${}^\varphi\nabla_{\partial_t}\varphi_*\partial_t\big|_{(0,0)} = 0$ and

$$
{}^\varphi\nabla_{\partial_t}\varphi_*\partial_s\big|_{(0,0)} = (\partial_t(\varphi_*\partial_s)^i)(\partial_i)_\varphi\big|_{(0,0)} = v.
$$

We now claim:

Claim 2.67. Under the assumption ${}^\varphi\nabla_{\partial_t}\varphi_*\partial_t \equiv 0$,

$$
\begin{aligned}
&{}^\varphi\nabla_{\partial_t}^{(k)}\varphi_*\partial_s = \\
&\quad \sum_{\ell=0}^{k-2} \binom{k-2}{\ell} (\nabla^{(k-2-\ell)}R)_\varphi\big(\underbrace{\varphi_*\partial_t, \ldots, \varphi_*\partial_t}_{k-2-\ell \text{ times}}, {}^\varphi\nabla_{\partial_t}^{(\ell)}(\varphi_*\partial_s), \varphi_*\partial_t\big)\varphi_*\partial_t.
\end{aligned}
$$

Proof of Claim. The case $k = 2$ is proved as follows:

$$
\begin{aligned}
{}^\varphi\nabla_{\partial_t}^{(2)}\varphi_*\partial_s &= {}^\varphi\nabla_{\partial_t}({}^\varphi\nabla_{\partial_s}\varphi_*\partial_t) && \text{by Proposition 2.60} \\
&= {}^\varphi\nabla_{\partial_s}({}^\varphi\nabla_{\partial_t}\varphi_*\partial_t) + {}^{\varphi\nabla}R(\partial_s,\partial_t)(\varphi_*\partial_t) && \text{by definition of } {}^{\varphi\nabla}R \\
&= {}^{\varphi\nabla}R(\partial_s,\partial_t)(\varphi_*\partial_t) && \text{by assumption} \\
&= {}^{\nabla}R(\varphi_*\partial_s,\varphi_*\partial_t)(\varphi_*\partial_t) && \text{by Proposition 2.59}
\end{aligned}
$$

as required. For the inductive step we suppose the identity is true for $k = j$, and differentiate. Since ${}^\varphi\nabla_{\partial_t}\varphi_*\partial_t = 0$, we find:

$$
\begin{aligned}
&{}^\varphi\nabla_{\partial_t}\left(\sum_{\ell=0}^{j-2}\binom{j-2}{\ell}(\nabla^{(j-2-\ell)}R)_\varphi\big(\underbrace{\varphi_*\partial_t,\ldots,\varphi_*\partial_t}_{j-2-\ell \text{ times}}, {}^\varphi\nabla_{\partial_t}^{(\ell)}\varphi_*\partial_s, \varphi_*\partial_t\big)\varphi_*\partial_t\right) \\
&= \sum_{\ell=0}^{j-2}\binom{j-2}{\ell}\,{}^\varphi\nabla_{\partial_t}(\nabla^{(j-2-\ell)}R)_\varphi\big(\underbrace{\varphi_*\partial_t,\ldots,\varphi_*\partial_t}_{j-2-\ell \text{ times}}, {}^\varphi\nabla_{\partial_t}^{(\ell)}\varphi_*\partial_s, \varphi_*\partial_t\big)\varphi_*\partial_t
\end{aligned}
$$

$$+\sum_{\ell=0}^{j-2}\binom{j-2}{\ell}(\nabla^{(j-2-\ell)}R)_\varphi\big(\underbrace{\varphi_*\partial_t,\ldots,\varphi_*\partial_t}_{j-2-\ell\text{ times}},{}^\varphi\nabla^{(\ell+1)}_{\partial_t}\varphi_*\partial_s,\varphi_*\partial_t\big)\varphi_*\partial_t$$

$$=\sum_{\ell=0}^{j-2}\binom{j-2}{\ell}(\nabla^{(j-1-\ell)}R)_\varphi\big(\underbrace{\varphi_*\partial_t,\ldots,\varphi_*\partial_t}_{j-1-\ell\text{ times}},{}^\varphi\nabla^{(\ell)}_{\partial_t}\varphi_*\partial_s,\varphi_*\partial_t\big)\varphi_*\partial_t$$

$$+\sum_{\ell=0}^{j-2}\binom{j-2}{\ell}(\nabla^{(j-2-\ell)}R)_\varphi(\underbrace{\varphi_*\partial_t,\ldots,\varphi_*\partial_t}_{j-2-\ell\text{ times}},{}^\varphi\nabla^{(\ell+1)}_{\partial_t}\varphi_*\partial_s,\varphi_*\partial_t)\varphi_*\partial_t$$

$$=\sum_{\ell=0}^{j-1}\left(\binom{j-2}{\ell}+\binom{j-2}{\ell-1}\right)(\nabla^{(j-1-\ell)}R)_\varphi\big(\underbrace{\varphi_*\partial_t,\ldots,\varphi_*\partial_t}_{j-1-\ell\text{ times}},{}^\varphi\nabla^{(\ell)}_{\partial_t}\varphi_*\partial_s,\varphi_*\partial_t\big)\varphi_*\partial_t$$

$$=\sum_{\ell=0}^{j-1}\binom{j-1}{\ell}(\nabla^{(j-1-\ell)}R)_\varphi\big(\underbrace{\varphi_*\partial_t,\ldots,\varphi_*\partial_t}_{j-1-\ell\text{ times}},{}^\varphi\nabla^{(\ell)}_{\partial_t}\varphi_*\partial_s,\varphi_*\partial_t\big)\varphi_*\partial_t$$

completing the induction. Here we used the identity ${}^\varphi\nabla_{\partial_t}(\nabla^{(j-2-\ell)}R)_\varphi = \nabla_{\varphi_*\partial_t}\left(\nabla^{(j-2-\ell)}R\right)$ to get from the first equality to the second – this is the characterisation of the pullback connection in Theorem 2.56. □

Finally, we compute the Taylor expansion of $t \mapsto g_{ij}(\varphi(0,t))$ around $t=0$ (so that $g_{ij}(\gamma_u(t)) = g_{ij}(0) + g_{ij}^{(1)}(0) + \frac{1}{2}g_{ij}^{(2)}(0) + \ldots$). The 0-order term $g_\varphi(\partial_i,\partial_j)\big|_{(0,0)} = g_{ij}(\gamma_u(0)) = g_{ij}(p) = \delta_{ij}$ as we are working in normal coordinates. The 1st order vanishes and by (2.23) we find that $12\, g_{ij}^{(2)}(0)v^iv^j = 8\, g_\varphi$ $(R_\varphi(v,u)u,v)$, so $g_{ij}^{(2)}(0) = -\frac{2}{3}R_{ikj\ell}u^ku^\ell$. The theorem now follows. □

2.9 Integration and Divergence Theorems

If (M,g) is an oriented Riemannian manifold with boundary and $\widetilde{g}$ is the induced Riemannian metric on ∂M, then we define the volume form of $\widetilde{g}$ by $d\sigma_{\widetilde{g}} = \iota_\nu d\mu_g|_{\partial M}$. In particular if X is a smooth vector field, we have $\iota_X d\mu_g|_{\partial M} = \langle X,\nu\rangle_g\, d\sigma_{\widetilde{g}}$. In light of this, we define the divergence $\mathrm{div}X$ to be the quantity that satisfies:

$$d(\iota_X d\mu) = \mathrm{div}X\, d\mu. \tag{2.24}$$

Theorem 2.68 (Divergence theorem). *Let (M,g) be a compact oriented Riemannian manifold. If X is a vector field, then*

$$\int_M \mathrm{div}X\, d\mu = \int_{\partial M}\langle X,\nu\rangle_g\, d\sigma.$$

In particular, if M is closed then $\int_M \mathrm{div}X\, d\mu = 0$.

Proof. Define the $(n-1)$-form α by $\alpha = \iota_X d\mu$. So by Stokes' theorem,

$$\int_M \operatorname{div} X \, d\mu = \int_M d\alpha = \int_{\partial M} \alpha = \int_{\partial M} \iota_X d\mu = \int_{\partial M} \langle X, \nu \rangle \, d\sigma. \qquad \square$$

From the divergence theorem we have the following useful formulas:

Proposition 2.69 (Integration by parts). *On a Riemannian manifold (M,g) with $u, v \in C^\infty(M)$ the following holds:*

(a) On a closed manifold,

$$\int_M \Delta u \, d\mu = 0.$$

(b) On a compact manifold,

$$\int_M (u\Delta v - v\Delta u) d\mu = \int_{\partial M} \left(u\frac{\partial u}{\partial \nu} - v\frac{\partial u}{\partial \nu} \right) d\sigma.$$

In particular, on a closed manifold $\int_M u\Delta v \, d\mu = \int_M v\Delta u \, d\mu$.

(c) On a compact manifold,

$$\int_M u\Delta v \, d\mu + \int_M \langle \nabla u, \nabla v \rangle \, d\mu = \int_{\partial M} \frac{\partial v}{\partial \nu} u \, d\sigma.$$

In particular, on a closed manifold $\int_M \langle \nabla u, \nabla v \rangle \, d\mu = -\int_M u\Delta v \, d\mu$.

2.9.1 Remarks on the Divergence Expression

We seek a local expression for the divergence, defined by (2.24), and show it is equivalent to the trace of the covariant derivative. That is,

$$\operatorname{div} X = \operatorname{tr} \nabla X = \operatorname{tr} (\nabla X)(\,\cdot\,,\,\cdot\,) = (\nabla_i X)(dx^i) \tag{2.25}$$

Lemma 2.70. *The divergence* $\operatorname{div} X$ *of a vector field X, defined by* (2.24), *can be expressed in local coordinates by*

$$\operatorname{div}(X^i \partial_i) = \frac{1}{\sqrt{\det g}} \partial_i (X^i \sqrt{\det g}). \tag{2.26}$$

Proof. By Cartan's formula[7] and (2.24) we have

$$\operatorname{div}(X) d\mu = d \circ \iota_X d\mu = (d \circ \iota_X + \iota_X \circ d) d\mu = \mathcal{L}_X d\mu.$$

[7] Which states that $\mathcal{L}_X \omega = \iota_X(d\omega) + d(\iota_X \omega)$, for any smooth vector field X and any smooth differential form ω.

From the left-hand side we find that $(\mathrm{div}X\, d\mu)(\partial_1, \ldots, \partial_n) = \mathrm{div}X \sqrt{\det g}$ and from the right-hand side we find that

$$\begin{aligned}(\mathcal{L}_X d\mu)(\partial_1, \ldots, \partial_n) &= \mathcal{L}_X(\sqrt{\det g}) - d\mu(\ldots, \mathcal{L}_X \partial_i, \ldots) \\ &= X(\sqrt{\det g}) + d\mu(\ldots, (\partial_i X^j)\partial_j, \ldots) \\ &= X(\sqrt{\det g}) + (\partial_i X^j)\delta^i_j \sqrt{\det g} \\ &= \partial_i(X^i \sqrt{\det g})\end{aligned}$$

□

Claim 2.71. The definitions of div$X$ given by (2.24) and (2.25) coincide.

Proof. As $(\nabla_X dx^j)(\partial_i) = -dx^j(\nabla_X \partial_i)$, equation (2.25) implies that

$$(\nabla_i X)(dx^i) = \partial_i X^i - X(\nabla_i dx^i) = \partial_i X^i + \Gamma^i_{ij} X^j \tag{2.27}$$

On the other hand, (2.26) implies that $\mathrm{div}X = \partial_i X^i + \frac{1}{\sqrt{\det g}} X^i \partial_i(\sqrt{\det g})$, where by the chain rule

$$\begin{aligned}\partial_i(\sqrt{\det g}) &= \frac{1}{2}\frac{1}{\sqrt{\det g}}\frac{\partial \det g}{\partial g_{pq}}\frac{\partial g_{pq}}{\partial x^i} = \frac{1}{2}\sqrt{\det g}\, g^{pq}\frac{\partial g_{pq}}{\partial x^i} \\ &= g^{pq}\Gamma^\ell_{ip} g_{\ell q}\sqrt{\det g} = \Gamma^p_{ip}\sqrt{\det g}.\end{aligned}$$

□

Chapter 3
Harmonic Mappings

When considering maps between Riemannian manifolds it is possible to associate a variety of invariantly defined 'energy' functionals that are of geometrical and physical interest. The core problem is that of finding maps which are 'optimal' in the sense of minimising the energy functional in some class; one of the techniques for finding minimisers (or more generally critical points) is to use a gradient descent flow to deform a given map to an extremal of the energy.

The first major study of Harmonic mappings between Riemannian manifolds was made by Eells and Sampson [ES64]. They showed, under suitable metric and curvature assumptions on the target manifold, gradient lines do indeed lead to extremals.

We motivate the study of Harmonic maps by considering a simple problem related to geodesics. Following this we discuss the convergence result of Eells and Sampson. The techniques and ideas used for Harmonic maps provide some motivation for those use later for Ricci flow, and will appear again explicitly when we discuss the short-time existence for Ricci flow.

3.1 Global Existence of Geodesics

By Lemma 2.23, geodesics enjoy local existence and uniqueness as solutions of an initial value problem. In this section we examine a global existence problem:

Theorem 3.1 (Global Existence). *On a compact Riemannian manifold M, every homotopy class of closed curves contains a curve which is geodesic.*

To prove this, we will use methods from Geometric analysis. Specifically, we want to start with some curve (in the homotopy class under consideration) and let it evolve according to a particular PDE that decreases its energy until, in the limit, the curve becomes energy minimising. This is achieved by gradient descent techniques commonly known as the heat flow method.

B. Andrews and C. Hopper, *The Ricci Flow in Riemannian Geometry*, Lecture Notes in Mathematics 2011, DOI 10.1007/978-3-642-16286-2_3,

Firstly, we say two closed curves[1] $\gamma_1, \gamma_2 : S^1 \to M$ are equivalent if there exists a homotopy between them. That is, they are equivalent if there exists a continuous map $c : S^1 \times [0,1] \to M$ such that $c(t,0) = \gamma_0(t)$, $c(t,1) = \gamma_1(t)$ for all $t \in S^1$.

Now consider a mapping $u = u(s,t) : S^1 \times [0,\infty) \to M$ that evolve (in local coordinates) according to the PDE:

$$\begin{aligned} \frac{\partial u^i}{\partial t} &= \frac{\partial^2 u^i}{\partial s^2} + \Gamma^i_{jk}(u(s,t))\frac{\partial u^j}{\partial s}\frac{\partial u^k}{\partial s} \qquad && s \in S^1, t \geq 0 \\ u(s,t) &= \gamma(s) && s \in S^1 \end{aligned} \tag{3.1}$$

where S^1 is parametrised by s, t denotes the 'time' and $\gamma : S^1 \to M$ is arbitrary curve in the given homotopy class. Note that the corresponding steady state equation is the geodesic equation (2.2). Although the equation is written in local coordinates, we will see in Sect. 3.2 that it does in fact – in a more general context – make sense independent of the choice of those coordinates.

Proof (Theorem 3.1, sketch only). We focus our interests on the geometric aspects of the proof and where appropriate, quote the necessary PDE results.

Step 1: By general existence and uniqueness theory, there exists a solution to (3.1), at least on a short time interval $[0,T)$. Consequently, the maximum time interval of existence is nonempty and open.

Step 2: Using PDE regularity theory, we show that a solution $u(s,t)$ has bounded spatial derivatives independent of time t. To do this compute

$$\begin{aligned} \left(\partial_s^2 - \partial_t\right) g_{ij}\partial_s u^i \partial_s u^j = {} & g_{ij,k}(\partial_s^2 u^k - \partial_t u^k)\partial_s u^i \partial_s u^j + 4 g_{ij,k}\partial_s u^k \partial_s^2 u^i \partial_s u^j \\ & + 2 g_{ij}\partial_s^2 u^i \partial_s^2 u^j + 2 g_{ij}(\partial_s^3 u^i - \partial_s \partial_t u^i)\partial_s u^j \\ & + g_{ij,k\ell}\partial_s u^i \partial_s u^j \partial_s u^k \partial_s u^\ell . \end{aligned}$$

From (3.1) note that:

$$\partial_s^2 u^k - \partial_t u^k = -\Gamma^k_{ij}\partial_s u^i \partial_s u^j \tag{3.2a}$$

$$\partial_s^3 u^i - \partial_s \partial_t u^i = -\Gamma^i_{jk,\ell}\partial_s u^\ell \partial_s u^j \partial_s u^k - 2\Gamma^i_{jk}\partial_s^2 u^j \partial_s u^k . \tag{3.2b}$$

By working in normal coordinates at a point (so that the first derivatives of the metric g_{ij} and the Christoffel symbol Γ^k_{ij} vanish), insert (3.2a) and (3.2b) (with the help of (2.9)) into the above computation, so that

$$\begin{aligned} \left(\partial_s^2 - \partial_t\right) g_{ij}\partial_s u^i \partial_s u^j &= 2 g_{ij}\partial_s^2 u^i \partial_s^2 u^j + \cancel{g_{ij,k\ell}}\partial_s u^i \partial_s u^j \partial_s u^k \partial_s u^\ell \\ &\quad - (\cancel{g_{ij,k\ell}} + \cancel{g_{ik,j\ell}} - \cancel{g_{jk,i\ell}})\partial_s u^i \partial_s u^j \partial_s u^k \partial_s u^\ell \\ &= 2 g_{ij}\partial_s^2 u^i \partial_s^2 u^j . \end{aligned} \tag{3.3}$$

[1] For technical convenience, we parametrise closed curves on S^1 as it is unnecessary to stipulate end conditions, like $\gamma(0) = \gamma(1)$, for closed loops.

Hence

$$\left(\frac{\partial^2}{\partial s^2} - \frac{\partial}{\partial t}\right) g_{ij} \frac{\partial u^i}{\partial s} \frac{\partial u^j}{\partial s} \geq 0.$$

Therefore $\|\partial_s u\|^2 = g_{ij}\partial_s u^i \partial_s u^j$ is a subsolution of (3.1). By the parabolic maximum principle

$$\sup_{s \in S^1} g_{ij}(u(s,t)) \frac{\partial u^i}{\partial s}(s,t) \frac{\partial u^j}{\partial s}(s,t)$$

is a non-increasing function of t. In particular $g_{ij}\partial_s u^i \partial_s u^j$ is bounded by some constant C independent of t and s. From this, regularity theory tells us that $\partial_s u^i$ stays bounded. So by standard bootstrapping methods we gain time-independent control of higher derivatives.

Step 3: When a solution exists on $[0,T)$, $u(s,t)$ will converge to a smooth curve $u(s,T)$ as $t \to T$. By regularity, this curve can be taken as a new initial value. In which case the solution can be continued beyond T. Therefore the maximum interval is closed and so, in conjunction with Step 1, the solution exists for all times $t > 0$.

Step 4: We directly show $t \mapsto E(u(\cdot,t))$ is a decreasing function. To do this, note that:

$$\begin{aligned}
\frac{d}{dt}E(u(\cdot,t)) &= \frac{1}{2}\frac{\partial}{\partial t}\int_{S^1} g_{ij}\partial_s u^i \partial_s u^j \\
&= \frac{1}{2}\int_{S^1} \left(g_{ij,k}\partial_t u^k \partial_s u^i \partial_s u^j + 2g_{ij}\partial_s\partial_t u^i \partial_s u^j\right) \\
&= \frac{1}{2}\int_{S^1} \left((g_{ij,k} - 2g_{ik,j})\partial_s u^i \partial_s u^j \partial_t u^k - 2g_{ij}\partial_t u^i \partial_s^2 u^j\right),
\end{aligned}$$

where the last equality follows from integration by parts. Using (3.1) observe that

$$g_{ip}\partial_t u^i = g_{ip}\partial_s^2 u^i + \frac{1}{2}(g_{jp,k} + g_{pk,j} - g_{jk,p})\partial_s u^j \partial_s u^k.$$

Multiply this by $\partial_t u^p$ on both sides and rearrange to get

$$(g_{jk,p} - 2g_{jp,k})\partial_s u^j \partial_s u^k \partial_t u^p = g_{ip}\partial_s^2 u^i \partial_t u^p - 2g_{ip}\partial_t u^i \partial_t u^p.$$

Thus

$$\frac{d}{dt}E(u(\cdot,t)) = -\int_{S^1} g_{ij}\partial_t u^i \partial_t u^j = -\int_{S^1} \|\partial_t u\|^2 \leq 0.$$

Since E is also non-negative (i.e. bounded below) we can find a sequence $t_n \to \infty$ for which $u(\cdot,t_n)$ will converge to a curve that satisfies the geodesic equation (2.2).

Step 5: Finally, by a similar computation (again in normal coordinates), we find that

$$\frac{d^2}{dt^2}E(u(\cdot,t)) = 2\int_{S^1} g_{ij}\partial_s\partial_t u^i \partial_s\partial_t u^j \geq 0.$$

Thus the energy $E(u(\cdot,t))$ is a convex function in t. As we already have $\frac{d}{dt}E(u(\cdot,t_n)) \to 0$ for some sequence $t_n \to \infty$, we conclude that $\frac{d}{dt}E(u(\cdot,t)) \to 0$ for $t \to \infty$. Thus, by our pointwise estimate $\partial_t u(s,t) \to 0$ as $t \to \infty$, the curve $u(s) = \lim_{t\to\infty} u(s,t)$ exists and is a geodesic. □

It is a rather natural step to generalise the above problem to maps between any Riemannian manifolds. To do so, one needs an energy functional that depends only on the intrinsic geometry of the domain manifold, target manifold and the map between them. Critical points of such an energy are called harmonic maps. Upon such a abstraction, we would like to know if the higher dimensional analogue to Theorem 3.1 remains true. That is:

Given Riemannian manifolds (M,g), (N,h) and a homotopy class of maps between them, does there exist a harmonic map in that homotopy class?

If M and N are compact, there are positive and negative answers to this question. There is a positive answer due to [ES64] if we assume the target manifold N has non-positive sectional curvature. In contrast, [EW76] showed the answer to be negative for maps of degree ± 1 from the 2-torus to the 2-sphere (there are of course many other counterexamples).

As Harmonic mappings seem to be one of the most natural problems one can pose, there are (not surprisingly) many varied examples. For instance:

- Identity and constant maps are harmonic.
- Geodesics as maps $S^1 \to M$ are harmonic.
- Every minimal isometric immersion is a harmonic map.
- If the target manifold $N = \mathbb{R}^n$, then it follows from the Dirichlet principle that f is a harmonic map if and only if it is a harmonic function in the usual sense (i.e. a solution of the Laplace equation).
- Holomorphic maps between Kähler manifolds are harmonic.
- Minimal submanifolds (or more generally submanifolds with parallel mean curvature vector) in Euclidean spaces have harmonic Gauss maps. The Gauss map takes a point in the submanifold to its tangent plane at that point, thought of as a point in the Grassmannian of subspaces of that dimension.
- Harmonic maps from surfaces depend only on the conformal structure of the source manifold – thus by the uniformisation theorem we can work with a constant curvature metric on the source manifold (or locally with a flat metric). The resulting equations have many nice properties – in

particular harmonic maps from surfaces into symmetric spaces have an integrable structure which leads to many explicit solutions.
- In theoretical physics, harmonic maps are also known as σ-models (cf. Sect. 11.6).

In what follows, we give a proof of the convergence result of Eells and Sampson [ES64] (Theorem 3.9, here) with improvements by Hartman [Har67].

3.2 Harmonic Map Heat Flow

Consider a C^∞-map $f : (M, g) \to (N, h)$ between Riemannian manifolds (M, g) and (N, h). Let (x^i) be a local chart on M about $p \in M$ and let (y^α) be a local chart on N about $f(p)$ with $f^\alpha = y^\alpha \circ f$. By Sect. 2.8.2, $f_* \in \Gamma(T^*M \otimes f^*TN)$ so $f_* = (f_*)_i^\alpha dx^i \otimes (\partial_\alpha)_f = \frac{\partial f^\alpha}{\partial x^i} dx^i \otimes (\frac{\partial}{\partial y^\alpha})_f$. Let $\langle \cdot, \cdot \rangle$ be the inner product on the bundle $T^*M \otimes f^*TN$ (in accordance with Sect. 2.4.6) so that

$$\begin{aligned}\langle f_*, f_* \rangle_{T^*M \otimes f^*TN} &= \frac{\partial f^\alpha}{\partial x^i} \frac{\partial f^\beta}{\partial x^j} \langle dx^i \otimes (\partial_\alpha)_f, dx^j \otimes (\partial_\beta)_f \rangle \\ &= g^{ij} (h_{\alpha\beta})_f \frac{\partial f^\alpha}{\partial x^i} \frac{\partial f^\beta}{\partial x^j}.\end{aligned}$$

In light of this, define the *energy density* $e(f)$ of the map f to be

$$e(f) = \frac{1}{2} \|f_*\|^2 = \frac{1}{2} g^{ij} (h_{\alpha\beta})_f \frac{\partial f^\alpha}{\partial x^i} \frac{\partial f^\beta}{\partial x^j} \tag{3.4}$$

and the *energy* E to be

$$E(f) = \int_M e(f) d\mu(g), \tag{3.5}$$

where $d\mu(g)$ is the volume form on M with respect to the metric g. The energy can be considered as a generalisation of the classical integral of Dirichlet; for if $N = \mathbb{R}$ (so that $f : M \to \mathbb{R}$) then E corresponds to the Dirichlet's energy $\frac{1}{2} \int_M |\nabla f|^2 d\mu(g)$.

As we have the connections ∇^M on TM and ${}^f\nabla^N$ on f^*TN, there is a connection (denoted simply by ∇) on each tensor bundle constructed from these (i.e. on $T^*M \otimes f^*TN$). Moreover, (2.12) gives a useful expression for the curvature of the connection constructed on $T^*M \otimes f^*TN$:

$$\begin{aligned}(R(U, V) f_*) (W) &= R_{{}^f\nabla^N}(U, V)(f_* W) - f_* \big(R_{\nabla^M}(U, V) W\big) \\ &= R^N (f_* U, f_* V) (f_* W) - f_* \left(R^M(U, V) W\right)\end{aligned} \tag{3.6}$$

where the second equality follows from Proposition 2.59.

3.2.1 Gradient Flow of E

For this we take a variation of the map f, i.e. a smooth map $f: M \times I \to N$ where $I \subset \mathbb{R}$ is an interval of time.

We wish to compute on TM at each time, so we define the 'spatial tangent bundle' to be the vector subbundle $\mathfrak{S} \subset T(M \times I)$ consisting of vectors tangent to the M factor, that is $\mathfrak{S} = \{v \in T(M \times I) : dt(v) = 0\}$ (ref. Example 2.62). On this bundle $\mathfrak{S} \to M \times I$ we place the time-independent metric and connection ∇ given by g and its Levi–Civita connection ${}^M\nabla$ – which must be augmented to include the time direction by defining $\nabla_{\partial_t} u = [\partial_t, u]$, for any $u \in \Gamma(\mathfrak{S})$. Further to this, we also extend the connection ∇ to $T(M \times I)$ by zero so that $\nabla \partial_t \equiv 0$.[2]

The map $f: M \times I \to N$ induces the pullback bundle f^*TN over $M \times I$, on which we can place the restriction metric and the pullback connection. We now look to compute

$$\frac{d}{dt} E(f) = \frac{1}{2} \int_M \frac{d}{dt} \langle f_*, f_* \rangle \, d\mu(g) = \int_M \langle f_*, \nabla_{\partial_t} f_* \rangle \, d\mu(g),$$

where ∇ is the connection on $\mathfrak{S}^* \otimes f^*TN$ over $M \times I$. By (2.7) we have $(\nabla_{\partial_t} f_*)(\partial_i) = {}^f\nabla_{\partial_t}(f_* \partial_i) - f_*(\cancel{\nabla_{\partial_t} \partial_i})$. Using this equation with Proposition 2.60 gives

$$(\nabla_{\partial_t} f_*)(\partial_i) = {}^f\nabla_{\partial_t}(f_* \partial_i) = {}^f\nabla_{\partial_i}(f_* \partial_t), \tag{3.7}$$

since $[\partial_t, \partial_i] = 0$. We denote by f_j^α the components of f_* in a local frame (for instance $f_* \partial_t = f_0^\alpha (\partial_\alpha)_f$), and by $\nabla_i f_j^\alpha$ the components of ∇f_* (cf. Sect. 2.5.2.1). Since $\nabla g = 0$ and $\nabla h = 0$ by compatibility, we have that

$$\begin{aligned} \nabla_j \big(g^{ij} (h_{\alpha\beta})_f f_i^\alpha f_0^\beta \big) &= g^{ij} (h_{\alpha\beta})_f \nabla_j f_i^\alpha f_0^\beta + g^{ij} (h_{\alpha\beta})_f f_i^\alpha \nabla_j f_0^\beta \\ &= \langle g^{ij} \nabla_j f_i, f_* \partial_t \rangle + \langle f_*, \nabla_{\partial_t} f_* \rangle \end{aligned}$$

where we used (3.7) in the last term and the fact that $\nabla_j \partial_t = 0$ by extension. So by the Divergence theorem with (2.25), we find that

$$\begin{aligned} \frac{d}{dt} E(f) &= \int_M \langle f_*, \nabla_{\partial_t} f_* \rangle \, d\mu \\ &= \int_M \nabla_j \big(g^{ij} h_f (f_* \partial_i, f_* \partial_t) \big) d\mu - \int_M \langle f_* \partial_t, g^{ij} \nabla_i f_j \rangle \, d\mu \\ &= - \int_M \langle f_* \partial_t, \Delta_{g,h} f \rangle \, d\mu, \end{aligned}$$

[2] This kind of construction is a natural one when dealing with variations of geometric structures; we will develop this in a more general context in Sect. 6.3.1.

where we define:

Definition 3.2. The harmonic map Laplacian

$$\Delta_{g,h} f = \mathrm{tr}_g \nabla f_* := g^{ij} (\nabla_{\partial_i} f_*)(\partial_j).$$

Note that $f_* \partial_t$ is the 'variation of f'. Hence the gradient of E, with respect to the inner product on f^*TN, is $-\Delta_{g,h} f$ and the gradient descent flow is:

$$f_* \partial_t = \Delta_{g,h} f. \tag{3.8}$$

Furthermore we say f is *harmonic* if $\Delta_{g,h} f = 0$, in which case f is a steady state solution of (3.8). We also note that an immediate consequence of (3.8) is that

$$\frac{d}{dt} E(f) = -\int_M |\Delta_{g,h} f|^2 d\mu = -\|\Delta_{g,h} f\|^2_{L^2(M)} \leq 0, \tag{3.9}$$

so the energy decays.

Remark 3.3. The idea now is quite simple: We want to deform a given $f_0 : M \to N$ to a harmonic map by evolving it along the gradient flow (3.8) so that $f(\cdot, t)$ converges to a harmonic map homotopic to f_0 as $t \to \infty$. This is the so-called *heat flow method.*

3.2.2 Evolution of the Energy Density

The key computation is the formula for the evolution of the energy density $e(f)$ under harmonic map heat flow (3.8). Here we compute this using the machinery of pullback connections.

First we compute an evolution equation for f_*:

$$\begin{aligned}
(\nabla_{\partial_t} f_*)(\partial_i) &= {}^f\nabla_{\partial_t}(f_* \partial_i) - f_* \cancel{(\nabla_{\partial_t} \partial_i)} \\
&= {}^f\nabla_{\partial_i}\left(g^{k\ell} \nabla_k f_\ell\right) \\
&= \nabla_{\partial_i}\left(g^{k\ell} \nabla_k f_\ell\right) \\
&= g^{k\ell} (\nabla_i \nabla_k f_*)(\partial_\ell) \\
&= g^{k\ell}\big((\nabla_k \nabla_i f_*)(\partial_\ell) + (R(\partial_k, \partial_i) f_*)(\partial_\ell)\big).
\end{aligned}$$

In the first term we observe by symmetry (i.e. Proposition 2.60) that

$$(\nabla_i f_*)(\partial_l) = {}^f\nabla_\ell (f_* \partial_i) - f_*(\nabla_\ell \partial_i) = (\nabla_l f_*)(\partial_i).$$

Using this together with the formula (3.6) for the second term implies that

$$(\nabla_{\partial_t} f_*)(\partial_i) = (\Delta f_*)(\partial_i) + g^{k\ell} R^N(f_*\partial_k, f_*\partial_i)(f_*\partial_\ell) \\ - g^{k\ell} f_* \left(R^M(\partial_k, \partial_i)\partial_\ell\right)$$

where $\Delta = g^{k\ell}\nabla_k \nabla_\ell$. It follows (noting that the metric on $T^*M \otimes f^*TN$ is parallel since g and h are) that

$$\begin{aligned} \frac{\partial}{\partial t} e &= \langle f_*, \Delta f_* \rangle + g^{k\ell} g^{ij} R^N(f_*\partial_k, f_*\partial_i, f_*\partial_\ell, f_*\partial_j) \\ &\quad - g^{k\ell} h_f \left(f_*(({}^M\mathrm{Ric})_k{}^p \partial_p), f_*\partial_\ell \right) \\ &= \Delta e - \|\nabla f_*\|^2 + g^{k\ell} g^{ij} R^N(f_*\partial_k, f_*\partial_i, f_*\partial_\ell, f_*\partial_j) \\ &\quad - g^{k\ell} h_f \left(f_*(({}^M\mathrm{Ric})_k{}^p \partial_p), f_*\partial_\ell \right). \end{aligned} \tag{3.10}$$

Remark 3.4. We note that if M is compact[3] then $g^{k\ell} h_f(f_*(\mathrm{Ric}^M(\partial_k), f_*\partial_\ell) \geq -Ce$, and if the target manifold N has non-positive curvature then the term $g^{k\ell} g^{ij} R^N(f_*\partial_k, f_*\partial_i, f_*\partial_\ell, f_*\partial_j) \leq 0$. Thus under these assumptions we have

$$\frac{\partial e}{\partial t} \leq \Delta e + Ce \tag{3.11}$$

for some constant C.

3.2.3 Energy Density Bounds

The problem here is to derive a bound on the energy density $e = e(f)$, having deduced the inequality (3.11). The idea is to use this inequality together with the bound on the Dirichlet energy (that is, a bound on the integral of e) to deduce the bound on e.

The argument is a variant on a commonly used technique called Moser iteration [Mos60], which uses the Sobolev inequality to 'bootstrap' up from an L^p bound to a sup bound. The argument is a little simpler in the elliptic case (for instance see [PRS08, p. 118] or [GT83]).

[3] In which case Ric^M is bounded, so $\mathrm{Ric}^M \geq -\frac{C}{2} g$ for some constant C.

To do this we need the following two ingredients. First, we need a computation for the time derivative of an L^p norm of e:

$$\begin{aligned}\frac{d}{dt}\int_M e^{2p} &\leq 2p\int_M e^{2p-1}\left(\Delta e + Ce\right)\\ &\leq -2p(2p-1)\int_M e^{2p-2}\left|\nabla e\right|^2 + 2pC\int_M e^{2p}\\ &= -\frac{2(2p-1)}{p}\int_M \left|\nabla(e^p)\right|^2 + 2pC\int_M e^{2p}\\ &\leq -2\left\|e^p\right\|^2_{W^{1,2}} + 2p\tilde{C}\int_M e^{2p}.\end{aligned}$$

The 'problem' term on the right can be absorbed by multiplying by an exponential, yielding

$$\frac{d}{dt}\left(e^{-\tilde{C}t}\|e\|_{2p}\right) \leq -\frac{1}{p}e^{-\tilde{C}t}\|e\|_{2p}^{1-2p}\|e^p\|^2_{W^{1,2}}. \tag{3.12}$$

The second ingredient is the Gagliardo–Nirenberg inequality. It is a standard PDE result (for instance see [Eva98, p. 263]) which says that there exists a constant C_1 such that for any $W^{1,2}$ function f on M,

$$\|f\|_2 \leq C_1\|f\|_{W^{1,2}}^{\frac{n}{n+2}}\|f\|_1^{\frac{2}{n+2}}. \tag{3.13}$$

The idea here is that if the L^1 norm is controlled, then we can use the good $W^{1,2}$ norm to yield a power bigger than 1 of the L^2 norm. Substituting this into (3.12) gives

$$\frac{d}{dt}\left(e^{-\tilde{C}t}\|e\|_{2p}\right) \leq -\frac{C_1^2}{p}e^{-\tilde{C}t}\|e\|_{2p}^{1+\frac{4p}{n}}\|e\|_p^{-\frac{4p}{n}},$$

which on rearrangement yields

$$\frac{d}{dt}\left(e^{-\tilde{C}t}\|e\|_{2p}\right)^{-\frac{4p}{n}} \geq \frac{4C_1^2}{n}\left(e^{-\tilde{C}t}\|e\|_p\right)^{-\frac{4p}{n}}. \tag{3.14}$$

From here we will use an induction argument to bound higher and higher L^p norms with a bound independent of p. This will imply a sup bound since $\|e\|_\infty = \lim_{p\to\infty}\|e\|_p$.

The first application of this is straightforward, since we have the energy bound $\|e\|_1 \leq E$. Plugging this into (3.14) with $p = 1$ gives

$$\begin{aligned}\frac{d}{dt}\left(e^{-\tilde{C}t}\|e\|_2\right)^{-\frac{4}{n}} &\geq \frac{4C_1^2}{n}\left(e^{-\tilde{C}t}E\right)^{-\frac{4}{n}}\\ &\geq \frac{4C_1^2}{n}E^{-4/n}t,\end{aligned}$$

since the exponential term is always at least 1. This is equivalent to a L^2 bound on e decaying like $t^{-4/n}$. This is the initial step, now we use induction to prove the following:

Theorem 3.5. *For each* $k \in \mathbb{N}$,

$$\left(e^{-\tilde{C}t}\|e\|_{2^k}\right)^{-\frac{4}{n}} \geq \left(\frac{4C_1^2 t}{n}\right)^{2(1-2^{-k})} B_k^{-2} E^{-4/n},$$

where $B_1 = 1$ *and* $B_{k+1} \leq 2^{\frac{k+1}{2^k}} B_k$.

The induction is easy from (3.14) with $p = 2^k$:

$$\begin{aligned}\frac{d}{dt}\left(e^{-\tilde{C}t}\|e\|_{2^{k+1}}\right)^{-\frac{2^{k+2}}{n}} &\geq \frac{4C_1^2}{n}\left(e^{-\tilde{C}t}\|e\|_{2^k}\right)^{-\frac{2^{k+2}}{n}} \\ &\geq \frac{4C_1^2}{n}\left(\frac{4C_1^2}{n}\right)^{2(2^k-1)} B_k^{-2^{k+1}} E^{-\frac{2^{k+2}}{n}} t^{2(2^k-1)}.\end{aligned}$$

Integration gives

$$\begin{aligned}\left(e^{-\tilde{C}t}\|e\|_{2^{k+1}}\right)^{-\frac{2^{k+2}}{n}} &\geq \frac{4C_1^2}{n}\left(\frac{4C_1^2}{n}\right)^{2(2^k-1)} B_k^{-2^{k+1}} E^{-\frac{2^{k+2}}{n}} \frac{t^{2^{k+1}-1}}{2^{k+1}-1} \\ &\geq \left(\frac{4C_1^2 t}{n}\right)^{2^{k+1}-1} B_k^{-2^{k+1}} E^{-\frac{2^{k+2}}{n}} 2^{-(k+1)}.\end{aligned}$$

Taking the power 2^{-k} gives

$$\begin{aligned}\left(e^{-\tilde{C}t}\|e\|_{2^{k+1}}\right)^{-\frac{4}{n}} &\geq \left(\frac{4C_1^2 t}{n}\right)^{2(1-2^{-(k+1)})} B_k^{-2} E^{-4/n} 2^{-\frac{k+1}{2^k}} \\ &\geq \left(\frac{4C_1^2 t}{n}\right)^{2(1-2^{-k})} B_{k+1}^{-2} E^{-4/n}\end{aligned}$$

for $B_{k+1} = 2^{\frac{k+1}{2^k}}$. This completes the induction and the bound for e.

3.2.4 Higher Regularity

Once e is controlled, so are all higher derivatives of f. This is a general result needing no special curvature assumptions:

Proposition 3.6. *Let* (M, g) *and* (N, h) *be compact Riemannian manifolds. Let* $f : M \times [0, T) \to N$ *be a smooth solution of the harmonic map heat flow*

(3.8) *with bounded energy density* e. *Then there exist constants* C_k, $k \geq 1$ *depending on* $\sup_{M\times[0,T)} e$, $|R^M|_{C^k}$ *and* $|R^N|_{C^k}$ *such that*

$$\|\nabla^{(k)} f_*\| \leq C_k \left(1 + t^{-k/2}\right).$$

For example, the evolution of ∇f_* is given by

$$\begin{aligned}\nabla_t (\nabla f_*) = \Delta \nabla f_* + R^N * (f_*)^2 * \nabla f_* \\ + R^M * \nabla f_* + \nabla R^N * (f_*)^4 + \nabla R^M * f_*,\end{aligned}$$

so that $\frac{\partial}{\partial t}\|\nabla f_*\|^2 \leq \Delta\|\nabla f_*\|^2 - 2\|\nabla^2 f_*\|^2 + C_1\|\nabla f_*\|^2 + C_2$, where C_1 and C_2 depend on $|R^M|$, $|R^N|$, $|\nabla R^M|$, $|\nabla R^N|$ and e. This gives

$$\begin{aligned}\frac{\partial}{\partial t}\left(t\|\nabla f_*\|^2 + \|f_*\|^2\right) \leq \Delta\left(t\|\nabla f_*\|^2 + \|f_*\|^2\right) \\ + (tC_1 - 1)\|\nabla f_*\|^2 + (tC_2 + C_3),\end{aligned}$$

which gives $t\|\nabla f_*\|^2 \leq C$ for $0 < t < 1/C_1$. The same argument applied on later time intervals gives a bound for any positive time, proving the case $k = 1$. For a similar argument applied to the Ricci flow see Theorem 8.1.

3.2.5 Stability Lemma of Hartman

Let $f(x, t, s)$ smooth family of solutions to (3.8) depending on a parameter s and a 'time' $t \in I$. We want to prove to following result by Hartman [Har67, p. 677].

Lemma 3.7 (Hartman). *Let* $F(x, s) : M \times [0, s_0] \to N$ *be of class* C^1 *and for fixed* s, *let* $f(x, t, s)$ *be a solution of (3.8) on* $0 \leq t \leq T$ *such that* $f(x, 0, s) = F(x, s)$. *Then for all* $s \in [0, s_0]$,

$$\sup_{M\times\{t\}\times\{s\}} h_f(f_*\partial_s, f_*\partial_s)$$

is non-increasing in t.

Proof. Since we now have a map with two parameters (s representing the variation through the family of harmonic map heat flows, and t representing the time parameter of the heat flows) we are now working on bundles over $M \times I_1 \times I_2$, where I_1 and I_2 are real intervals. The bundles we need are again the spatial tangent bundle $\mathfrak{S}$, defined by $\{v \in T(M \times I_1 \times I_2) : dt(v) = ds(v) = 0\}$, and the pullback bundle f^*TN. On the former we have the 'time-independent' metric g, and a connection given by the Levi–Civita connection in spatial directions, together with the prescription $\nabla_{\partial_t}\partial_i = 0$ and $\nabla_s\partial_i = 0$ (note that this choice gives a compatible metric and connection on $\mathfrak{S}$, for which ∇_{∂_s} and ∇_{∂_t} commute with each other and with ∇_{∂_i}). On f^*TN we have as usual the pullback metric and connection.

We compute an evolution equation for $f_*\partial_s$:

$$\begin{aligned}\nabla_{\partial_t}(f_*\partial_s) &= \nabla_{\partial_s}(f_*\partial_t)\\ &= \nabla_s\left(g^{k\ell}(\nabla_k f_*)(\partial_\ell)\right)\\ &= g^{k\ell}\big((\nabla_k\nabla_s f_*)(\partial_\ell) + (R(\partial_k,\partial_s)f_*)(\partial_\ell)\big)\\ &= g^{k\ell}(\nabla_k\nabla_\ell f_*)(\partial_s) + g^{k\ell}R^N(f_*\partial_k, f_*\partial_s)(f_*\partial_\ell)\\ &= \Delta(f_*\partial_s) + g^{k\ell}R^N(f_*\partial_k, f_*\partial_s)(f_*\partial_\ell)\end{aligned}$$

since $R^M(\partial_k,\partial_s) = 0$ and $\nabla_\ell\partial_s = 0$. Therefore the evolution of $\|f_*\partial_s\|^2$ is given by

$$\begin{aligned}\frac{\partial}{\partial t}\|f_*\partial_s\|^2 &= 2\langle f_*\partial_s, \Delta(f_*\partial_s)\rangle + 2g^{k\ell}R^N(f_*\partial_k, f_*\partial_s, f_*\partial_\ell, f_*\partial_s)\\ &\le \Delta\|f_*\partial_s\|^2 - 2\|\nabla f_*\partial_s\|^2,\end{aligned}$$

since the curvature term can be written as a sum of sectional curvatures (cf. Proposition 2.49).

By letting $Q = \|f_*\partial_s\|^2$ we note, for fixed s, that $Q(x,s,t)$ satisfies the parabolic differential inequality $\partial_t Q - \Delta Q \le 0$. Hence the maximum principle implies that if $0 \le \tau \le t$ then $\max_x Q(x,s,t) \le \max_x Q(x,s,\tau)$ for every fixed $s \in [0,s_0]$. Consequently $\max_{x,s} Q(x,s,t) \le \max_{x,s} Q(x,s,\tau)$. Hence the desired quantity is non-increasing. □

An important application of Hartman's Lemma is to prove that the distance between homotopic solutions of the flow cannot increase. We define the distance between two homotopic maps f_0 and f_1 as follows: If $H : M \times [0,1] \to N$ is a smooth homotopy from f_0 to f_1, so that $H(x,0) = f_0(x)$ and $H(x,1) = f_1(x)$, then the *length* of H (in analogy to Sect. 2.4.1.1) is defined to be

$$L(H) = \sup_{x\in M}\int_0^1 \left\|\frac{\partial H}{\partial s}(x,s)\right\| ds.$$

We define $\tilde{d}(f_0,f_1)$ to be the infimum of the lengths over all homotopies from f_0 to f_1. When N is non-positively curved the infimum is attained by a smooth homotopy H in which $s \mapsto H(x,s)$ is a geodesic for each $x \in M$, and in this case $L(H) = \sup\{|\partial_s H(x,s)| : x \in M\}$ for each $s \in [0,1]$.

Corollary 3.8 **([Har67, Sect. 8]).** *If N is non-positively curved and $f_0(x,t)$ and $f_1(x,t)$ are solutions of the harmonic map heat flow with homotopic initial data, then $t \mapsto \tilde{d}(f_0(\cdot,t), f_1(\cdot,t))$ is non-increasing.*

Proof. Fix $t_0 \geq 0$ and let H be the minimising homotopy from $f_0(\cdot, t_0)$ to $f_1(\cdot, t_0)$. There exists $\delta > 0$ such that the flows $f(x,t,s)$ with $f(x,t_0,s) = H(x,s)$ exist for $t_0 \leq t \leq t_0 + \delta$. By Lemma 3.7, $\sup\{|\partial_s f(x,t,s)| : x \in M\}$ is non-increasing in t for each s, and therefore for $t_0 \leq t \leq t_0 + \delta$,

$$\begin{aligned}
\tilde{d}(f_0(t), f_1(t)) \leq L[H(\cdot,t,\cdot)] &= \sup_{x \in M} \int_0^1 \left\| \frac{\partial H}{\partial s}(x,t,s) \right\| ds \\
&\leq \int_0^1 \sup_{x \in M} \left\| \frac{\partial H}{\partial s}(x,t,s) \right\| ds \\
&\leq \int_0^1 \sup_{x \in M} \left\| \frac{\partial H}{\partial s}(x,t_0,s) \right\| ds \\
&= \tilde{d}(f_0(t_0), f_1(t_0)).
\end{aligned}$$

□

3.2.6 Convergence to a Harmonic Map

Theorem 3.9 (Eells and Sampson). *If N is a non-positively curved compact manifold, then $f(\cdot,t)$ converges in $C^\infty(M,N)$ to a limit $\bar{f}$ in the same homotopy class as f_0 with $\Delta_{g,h}\bar{f} = 0$.*

Proof. The energy decay formula (3.9) implies that

$$\int_0^\infty \|\Delta_{g,h} f(\cdot,t)\|^2_{L^2(M)} < \infty.$$

Therefore there exists a sequence $t_n \to \infty$ such that $\|\Delta_{g,h} f(\cdot,t_n)\|_{L^2(M)} \to 0$. Since all higher derivatives are bounded by Proposition 3.6, this also implies $\|\Delta_{g,h} f(\cdot,t_n)\| \to 0$ uniformly. Also using the higher derivative bounds and the Arzela–Ascoli theorem, by passing to a subsequence we can ensure that $f(\cdot,t_n)$ converges in C^∞ to a limit $\bar{f}$ which therefore satisfies $\Delta_{g,h}\bar{f} = 0$. The function $\bar{f}$ is in the same homotopy class as f_0, since for large n we have $d_N(f(x,t_n), \bar{f}(x)) < \text{inj}(N)$ and there is a unique minimising geodesic from $f(x,t_n)$ to $\bar{f}(x)$, which depends smoothly on x. Following these geodesics defines a homotopy from $f(\cdot,t_n)$ to $\bar{f}$.

Stronger convergence follows from Corollary 3.8 (essentially the argument as [Har67, Sect. 4]): Since $\tilde{d}(f(\cdot,t),\bar{f})$ is non-increasing, and $\tilde{d}(f(\cdot,t_n),\bar{f}) \to 0$, we have $\tilde{d}(f(\cdot,t),\bar{f}) \to 0$, which implies $f(\cdot,t)$ converges to $\bar{f}$ uniformly. Convergence in C^∞ follows since all higher derivatives are bounded. □

3.2.7 Further Results

3.2.7.1 Uniqueness

One can also show that the harmonic map is essentially unique in its homotopy class. This can be done using the following result by Hartman [Har67, p. 675].

Theorem 3.10. *Let $f_1(x)$ and $f_2(x)$ be two homotopic harmonic maps from M into the non-positively curved manifold N. For fixed x, let $F(x,s)$ be the unique geodesic from $F(x,0) = f_1(x)$ to $F(x,1) = f_2(x)$ in the homotopy class determined by the homotopy between f_1 and f_2, and let the parameter $s \in [0,1]$ be proportional to arc length.*

Then $F(\cdot, s)$ is harmonic for each $s \in [0,1]$, and $E(F(\cdot,s)) = E(f_1) = E(f_2)$. Furthermore, the length of the geodesic $F(x,\cdot)$ is independent of x.

Remark 3.11. This cannot be improved since a torus $S^1 \times S^1$ has non-positive (zero!) curvature, and has an infinite family of homotopic harmonic maps from S^1 given by $z \mapsto (z, z_0)$ for fixed $z_0 \in S^1$.

If the sectional curvatures of N are strictly negative then this cannot occur:

Theorem 3.12. *If N has negative sectional curvature, then a harmonic map $f : M \to N$ is unique in its homotopy class, unless it is constant or maps M onto a closed geodesic. In the latter case, non-uniqueness can only occur by rotations of this geodesic.*

3.2.7.2 Dirichlet and Neumann Problems

It is interesting to note that Richard S. Hamilton, with some advice and encouragement from James Eells Jr., looked at solving the Dirichlet problem for harmonic mappings into non-positively curved manifolds. In [Ham74] he was able to prove the following.

Theorem 3.13. *Suppose N is a compact manifold with nonempty boundary ∂M, and N is complete with non-positive sectional curvature and convex (or empty) boundary. If $f_0 : M \to N$ is continuous, then the parabolic system*

$$\begin{aligned} \frac{\partial f}{\partial t}(x,t) &= \Delta_{g,h} f(x,t) && (x,t) \in M \times (0,\infty) \\ f(x,0) &= f_0(x) && x \in M \\ f(y,t) &= f_0(y) && y \in \partial M \end{aligned}$$

has a smooth solution $f(x,t)$ for all t. As $f(x,t)$ converges to the unique harmonic map homotopic to f_0 with the same boundary values as f_0 on ∂M.

Hamilton also treated natural Neumann and mixed boundary problems with similar results.

Chapter 4
Evolution of the Curvature

The Ricci flow is introduced in this chapter as a geometric heat-type equation for the metric. In Sect. 4.4 we derive evolution equations for the curvature, and its various contractions, whenever the metric evolves by Ricci flow. These equations, particularly that of Theorem 4.14, are pivotal to our analysis throughout the coming chapters. In Sect. 4.5.3 we discuss a historical result concerning the convergence theory for the Ricci flow in n-dimensions. This will motivational much of the Böhm and Wilking analysis discussed in Chap. 11.

4.1 Introducing the Ricci Flow

The Ricci flow was first introduced by Richard Hamilton in the early 1980s. Inspired by Eells and Sampson's work on Harmonic map heat flow [ES64] (where they take maps between manifolds and try to 'make them better'), Hamilton speculated that it should be possible to take other geometric objects, for instance the metric g_{ij}, and try to 'improve it' by means of a heat-type equation.

In looking for a suitable parabolic equation, one would like $\frac{\partial g_{ij}}{\partial t}$ to equal a Laplacian-type expression involving second-order derivatives of the metric. Computing derivatives of the metric with the Levi–Civita connection is of no help, as they vanish in normal coordinates. However, computing derivatives with respect to a fixed background connection does give something non-trivial. In particular, the Ricci tensor in Lemma 2.47 has an expression which involves second derivatives the components of the metric, thus it is a natural candidate for such a $(2,0)$-tensor (specifically, the last term in the bracket in Lemma 2.47 is $-\frac{1}{2}$ times a 'Laplacian' of the metric computed using coordinate second derivatives). Taking this factor into account, we are led to the evolution equation

$$\frac{\partial}{\partial t}g_{ij} = -2\,\mathrm{Ric}_{ij} \tag{4.1}$$

B. Andrews and C. Hopper, *The Ricci Flow in Riemannian Geometry*, Lecture Notes in Mathematics 2011, DOI 10.1007/978-3-642-16286-2_4,

known simply as the Ricci flow. Further motivation comes from the expression for the metric in exponential coordinates given in Theorem 2.65, as it implies that the Laplacian of the metric computed at the origin in exponential coordinates is equal to $-\frac{2}{3}\mathrm{Ric}_{ij}$.

In contrast to other natural geometric equations (such as the minimal surface equation and its parabolic analogue the mean curvature flow; the harmonic map equation and the associated flow; and many other examples), the Ricci flow was not initially derived as the gradient flow of a geometric functional. However despite this, Perelman's work [Per02] shows there is a natural energy functional lying behind the Ricci flow. We discuss this in Chap. 10.

In light of (4.1), it is clear that we are no longer interested in just a single manifold (M, g), but rather – assuming a suitable local existence theory – a manifold with a one-parameter family of metrics $t \mapsto g(t)$ parametrised by a 'time' t. We adopt this point of view here, taking the metrics $g(t)$ as sections of the fixed bundle $\mathrm{Sym}^2 T^*M$. The rate of change in time of the metric makes sense since at each point $p \in M$ we are simply differentiating $g(t)$ in the vector space given by the fibre of this bundle at p. Later (in Chap. 6) we will see that there are advantages in instead working with bundles defined over $M \times \mathbb{R}$, where a more geometrically meaningful notion of the time derivative can be used.

4.1.1 Exact Solutions

In order to get a feel for the evolution equation, we present some simple solutions of the Ricci flow.

4.1.1.1 Einstein Metrics

If the initial metric is Ricci flat, that is Ric=0, then clearly the metric remains stationary for all subsequent times. Concrete examples of this are the Euclidean space $\mathbb{R}^n$ and the flat torus $\mathbb{T}^n = S^1 \times \cdots \times S^1$.

If the initial metric is Einstein, that is $\mathrm{Ric}(g_0) = \lambda g_0$ for some $\lambda \in \mathbb{R}$, then a solution $g(t)$ with $g(0) = g_0$ is given by $g(t) = (1 - 2\lambda t)g_0$. The cases $\lambda > 0$, $\lambda = 0$ and $\lambda < 0$ correspond to shrinking, steady and expanding solutions. The simplest shrinking solution is that of the unit sphere (S^n, g_0). Here $\mathrm{Ric}(g_0) = (n-1)g_0$, so $g(t) = (1-2(n-1)t)g_0$. Thus the sphere will collapse to a point in finite time $T = 1/2(n-1)$. By contrast, if the initial metric g_0 were hyperbolic, $\mathrm{Ric}(g_0) = -(n-1)g_0$ so the evolution $g(t) = (1+2(n-1)t)g_0$ will expand the manifold homothetically for all time. In this case, the solution only goes back to $T = -1/2(n-1)$ upon which the metric explodes out of a single point.

4.1.1.2 Quotient Metrics

If the initial Riemannian manifold $N = M/\mathcal{G}$ is a quotient of a Riemannian manifold M by a discrete group of isometries $\mathcal{G}$, it will remain so under the Ricci flow – as the flow on M preserves the isometry group. For example, the projective space $\mathbb{R}P^n = S^n/\mathbb{Z}_2$ of constant curvature shrinks to a point, as does its cover S^n.

4.1.1.3 Product Metrics

If we take the product metric (cf. Sect. 2.4.2) on a product manifold $M \times N$ initially, the metric will remain a product metric under the Ricci flow. Hence the metric on each factor evolves by the Ricci flow independently of the other.

For example, on $S^2 \times S^1$ the S^2 shrinks to a point while S^1 stays fixed so that the manifold collapses to a circle. Moreover, if we take any Riemannian manifold (M, g) and evolve the metric g by the Ricci flow, then it will also evolve on the product manifold $M \times \mathbb{R}^n$, $n \geq 1$, with product metric $g \oplus (dx_1^2 + \cdots + dx_n^2)$ – since the flow is stationary on the $\mathbb{R}^n$ part.

4.1.2 Diffeomorphism Invariance

The curvature of a manifold M can be thought of as the obstruction to being locally isometric to Euclidean space. Indeed, a Riemannian manifold is flat if and only if its curvature tensor vanishes identically. Moreover, recall that:

Theorem 4.1. *The Riemann curvature tensor R is invariant under local isometries: If $\varphi : (M, g) \to (\widetilde{M}, \widetilde{g})$ is a local isometry, then $\varphi^* \widetilde{R} = R$.*

So if $\varphi : M \to \widetilde{M}$ is a time-*independent* diffeomorphism such that $g(t) = \varphi^* \widetilde{g}(t)$ and $\widetilde{g}$ is a solution of the Ricci flow, we see that

$$\begin{aligned}\frac{\partial}{\partial t} g &= \varphi^* \Big(\frac{\partial}{\partial t} \widetilde{g} \Big) \\ &= \varphi^* \big(-2 \operatorname{Ric}(\widetilde{g}) \big) \\ &= -2 \operatorname{Ric}(\varphi^* \widetilde{g}) \\ &= -2 \operatorname{Ric}(g),\end{aligned}$$

where the second last equality is due to Theorem 4.1. Hence g is also a solution to the Ricci flow. Therefore we conclude that (4.1) is invariant under the full diffeomorphism group.[1]

---

[1] Note that if φ is time-*dependent*, an extra Lie derivative term is introduced into the equation (cf. Sect. 5.4).

4.1.2.1 Preservation of Symmetries

The Ricci flow preserves any symmetries that are present in the initial metric. To see this, note that each symmetry is an isometric diffeomorphism of the initial metric; so the pullback of a solution of the Ricci flow by this diffeomorphism gives a solution of Ricci flow. Since the symmetry is an isometry of the initial metric, these are solutions of Ricci flow with the same initial data, and so are identical (assuming the uniqueness result proved in Chap. 5). Therefore the symmetry is an isometry of the metric at any positive time.

4.1.3 Parabolic Rescaling of the Ricci Flow

Aside from the geometric symmetries of diffeomorphism invariance, the Ricci flow has additional scaling properties that are essential for blow-up analysis of singularities. The time-independent diffeomorphism invariance allows for changes in the spatial coordinates. We can also translate in the time coordinates: If $g(x,t)$ satisfies Ricci flow, then so does $g(x, t-t_0)$ for any $t_0 \in \mathbb{R}$. Further, the Ricci flow has a *scale invariance*: If g is a solution of Ricci flow, and $\lambda > 0$, then g_λ is also a solution, where

$$g_\lambda(x,t) = \lambda^2 g\left(x, \frac{t}{\lambda^2}\right).$$

The main use of this rescaling will be to analyse singularities that develop under the Ricci flow. In such a case the curvature tends to infinity, so we perform a rescaling to produce metrics with bounded curvature and try to produce a smooth limit of these. We will discuss the machinery required for this in Chap. 9, and return to the blow-up procedure in Sect. 9.5.

4.2 The Laplacian of Curvature

We devote this section to the derivation of an expression for the Laplacian of $R_{ijk\ell}$. This will be used to derive an evolution equation for the curvature in Sect. 4.4, where the Ricci flow (4.1) implies a heat-type equation for the Riemannian curvature R. In order to do this, we need to introduce quadratic $B_{ijk\ell}$ terms which will help simplify our computation.

4.2.1 Quadratic Curvature Tensor

Various second order derivatives of the curvature tensor are likely to differ by terms quadratic in the curvature tensor. To this end we introduce the $(4,0)$-tensor $B \in \mathscr{T}_0^4(M)$ defined by

$$B(X,Y,W,Z) = \langle R(X,\cdot,Y,\star), R(W,\cdot,Z,\star)\rangle\,,$$

where the inner product $\langle,\rangle$ is given by (2.3). In components this becomes

$$B_{ijk\ell} = g^{pr}g^{qs}R_{piqj}R_{rks\ell} = R^{p\ \,q}_{\ i\ \,j}R_{pkq\ell}. \tag{4.2}$$

This tensor has some of the symmetries of the curvature tensor, namely

$$B_{ijk\ell} = B_{ji\ell k} = B_{k\ell ij}.$$

However other symmetries of the curvature tensor may fail to hold for $B_{ijk\ell}$.

4.2.2 Calculating the Connection Laplacian $\Delta R_{ijk\ell}$

Using the definition of the Laplacian explicated in Sect. 2.6, we compute the connection Laplacian acting on the $\mathscr{T}_0^4(M)$ tensor bundle.

Proposition 4.2. *On a Riemannian manifold (M,g), the Laplacian of the curvature tensor R satisfies*

$$\begin{aligned}\Delta R_{ijk\ell} &= \nabla_i\nabla_k R_{j\ell} - \nabla_j\nabla_k R_{i\ell} + \nabla_j\nabla_\ell R_{ik} - \nabla_i\nabla_\ell R_{jk}\\ &\quad - 2\big(B_{ijk\ell} - B_{ij\ell k} - B_{i\ell jk} + B_{ikj\ell}\big) + g^{pq}\big(R_{qjk\ell}R_{pi} + R_{iqk\ell}R_{pj}\big).\end{aligned}$$

Proof. Using the second Bianchi identity – together with the linearity of ∇ over the space of tensor fields – we find that

$$\begin{aligned}0 &= (\nabla_p\nabla_q R)(\partial_i,\partial_j,\partial_k,\partial_\ell)\\ &\quad + (\nabla_p\nabla_i R)(\partial_j,\partial_q,\partial_k,\partial_\ell) + (\nabla_p\nabla_j R)(\partial_q,\partial_i,\partial_k,\partial_\ell).\end{aligned}$$

Combining this identity with (2.10) implies that the Laplacian of R takes the form

$$\begin{aligned}\Delta R_{ijk\ell} &= (\mathrm{tr}_g \nabla^2 R)_{ijk\ell}\\ &= g^{pq}(\nabla_{\partial_p}\nabla_{\partial_q} R)(\partial_i,\partial_j,\partial_k,\partial_\ell)\\ &= g^{pq}\left(-(\nabla_p\nabla_i R)(\partial_j,\partial_q,\partial_k,\partial_\ell) - (\nabla_p\nabla_j R)(\partial_q,\partial_i,\partial_k,\partial_\ell)\right)\\ &= g^{pq}\big(\nabla_p\nabla_i R_{qjk\ell} - \nabla_p\nabla_j R_{qik\ell}\big).\end{aligned}$$

From this it suffices to express $g^{pq}\nabla_p\nabla_i R_{qjk\ell}$ in terms of lower order terms by commuting derivatives and contracting with the metric.

Firstly, Proposition 2.44 implies that

$$\nabla_p\nabla_i R_{qjk\ell} = \nabla_i\nabla_p R_{qjk\ell} + \big(R(\partial_i,\partial_p)R\big)(\partial_q,\partial_j,\partial_k,\partial_\ell). \tag{4.3}$$

From the second Bianchi identity, the first term on the right-hand side of (4.3) becomes

$$\nabla_i\nabla_p R_{qjk\ell} = \nabla_i\nabla_k R_{jq\ell p} - \nabla_i\nabla_\ell R_{jqkp}.$$

Contracting with the metric and invoking Proposition 2.48, (2.15), gives

$$\begin{aligned} g^{pq}\nabla_i\nabla_p R_{qjk\ell} &= g^{pq}\nabla_i\nabla_k R_{jq\ell p} - g^{pq}\nabla_i\nabla_\ell R_{jqkp} \\ &= \nabla_i\nabla_k R_{j\ell} - \nabla_i\nabla_\ell R_{jk}. \end{aligned} \tag{4.4}$$

Turning our attention to the second term on the right-hand side of (4.3), we find by Proposition 2.43 that

$$\begin{aligned} &\big(R(\partial_i,\partial_p)R\big)(\partial_q,\partial_j,\partial_k,\partial_\ell) \\ &\quad = \cancel{R(\partial_i,\partial_p)(R_{qjk\ell})} \\ &\qquad - R(R(\partial_i,\partial_p)\partial_q,\partial_j,\partial_k,\partial_\ell) - \cdots - R(\partial_q,\partial_j,\partial_k,R(\partial_i,\partial_p)\partial_\ell) \\ &\quad = -R_{ipq}{}^{n}R_{njk\ell} - R_{ipq}{}^{n}R_{qnk\ell} - R_{ipq}{}^{n}R_{qjn\ell} - R_{ipq}{}^{n}R_{qjkn} \\ &\quad = g^{mn}\big(R_{piqm}R_{njk\ell} + R_{pijm}R_{qnk\ell} + R_{pikm}R_{qjn\ell} + R_{pi\ell m}R_{qjkn}\big). \end{aligned}$$

Therefore

$$\begin{aligned} &g^{pq}\big(R(\partial_i,\partial_p)R\big)(\partial_q,\partial_j,\partial_k,\partial_\ell) \\ &\quad = g^{pq}g^{mn}\big(R_{piqm}R_{njk\ell} + R_{pijm}R_{qnk\ell} + R_{pikm}R_{qjn\ell} + R_{pi\ell m}R_{qjkn}\big). \end{aligned}$$

We now look to simplify this expression by observing that the first term contracts to

$$g^{pq}g^{mn}R_{piqm}R_{njk\ell} = g^{mn}R_{ipm}{}^{p}R_{njk\ell} = g^{mn}R_{njk\ell}R_{im} = g^{pq}R_{pjk\ell}R_{iq},$$

the second term contracts to

$$\begin{aligned} g^{pq}g^{mn}R_{pijm}R_{qnk\ell} &= g^{pq}g^{mn}R_{pijm}(-R_{nkq\ell} - R_{kqn\ell}) \\ &= g^{pq}g^{mn}(R_{pimj}R_{q\ell nk} - R_{pimj}R_{qkn\ell}) \\ &= R^{q}{}_{i}{}^{n}{}_{j}R_{q\ell nk} - R^{q}{}_{i}{}^{n}{}_{j}R_{qkn\ell} \\ &= B_{ij\ell k} - B_{ijk\ell}, \end{aligned}$$

and the third and forth terms contract to

$$\begin{aligned} g^{pq}g^{mn}(R_{pikm}R_{qjn\ell} + R_{pi\ell m}R_{qjkn}) &= g^{pq}g^{mn}(-R_{pimk}R_{qjn\ell} + R_{pim\ell}R_{qjnk}) \\ &= -R^{q\ n}_{\ i\ k}R_{qjn\ell} + R^{q\ n}_{\ i\ \ell}R_{qjnk} \\ &= -B_{ikj\ell} + B_{i\ell jk}. \end{aligned}$$

Combining these results gives

$$\begin{aligned} &g^{pq}\big(R(\partial_i, \partial_p)R\big)(\partial_q, \partial_j, \partial_k, \partial_\ell) \\ &\qquad = g^{pq}R_{pjk\ell}R_{iq} - (B_{ijk\ell} - B_{ij\ell k} - B_{i\ell jk} + B_{ikj\ell}). \end{aligned} \tag{4.5}$$

Finally, by putting (4.4) and (4.5) together we get

$$\begin{aligned} g^{pq}\nabla_p\nabla_i R_{qjk\ell} = {} & \nabla_i\nabla_k R_{j\ell} - \nabla_i\nabla_\ell R_{jk} \\ & - \ (B_{ijk\ell} - B_{ij\ell k} - B_{i\ell jk} + B_{ikj\ell}) + g^{pq}R_{pjk\ell}R_{iq}. \end{aligned}$$

Since $\Delta R_{ijk\ell} = g^{pq}\nabla_p\nabla_i R_{qjk\ell} - (i \leftrightarrow j)$ the desired formula now follows. □

4.3 Metric Variation Formulas

In this section we establish how one can formally take the time derivative of a metric and the associated Levi–Civita connection. Thereafter we derive various variational equations for the Levi–Civita connection, curvature tensor and various traces thereof.

4.3.1 Interpreting the Time Derivative

Consider a one-parameter family of smooth metrics $g = g(t) \in \Gamma(\mathrm{Sym}^2 T^*M)$ parametrised by 'time' t. We define the time derivative $\frac{\partial}{\partial t}g : \mathscr{X}(M) \times \mathscr{X}(M) \to C^\infty(M)$ of the metric g by letting

$$\Big(\frac{\partial}{\partial t}g\Big)(X, Y) := \frac{\partial}{\partial t}g(X, Y) \tag{4.6}$$

for any time *independent* vector fields $X, Y \in \mathscr{X}(M)$ (where $\frac{\partial}{\partial t}g(X, Y)$ is the time derivative of the smooth function $g(X, Y) \in C^\infty(M)$ given by the standard difference quotient). The metric in local coordinates can be expressed as $g(t) = g_{ij}(t)dx^i \otimes dx^j$, in which case (4.6) implies that

$$\frac{\partial}{\partial t}g = \dot{g}_{ij}(t)\, dx^i \otimes dx^j.$$

Therefore we regard the time derivative of the metric as the derivative of its the component functions with respect to a fixed basis.

Since the Levi–Civita connection ∇ can be written locally in terms of the metric (2.9), it too will be time dependent. So in a similar fashion, we define the time derivative of $\nabla = \nabla^{(t)}$ by letting

$$\Big(\frac{\partial}{\partial t}\nabla\Big)(X,Y) := \frac{\partial}{\partial t}\nabla_X Y \tag{4.7}$$

for time-independent vector fields X and Y. As ∇ satisfies the product rule by definition, it is not tensorial. However, we observe the following special properties of its time derivative:

Lemma 4.3. *The time derivative* $\frac{\partial}{\partial t}\nabla$ *of the Levi–Civita connection* ∇ *is tensorial.*

Proof. For $f \in C^\infty(M)$ and any time independent vector fields $X, Y \in \mathscr{X}(M)$, we see that

$$\Big(\frac{\partial}{\partial t}\nabla\Big)(X,fY) = \frac{\partial}{\partial t}\Big((Xf)Y + f\nabla_X Y\Big) = f\frac{\partial}{\partial t}\nabla_X Y = f\Big(\frac{\partial}{\partial t}\nabla\Big)(X,Y).$$

Thus $\frac{\partial}{\partial t}\nabla$ is a tensor by Proposition 2.19. □

Lemma 4.4. *If* X *is a time independent vector field and* $V = V(t)$ *is a time dependent vector field, then*

$$\frac{\partial}{\partial t}\nabla_X V = \Big(\frac{\partial}{\partial t}\nabla\Big)(X,V) + \nabla_X\frac{\partial V}{\partial t}.$$

Proof. Fix a time independent vector field X. As $\nabla_X = \nabla_X^{(t)} : \mathscr{X}(M) \to \mathscr{X}(M)$ we see that

$$\frac{\partial}{\partial t}\nabla_X V \lim_{\delta\to 0}\frac{\nabla^{(t+\delta)}(X,V(t+\delta)) - \nabla^{(t)}(X,V(t))}{\delta}.$$

Since

$$\begin{aligned}
&\nabla^{(t+\delta)}(X,V(t+\delta)) - \nabla^{(t)}(X,V(t))\\
&\qquad = \nabla^{(t+\delta)}(X,V(t+\delta)) - \nabla^{(t+\delta)}(X,V(t))\\
&\qquad\quad + \Big(\nabla^{(t+\delta)} - \nabla^{(t)}\Big)(X,V(t))\\
&\qquad = \nabla^{(t+\delta)}(X,V(t+\delta) - V(t)) + \big(\nabla^{(t+\delta)} - \nabla^{(t)}\big)(X,V(t)),
\end{aligned}$$

we conclude that

$$\begin{aligned}\frac{\partial}{\partial t}\nabla_X V &= \lim_{\delta\to 0}\frac{1}{\delta}\nabla^{(t+\delta)}(X, V(t+\delta) - V(t)) \\ &\quad + \lim_{\delta\to 0}\frac{1}{\delta}\left(\nabla^{(t+\delta)} - \nabla^{(t)}\right)(X, V(t)) \\ &= \nabla_X^{(t)}\left(\lim_{\delta\to 0}\frac{V(t+\delta) - V(t)}{\delta}\right) + \left(\frac{\partial}{\partial t}\nabla\right)(X, V).\end{aligned}$$

□

Furthermore, we would like to differentiate the Christoffel symbols $\Gamma_{ij}^k = dx^k(\nabla_{\partial_i}\partial_j)$ of the Levi–Civita connection ∇. To do this, we consider them as a map $\Gamma : \mathscr{X}(M) \times \mathscr{X}(M) \times \mathscr{T}_0^1(M) \to C^\infty(M)$ defined by $\Gamma(X, Y, \omega) := \omega(\nabla_X Y)$. With this we proceed in a similar fashion by defining $\frac{\partial}{\partial t}\Gamma$ to be

$$\left(\frac{\partial}{\partial t}\Gamma\right)(X, Y, \omega) := \frac{\partial}{\partial t}\Gamma(X, Y, \omega). \tag{4.8}$$

It is easy so see that $\frac{\partial}{\partial t}\Gamma$ is tensorial since $(\frac{\partial}{\partial t}\Gamma)(X, Y, \omega) = \frac{\partial}{\partial t}\omega(\nabla_X Y) = \omega(\frac{\partial}{\partial t}\nabla_X Y)$ and $\frac{\partial}{\partial t}\nabla$ is a tensor by Lemma 4.3.

4.3.2 Variation Formulas of the Curvature

We now derive evolution equations for geometric quantities under arbitrary metric variations. Note that $\frac{\partial}{\partial t}T = (\delta_{\dot g}T)(g) = (\delta_h T)(g)$ for a (k, ℓ)-tensor T, whenever the direction $h = \frac{\partial}{\partial t}g$ (where the first variation δ_h is taken in the sense of Appendix A).

Lemma 4.5. *Suppose $g(t)$ is a smooth one-parameter family of metrics on a manifold M such that $\frac{\partial}{\partial t}g = h$. Then*

$$\frac{\partial}{\partial t}g^{ij} = -g^{ik}g^{j\ell}h_{k\ell}. \tag{4.9}$$

Proof. As $g^{ik}g_{k\ell} = \delta_\ell^i$ we find that $0 = (\frac{\partial}{\partial t}g^{ik})g_{k\ell} + g^{ik}h_{k\ell}$. In which case $\frac{\partial}{\partial t}g^{ij} = -g^{ik}g^{j\ell}h_{k\ell} = -h^{ij}$. □

Proposition 4.6. *Suppose $g(t)$ is a smooth one-parameter family of metrics on a manifold M such that $\frac{\partial}{\partial t}g = h$. Then the Levi–Civita connection Γ_{ij}^k of g evolves by*

$$\frac{\partial}{\partial t}\Gamma_{ij}^k = \frac{1}{2}g^{k\ell}\Big((\nabla_j h)(\partial_i, \partial_\ell) + (\nabla_i h)(\partial_j, \partial_\ell) - (\nabla_\ell h)(\partial_i, \partial_j)\Big).$$

Proof. As $\frac{\partial}{\partial t}\Gamma$ and ∇h are tensorial, we are free to work in any local coordinate chart (x^i). In particular, by choosing normal coordinates about a point p (and evaluating the computation at that point), (2.9) implies that

$$\frac{\partial}{\partial t}\Gamma_{ij}^{k} = \frac{1}{2}\Big(\frac{\partial}{\partial t}g^{k\ell}\Big)\cancel{(\partial_j g_{i\ell} + \partial_i g_{j\ell} - \partial_\ell g_{ij})} + \frac{1}{2}g^{k\ell}\frac{\partial}{\partial t}\Big(\partial_j g_{i\ell} + \partial_i g_{j\ell} - \partial_\ell g_{ij}\Big).$$

Using Proposition 2.29 we observe that

$$\frac{\partial}{\partial t}\partial_j g_{i\ell} = \partial_j\big(h(\partial_i,\partial_\ell)\big) = (\nabla_j h)(\partial_i,\partial_\ell) + h(\cancel{\nabla_j\partial_i},\partial_\ell) + h(\partial_i,\cancel{\nabla_j\partial_\ell}),$$

from which the result now follows. □

Proposition 4.7. *Suppose $g(t)$ is a smooth one-parameter family of metrics on a manifold M such that $\frac{\partial}{\partial t}g = h$. Then the Riemannian curvature tensor evolves by*

$$\begin{aligned}\frac{\partial}{\partial t}R_{ijk}{}^{\ell} = \frac{1}{2}g^{\ell p}\Big(&\nabla_i\nabla_p h_{jk} + \nabla_j\nabla_k h_{ip} - \nabla_i\nabla_k h_{jp} - \nabla_j\nabla_p h_{ik}\\ &- R_{ijk}{}^{q}h_{qp} - R_{ijp}{}^{q}h_{qk}\Big).\end{aligned}$$

Proof. As the expression is tensorial, we are free to work in normal coordinates in a neighbourhood about a point p (and evaluating the expression at that point). Using Lemma 2.46 in these coordinates, we find that

$$\begin{aligned}\frac{\partial}{\partial t}R_{ijk}{}^{\ell} &= \frac{\partial}{\partial t}(\partial_j\Gamma_{ik}^{\ell} - \partial_i\Gamma_{jk}^{\ell})\\ &\quad + \Big(\Big(\frac{\partial}{\partial t}\Gamma_{ik}^{m}\Big)\cancel{\Gamma_{jm}^{\ell}} + \cancel{\Gamma_{ik}^{m}}\Big(\frac{\partial}{\partial t}\Gamma_{jm}^{\ell}\Big)\Big) - \cancel{(i\leftrightarrow j)}\\ &= \partial_j\Big(\frac{\partial}{\partial t}\Gamma_{ik}^{\ell}\Big) - \partial_i\Big(\frac{\partial}{\partial t}\Gamma_{jk}^{\ell}\Big).\end{aligned}$$

From this, together with Proposition 4.6, we have that

$$\begin{aligned}\partial_j\Big(\frac{\partial}{\partial t}\Gamma_{ik}^{\ell}\Big) &= \frac{1}{2}\cancel{\partial_j g^{\ell p}}\Big(\nabla_k h_{ip} + \nabla_i h_{kp} - \nabla_p h_{ik}\Big)\\ &\quad + \frac{1}{2}g^{\ell p}\Big(\partial_j(\nabla_k h_{ip} + \nabla_i h_{kp} - \nabla_p h_{ik})\Big)\\ &= \frac{1}{2}g^{\ell p}\Big(\nabla_j\nabla_k h_{ip} + \nabla_j\nabla_i h_{kp} - \nabla_j\nabla_p h_{ik}\Big),\end{aligned}$$

where the last equality – using Proposition 2.29 – is due to

$$\begin{aligned}\partial_j\big((\nabla h)(\partial_i,\partial_k,\partial_\ell)\big) &= \big(\nabla_j(\nabla h)\big)(\partial_i,\partial_k,\partial_\ell)\\ &\quad + (\nabla h)(\cancel{\nabla_j\partial_i},\ldots) + \cdots + (\nabla h)(\ldots,\cancel{\nabla_j\partial_\ell})\\ &= (\nabla_j\nabla_i h)(\partial_k,\partial_\ell).\end{aligned}$$

Therefore,

$$\begin{aligned}\frac{\partial}{\partial t} R_{ijk}{}^{\ell} &= \frac{1}{2} g^{\ell p} (\nabla_j \nabla_k h_{ip} + \nabla_j \nabla_i h_{kp} - \nabla_j \nabla_p h_{ik}) \\ &\quad - \frac{1}{2} g^{\ell p} (\nabla_i \nabla_k h_{jp} + \nabla_i \nabla_j h_{kp} - \nabla_i \nabla_p h_{jk}) \qquad (4.10) \\ &= \frac{1}{2} g^{\ell p} \Big(\nabla_i \nabla_p h_{jk} + \nabla_j \nabla_k h_{ip} - \nabla_i \nabla_k h_{jp} - \nabla_j \nabla_p h_{ik} \\ &\qquad + \big(R(\partial_i, \partial_j) h\big)(\partial_k, \partial_p) \Big) \qquad (4.11)\end{aligned}$$

since $\nabla_j \nabla_i h_{kp} - \nabla_i \nabla_j h_{kp} = (R(\partial_i, \partial_j) h)(\partial_k, \partial_p)$ by Proposition 2.44. The desired equation now follows since we find, by Proposition 2.43, that

$$\begin{aligned}\big(R(\partial_i, \partial_j) h\big)(\partial_k, \partial_p) &= \cancel{R(\partial_i, \partial_j) h_{kp}} - h(R(\partial_i, \partial_j)\partial_k, \partial_p) - h(\partial_k, R(\partial_i, \partial_j)\partial_p) \\ &= -R_{ijk}{}^{q} h_{qp} - R_{ijp}{}^{q} h_{qk}.\end{aligned}$$

□

Proposition 4.8. *Suppose $g(t)$ is a smooth family of metrics on a manifold M with $\frac{\partial}{\partial t} g = h$. Then the $(4,0)$-Riemann curvature tensor R evolves by*

$$\begin{aligned}\frac{\partial}{\partial t} R_{ijk\ell} &= \frac{1}{2} (\nabla^2_{i,\ell} h_{jk} + \nabla^2_{j,k} h_{i\ell} - \nabla^2_{i,k} h_{j\ell} - \nabla^2_{j,\ell} h_{ik}) \\ &\quad + \frac{1}{2} g^{pq} \big(R_{ijkp} h_{q\ell} + R_{ijp\ell} h_{qk}\big).\end{aligned}$$

Proof. As $R_{ijk\ell} = R_{ijk}{}^{m} g_{m\ell}$, we have

$$\frac{\partial}{\partial t} R_{ijk\ell} = R_{ijk}{}^{q} \frac{\partial}{\partial t} g_{q\ell} + g_{a\ell} \frac{\partial}{\partial t} R_{ijk}{}^{a}.$$

Now since

$$R_{ijk}{}^{q} \frac{\partial}{\partial t} g_{q\ell} = g^{pq} R_{ijkp} h_{q\ell},$$

and (by Proposition 4.7)

$$\begin{aligned}g_{a\ell} \frac{\partial}{\partial t} R_{ijk}{}^{a} &= \frac{1}{2} (\nabla^2_{i,\ell} h_{jk} + \nabla^2_{j,k} h_{i\ell} - \nabla^2_{i,k} h_{j\ell} - \nabla^2_{j,\ell} h_{ik}) \\ &\quad - \frac{1}{2} g^{pq} \big(R_{ijkp} h_{q\ell} + R_{ij\ell p} h_{qk}\big)\end{aligned}$$

the result naturally follows. □

Proposition 4.9. *Suppose $g(t)$ is a smooth one-parameter family of metrics on a manifold M such that $\frac{\partial}{\partial t} g = h$. Then the Ricci tensor* Ric *evolves by*

$$\frac{\partial}{\partial t} R_{ik} = \frac{1}{2} g^{pq} \big(\nabla^2_{q,k} h_{ip} - \nabla^2_{i,k} h_{qp} + \nabla^2_{q,i} h_{kp} - \nabla^2_{q,p} h_{ik}\big).$$

Proof. Recall from (4.10) that

$$\frac{\partial}{\partial t} R_{ijk}{}^{\ell} = \frac{1}{2} g^{\ell p} \big(\nabla^2_{j,k} h_{ip} + \nabla^2_{j,i} h_{kp} - \nabla^2_{j,p} h_{ik} \\ - \nabla^2_{i,k} h_{jp} - \nabla^2_{i,j} h_{kp} + \nabla^2_{i,p} h_{jk} \big).$$

From which we find that

$$\frac{\partial}{\partial t} R_{ik} = \frac{\partial}{\partial t} R_{ijk}{}^{j} = \frac{1}{2} g^{jp} \big(\nabla^2_{j,k} h_{ip} - \nabla^2_{i,k} h_{jp} + \nabla^2_{j,i} h_{kp} - \nabla^2_{j,p} h_{ik} \big) \\ + \frac{1}{2} g^{qp} \big(- \nabla^2_{i,q} h_{kp} + \nabla^2_{i,p} h_{qk} \big).$$ □

Proposition 4.10. *Suppose $g(t)$ is a smooth one-parameter family of metrics on a manifold M with $\frac{\partial}{\partial t} g = h$. Then the scalar curvature* Scal *evolves by*

$$\frac{\partial}{\partial t} \mathrm{Scal} = -\Delta \mathrm{tr}_g h + \delta^2 h - \langle h, \mathrm{Ric} \rangle ,$$

where $\delta^2 h = g^{ij} g^{pq} \nabla^2_{q,j} h_{pi}$ is the 'divergence term'.

Proof. Using (4.9) and Proposition 4.9 we find that

$$\begin{aligned}\frac{\partial}{\partial t} \mathrm{Scal} &= R_{ik} \frac{\partial}{\partial t} g^{ik} + g^{ik} \frac{\partial}{\partial t} R_{ik} \\ &= -g^{ij} g^{k\ell} h_{j\ell} R_{ik} + \frac{1}{2} g^{ik} g^{pq} \Big(\nabla^2_{q,k} h_{ip} - \nabla^2_{i,k} h_{qp} + \nabla^2_{q,i} h_{kp} - \nabla^2_{q,p} h_{ik} \Big) \\ &= -g^{ij} g^{pq} h_{jq} R_{ip} + g^{ij} g^{pq} \nabla^2_{q,j} h_{ip} - g^{ij} g^{pq} \nabla^2_{i,j} h_{pq} \\ &= -h_{jq} R^{jq} + g^{ij} g^{pq} \nabla^2_{q,j} h_{pi} - \Delta \mathrm{tr}_g h.\end{aligned}$$ □

Proposition 4.11. *Suppose $g(t)$ is a smooth one-parameter family of metrics on a manifold M with $\frac{\partial}{\partial t} g = h$. Then the volume form $d\mu(g(t))$ evolves by*

$$\frac{\partial}{\partial t} d\mu = \frac{1}{2} \mathrm{tr}\, h \, d\mu.$$

To prove this, recall that:

Definition 4.12. The *adjunct* of a square matrix A is defined as the transpose of the cofactor matrix of A, that is

$$\mathrm{adj}\, A = \begin{pmatrix} \det A_{11} & -\det A_{21} & \dots \\ -\det A_{12} & \det A_{22} & \dots \\ \vdots & \vdots & \ddots \end{pmatrix}$$

where A_{ij} is obtained from A by striking out the i-th row and the j-th column.

Also recall that:

Lemma 4.13. *If A is a square matrix then $A \operatorname{adj} A = \operatorname{adj} A\, A = \det A\, I_{n\times n}$. Moreover, if $\det A \neq 0$ then $A^{-1} = \frac{1}{\det A}\operatorname{adj} A$.*

So if A is a square matrix with (i,j)-th entry a_{ij}, we can expand $\det A$ along the i-th row to get: $\det A = \sum_{k=1}^{n}(-1)^{i+k}\det A_{ik}$. Therefore by Lemma 4.13, the partial derivative of $\det A$ with respect to the (i,j)-th entry are

$$\begin{aligned}\frac{\partial}{\partial a_{ij}}\det A &= (-1)^{i+j}\det A_{ij}\\ &= (\operatorname{adj} A)_{ji} = \det A\,(A^{-1})_{ji}.\end{aligned} \tag{4.12}$$

Proof (Proposition 4.11). In local coordinates (x^i) the volume form can be written as $d\mu = \sqrt{\det g}\,dx^1\wedge\ldots\wedge x^n$. So by (4.12) and the chain rule,

$$\begin{aligned}\frac{\partial}{\partial t}\sqrt{\det g} &= \frac{1}{2}\frac{1}{\sqrt{\det g}}\frac{\partial}{\partial t}\det g\\ &= \frac{1}{2}\frac{1}{\sqrt{\det g}}\frac{\partial \det g}{\partial g_{ij}}\frac{\partial g_{ij}}{\partial t}\\ &= \frac{1}{2}\sqrt{\det g}\,(g^{-1})_{ji}h_{ij}\\ &= \frac{1}{2}g^{ij}h_{ij}\sqrt{\det g} = \frac{1}{2}\operatorname{tr} h\sqrt{\det g}.\end{aligned}$$

Thus,

$$\frac{\partial}{\partial t}d\mu = \frac{\partial\sqrt{\det g}}{\partial t}dx^1\wedge\ldots\wedge x^n = \frac{1}{2}\operatorname{tr} h\,d\mu. \qquad \square$$

4.4 Evolution of the Curvature Under the Ricci Flow

Using the results of the previous sections, it is now a relatively easy task to derive the evolution equations of the curvature, and its various traces, under the Ricci flow.

Theorem 4.14. *Suppose $g(t)$ is a solution of the Ricci flow, the $(4,0)$-Riemannian tensor R evolves by*

$$\begin{aligned}\frac{\partial}{\partial t}R_{ijk\ell} &= \Delta R_{ijk\ell} + 2(B_{ijk\ell} - B_{ij\ell k} - B_{i\ell jk} + B_{ikj\ell})\\ &\quad - g^{pq}(R_{pjk\ell}R_{qi} + R_{ipk\ell}R_{qj} + R_{ijkp}R_{q\ell} + R_{ijp\ell}R_{qk}).\end{aligned}$$

Proof. By Proposition 4.8, with $\frac{\partial}{\partial t}g_{ij} = -2R_{ij}$, the time derivative of $R_{ijk\ell}$ satisfies

$$\begin{aligned}\nabla^2_{i,\ell}R_{jk} + \nabla^2_{j,k}R_{i\ell} - \nabla^2_{i,k}R_{j\ell} - \nabla^2_{j,\ell}R_{ik} \\ = -\frac{\partial}{\partial t}R_{ijk\ell} - g^{pq}\big(R_{ijkp}R_{q\ell} + R_{ijp\ell}R_{qk}\big).\end{aligned}$$

By Proposition 4.2, with indices k and ℓ switched, the Laplacian of R satisfies

$$\begin{aligned}&\nabla^2_{i,\ell}R_{jk} - \nabla^2_{j,\ell}R_{ik} + \nabla^2_{j,k}R_{i\ell} - \nabla^2_{i,k}R_{j\ell} \\ &\quad = \Delta R_{ij\ell k} + 2(B_{ij\ell k} - B_{ijk\ell} - B_{ikj\ell} + B_{i\ell jk}) \\ &\qquad - g^{pq}(R_{qj\ell k}R_{pi} + R_{iq\ell k}R_{pj}).\end{aligned}$$

Combining these equations gives

$$\begin{aligned}-\Delta R_{ijk\ell} = \Delta R_{ij\ell k} = &-\frac{\partial}{\partial t}R_{ijk\ell} - 2(B_{ij\ell k} - B_{ijk\ell} - B_{ikj\ell} + B_{i\ell jk}) \\ &+ g^{pq}(R_{qj\ell k}R_{pi} + R_{iq\ell k}R_{pj}) \\ &- g^{pq}(R_{ijkp}R_{q\ell} + R_{ijp\ell}R_{qk}).\end{aligned}$$

□

We can also derive, without too much effort, the evolution of the following quantities under the Ricci flow.

Corollary 4.15. *Under the Ricci flow, the connection coefficients evolve by*

$$\frac{\partial}{\partial t}\Gamma^k_{ij} = -g^{k\ell}\Big((\nabla_j \mathrm{Ric})(\partial_i,\partial_\ell) + (\nabla_i \mathrm{Ric})(\partial_j,\partial_\ell) - (\nabla_\ell \mathrm{Ric})(\partial_i,\partial_j)\Big).$$

Corollary 4.16. *Under the Ricci flow, the volume form of g evolves by*

$$\frac{\partial}{\partial t}d\mu = -\mathrm{Scal}\, d\mu.$$

Corollary 4.17. *Under the Ricci flow,*

$$\begin{aligned}\frac{\partial}{\partial t}R_{ik} &= \Delta R_{ik} + \nabla^2_{ik}\mathrm{Scal} - g^{pq}(\nabla^2_{q,i}R_{kp} + \nabla^2_{q,k}R_{ip}), \\ \frac{\partial}{\partial t}\mathrm{Scal} &= 2\Delta\mathrm{Scal} - 2g^{ij}g^{pq}\nabla^2_{q,j}R_{pi} + 2|\mathrm{Ric}|^2.\end{aligned}$$

The proof of these corollaries follow easily by substituting $\frac{\partial}{\partial t}g_{ij} = -2R_{ij}$ into Propositions 4.6, 4.11, 4.9 and 4.10 respectively.

We note that the formulas in the last corollary can be simplified as follows.

Corollary 4.18. *Under the Ricci flow,*

$$\frac{\partial}{\partial t}R_{ik} = \Delta R_{ik} + 2g^{pq}g^{rs}R_{pikr}R_{qs} - 2g^{pq}R_{ip}R_{qk}.$$

Proof. By (4.9) the time derivative of $R_{ik} = g^{j\ell}R_{ijk\ell}$ is $\frac{\partial}{\partial t}R_{ik} = g^{j\ell}\frac{\partial}{\partial t}R_{ijk\ell} - 2g^{jp}g^{\ell q}R_{pq}R_{ijk\ell}$. Substituting the expression for $\frac{\partial}{\partial t}R_{ijk\ell}$ in Theorem 4.14 (with $g^{j\ell}\Delta R_{ijk\ell} = \Delta R_{ik}$ from (2.15)) results in

$$\begin{aligned}\frac{\partial}{\partial t}R_{ik} &= \Delta R_{ik} + 2g^{j\ell}(B_{ijk\ell} - B_{ij\ell k} - B_{i\ell jk} + B_{ikj\ell})\\ &\quad - g^{j\ell}g^{pq}(R_{pjk\ell}R_{qi} + R_{ipk\ell}R_{qj} + R_{ijp\ell}R_{qk} + R_{ijkp}R_{q\ell}).\end{aligned}$$

As we find that

$$\begin{aligned}&2g^{j\ell}(B_{ijk\ell} - B_{ij\ell k} - B_{i\ell jk} + B_{ikj\ell})\\ &\quad = 2g^{j\ell}B_{ijk\ell} - 2g^{j\ell}(B_{i\ell jk} + B_{ij\ell k}) + 2g^{pr}g^{qs}R_{piqk}R_{rs}\\ &\quad = 2g^{j\ell}B_{ijk\ell} - 4g^{j\ell}B_{ij\ell k} + 2g^{pr}g^{qs}R_{piqk}R_{rs}\\ &\quad = 2g^{j\ell}(B_{ijk\ell} - 2B_{ij\ell k}) + 2g^{pr}g^{qs}R_{piqk}R_{rs}\end{aligned}$$

and

$$\begin{aligned}&g^{j\ell}g^{pq}(R_{pjk\ell}R_{qi} + R_{ipk\ell}R_{qj} + R_{ijp\ell}R_{qk} + R_{ijkp}R_{q\ell})\\ &\quad = 2g^{pq}R_{pi}R_{qk} + g^{j\ell}g^{pq}R_{ipk\ell}R_{qj} + g^{j\ell}g^{pq}R_{ijkp}R_{q\ell}\\ &\quad = 2g^{pq}R_{pi}R_{qk} + 2g^{pr}g^{qs}R_{piqk}R_{rs},\end{aligned}$$

it follows that

$$\frac{\partial}{\partial t}R_{ik} = \Delta R_{ik} + 2g^{j\ell}(B_{ijk\ell} - 2B_{ij\ell k}) + 2g^{pr}g^{qs}R_{piqk}R_{rs} - 2g^{pq}R_{pi}R_{qk}.$$

The desired result now follows from the following claim.

Claim 4.19. For any metric g_{ij}, the tensor $B_{ijk\ell}$ satisfies the identity

$$g^{j\ell}(B_{ijk\ell} - 2B_{ij\ell k}) = 0.$$

Proof of Claim. Using the Bianchi identities,

$$\begin{aligned}g^{j\ell}B_{ijk\ell} &= g^{j\ell}g^{pr}g^{qs}R_{piqj}R_{rks\ell}\\ &= g^{j\ell}g^{pr}g^{qs}R_{pqij}R_{rsk\ell}\\ &= g^{j\ell}g^{pr}g^{qs}(R_{piqj} - R_{pjqi})(R_{rks\ell} - R_{r\ell sk})\\ &= 2g^{j\ell}(B_{ijk\ell} - B_{ij\ell k}).\end{aligned}$$

□

Corollary 4.20. *Under the Ricci flow,*

$$\frac{\partial}{\partial t}\mathrm{Scal} = \Delta\mathrm{Scal} + 2|\mathrm{Ric}|^2.$$

Proof. From Corollary 4.17 it suffices to show that

$$2g^{ij}g^{pq}\nabla^2_{q,j}R_{pi} = g^{pq}\nabla_q\nabla_p\mathrm{Scal} = \Delta\mathrm{Scal}.$$

To do this we claim:

Claim 4.21. The identity

$$\frac{1}{2}\nabla_q\nabla_p\mathrm{Scal} = g^{ij}\nabla_q\nabla_j R_{pi}$$

holds true.

Proof of Claim. By the contracted second Bianchi identity (2.16),

$$\begin{aligned}\frac{1}{2}(\nabla\mathrm{Scal})(\partial_p) = \frac{1}{2}\nabla_p\mathrm{Scal} = g^{ij}\nabla_j R_{pi} &= (g^{-1}\otimes\nabla\mathrm{Ric})(\partial_i,\partial_j,\partial_j,\partial_p,\partial_i)\\ &= (\mathrm{tr}_{14}\,\mathrm{tr}_{23}\,g^{-1}\otimes\nabla\mathrm{Ric})(\partial_p)\\ &= (\mathrm{tr}\,g^{-1}\otimes\nabla\mathrm{Ric})(\partial_p).\end{aligned}$$

As $\nabla_q(\mathrm{tr}\,g^{-1}\otimes\nabla\mathrm{Ric}) = \mathrm{tr}\,g^{-1}\otimes\nabla_q(\nabla\mathrm{Ric})$ we find that

$$\begin{aligned}\frac{1}{2}\nabla_q\nabla_p\mathrm{Scal} &= \frac{1}{2}(\nabla_q(\nabla\mathrm{Scal}))(\partial_p)\\ &= (\mathrm{tr}\,g^{-1}\otimes\nabla_q(\nabla\mathrm{Ric}))(\partial_p)\\ &= g^{ij}\nabla_q\nabla_j R_{pi}.\end{aligned}$$

□

Remark 4.22. The evolution of the scalar curvature by Corollary 4.20 provides a simple illustration of the fact that the Ricci flow 'prefers' positive curvature. In this case the two components $\Delta\mathrm{Scal}$ and $2|\mathrm{Ric}|^2$ can be interpreted in the following way: The dissipative term $\Delta\mathrm{Scal}$ reflects the fact that a point in M with a higher average curvature than its neighbours will tend to revert to the mean. The nonlinear term $2|\mathrm{Ric}|^2$ reflects the fact that if one is in a positive curvature region (e.g. a region behaving like a sphere), then the metric will contract, thus increasing the curvature to be even more positive. Conversely, if one is in a negative curvature region (such as a region behaving like a saddle), then the metric will expand, thus weakening the negativity of curvature. In both cases the curvature is trending upwards, consistent with the non-negativity of $2|\mathrm{Ric}|^2$.

4.5 A Closer Look at the Curvature Tensor

So far we have managed to derive, in Theorem 4.14, a heat-type evolution equation for the curvature tensor under the Ricci flow. As we seek to deform the metric so it has constant sectional curvature, we need to look at the relationship between the algebraic properties of the curvature tensor and the global topological and geometry of the manifold.

In three dimensions the curvature is relatively simple algebraically, and in four dimensions still somewhat tractable, but in higher dimensions the curvature becomes very complicated and hard to study. To get a glimpse of this we will look closely at the algebraic structure, in particular that of the Weyl curvature tensor. To do this though, we first need to define the following product.

4.5.1 *Kulkarni–Nomizu Product*

Given two $(2,0)$-tensors we want to build a $(4,0)$-tensor that has the same symmetries as that of the algebraic Riemannian curvature tensor (that is, a $(4,0)$-tensor that satisfies symmetry properties (i)–(iv) in Sect. 2.7.4.1). To do this we need a map, called the Kulkarni–Nomizu product, of the form

$$\owedge : \mathrm{Sym}^2(M) \times \mathrm{Sym}^2(M) \to \mathrm{Curv}(M),$$

where $\mathrm{Sym}^2(M)$ is the bundle of symmetric $(2,0)$-tensors and $\mathrm{Curv}(M)$ is the bundle of curvature tensors. Thus given $\alpha, \beta \in \mathrm{Sym}^2(M)$ we require the product $\alpha \owedge \beta$ to satisfy the following symmetries:

(a) Antisymmetric in the first two arguments: $(\alpha \owedge \beta)_{ijk\ell} = -(\alpha \odot \beta)_{jik\ell}$
(b) Antisymmetric in the last two arguments: $(\alpha \owedge \beta)_{ijk\ell} = -(\alpha \owedge \beta)_{ij\ell k}$
(c) Symmetric paired arguments: $(\alpha \owedge \beta)_{ijk\ell} = (\alpha \owedge \beta)_{k\ell ij}$
(d) Satisfy the Bianchi identity: $(\alpha \owedge \beta)_{ijk\ell} + (\alpha \owedge \beta)_{jki\ell} + (\alpha \owedge \beta)_{kij\ell} = 0$

A natural way to build $\alpha \owedge \beta$ is to use the tensor product. By linearity we expect $\alpha \owedge \beta$ to be a sum of $\alpha \otimes \beta$ with components permuted in such a way that the symmetries matches that of the algebraic Riemann curvature tensor. Hence all that is needed is to find the correct permutations.

Now as α and β are symmetric, terms of the form $\alpha_{ij}\beta_{k\ell}$ are disallowed since this would contradict (a) and (b). So we need to mix i, j with k, ℓ across the tensors α and β. If we naïvely suppose one of the terms is $\alpha_{ik}\beta_{j\ell}$, then (c) implies we will also need $\alpha_{j\ell}\beta_{ik}$ and by (a), (b) we will also need $-\alpha_{jk}\beta_{i\ell}$ and $-\alpha_{i\ell}\beta_{jk}$. Therefore by defining

$$(\alpha \owedge \beta)_{ijk\ell} := \alpha_{ik}\beta_{j\ell} + \alpha_{j\ell}\beta_{ik} - \alpha_{jk}\beta_{i\ell} - \alpha_{i\ell}\beta_{jk} \tag{4.13}$$

or alternatively

$$\begin{aligned}(\alpha \owedge \beta)(v_1, v_2, v_3, v_4) = {} & (\alpha \otimes \beta)(v_1, v_3, v_2, v_4) + (\alpha \otimes \beta)(v_2, v_4, v_1, v_3) \\ & - (\alpha \otimes \beta)(v_1, v_4, v_2, v_3) - (\alpha \otimes \beta)(v_2, v_3, v_1, v_4),\end{aligned}$$

properties (a)–(c) are immediately satisfied; inspection also shows that the Bianchi identity holds as well. Thus we have constructed a $(4,0)$-tensor that has all of the symmetries of that of the algebraic curvature tensor.

4.5.2 Weyl Curvature Tensor

The Weyl curvature tensor is defined to be the *traceless component* of the Riemann curvature tensor. It can be obtained from the full curvature tensor R by subtracting out various traces. To find an exact expression, we seek to subtract a tensor C which is the sum of the scalar and traceless Ricci[2] parts of R, with the additional conditions that C must have the same traces and algebraic structures as that of R. By using the Kulkarni–Nomizu product, we consider the tensor C as taking the form

$$C = c_1 \mathrm{Scal}\, g \owedge g + c_2 \overset{\circ}{\mathrm{Ric}} \owedge g,$$

where c_1 and c_2 are scalars. As $\mathrm{tr}\, R = \mathrm{tr}\, C$, where the trace is taken over *any* pair of indices, we must have

$$\mathrm{Ric}_{ik} = g^{j\ell} R_{ijk\ell} = g^{j\ell} C_{ijk\ell}.$$

Now since

$$\begin{aligned} C_{ijk\ell} &= 2c_1 \mathrm{Scal}(g_{ik} g_{j\ell} - g_{i\ell} g_{jk}) \\ &\quad + c_2 (\overset{\circ}{\mathrm{Ric}}_{ik} g_{j\ell} + \overset{\circ}{\mathrm{Ric}}_{j\ell} g_{ik} - \overset{\circ}{\mathrm{Ric}}_{i\ell} g_{jk} - \overset{\circ}{\mathrm{Ric}}_{jk} g_{i\ell}), \end{aligned}$$

it follows that

$$\begin{aligned} \frac{1}{n} \mathrm{Scal}\, g_{ik} + \overset{\circ}{\mathrm{Ric}}_{ik} = \mathrm{Ric}_{ik} &= g^{j\ell} C_{ijk\ell} \\ &= 2c_1 (n-1) \mathrm{Scal}\, g_{ik} + c_2 (n-2)\, \overset{\circ}{\mathrm{Ric}}_{ik}. \end{aligned}$$

In which case the scalars $c_1 = \frac{1}{2n(n-1)}$ and $c_2 = \frac{1}{n-2}$.

Therefore the Weyl tensor W must be of the form

$$\mathrm{Weyl} = R - \frac{1}{n-2} \overset{\circ}{\mathrm{Ric}} \owedge g - \frac{\mathrm{Scal}}{2n(n-1)} g \owedge g. \tag{4.14}$$

---

[2] Where the traceless Ricci tensor $\overset{\circ}{\mathrm{Ric}} := \mathrm{Ric} - \frac{\mathrm{Scal}}{n} g$, since $g^{ik} \overset{\circ}{\mathrm{Ric}}_{ik} = \mathrm{Scal} - \frac{\mathrm{Scal}}{n} g^{ik} \mathrm{Ric}_{ik} = \mathrm{Scal} - \frac{\mathrm{Scal}}{n} \delta^i_i = 0$.

Alternatively, by using $\overset{\circ}{\mathrm{Ric}} = \mathrm{Ric} - \frac{\mathrm{Scal}}{n} g$, we could also write

$$\mathrm{Weyl} = R - \frac{1}{n-2}\mathrm{Ric} \owedge g + \frac{\mathrm{Scal}}{2(n-1)(n-2)} g \owedge g. \tag{4.15}$$

Since W is defined using the Kulkarni–Nomizu product it has the same symmetries as the curvature tensor, but by definition all of its traces vanish. Indeed, one can check by hand that $W_{ijk}{}^{j} = 0$, or indeed any other trace, as per definition. Also note that

$$|R|^2 = \Big|\frac{\mathrm{Scal}}{2n(n-1)} g \owedge g\Big|^2 + \Big|\frac{1}{n-2}\overset{\circ}{\mathrm{Ric}} \owedge g\Big|^2 + |\mathrm{Weyl}|^2$$

as the decomposition is orthogonal and that the metric g has constant sectional curvature if and only if $\overset{\circ}{\mathrm{Ric}} = 0$ and $\mathrm{Weyl} = 0$ (cf. Sect. 12.3.1).

4.5.3 Sphere Theorem of Huisken–Margerin–Nishikawa

In three dimensions the Weyl tensor vanishes, and the curvature can be understood solely in terms of the Ricci tensor (see Sect. 7.5.3 for one way of doing this). One of the first insights into the differentiable pinching problem, for dimensions $n \geq 4$, was made independently by Huisken [Hui85], Nishikawa [Nis86] and Margerin [Mar86]. By using the Ricci flow, they were able show that a manifold is diffeomorphic to a spherical space form, provided the norm of the Weyl curvature tensor and the norm of the traceless Ricci tensor are not too large compared to the scalar curvature at each point.

Theorem 4.23. *Let $n \geq 4$. If the curvature tensor of a smooth compact n-dimensional Riemannian manifold M of positive scalar curvature satisfies*

$$|\mathrm{Weyl}|^2 + \Big|\frac{1}{n-2}\overset{\circ}{\mathrm{Ric}} \owedge g\Big|^2 \leq \delta_n \frac{2}{n(n-1)}\mathrm{Scal}^2 \tag{4.16}$$

with $\delta_4 = \frac{1}{5}$, $\delta_5 = \frac{1}{10}$, and $\delta_n = \frac{2}{(n-2)(n+1)}$ for $n \geq 6$, then the Ricci flow has a solution $g(t)$ on a maximal finite time interval $[0, T)$, and $\frac{g(t)}{2(n-1)(T-t)}$ converges in C^∞ to a metric of constant curvature 1 as $t \to \infty$. In particular, M is diffeomorphic to a spherical space form.

Remark 4.24. The Riemann curvature tensor can be considered as an element of the vector space of symmetric bilinear forms acting on the space $\bigwedge^2 TM$ (see Sect. 12.2). The inequality (4.16) defines a cone in this vector space around the line of constant curvature tensors, in which the Riemann tensor must lie. In Chaps. 12 and 13 we will return to this idea of constructing cones in the space of curvature tensors.

The technique of the proof follows Hamilton's original paper [Ham82b]. The approach is to show that if the scalar-free part of the curvature is small compared to the scalar curvature initially, then it must remain so for all time. In which case we need to show, under the hypothesis (4.16) with an appropriate δ_n, there exists $\varepsilon > 0$ such that

$$\left|\mathrm{Weyl}\right|^2 + \left|\frac{1}{n-2}\overset{\circ}{\mathrm{Ric}} \owedge g\right|^2 \leq \delta_n \,(1-\varepsilon)^2 \,\frac{2}{n(n-1)}\mathrm{Scal}^2 \tag{4.17}$$

remains valid as long as the solution to the Ricci flow exists for times $t \in [0, T)$. This is achieved by working with

$$\overset{\circ}{R} := R - \frac{2\,\mathrm{Scal}}{n(n-1)} g \owedge g,$$

which measures the failure of g to have (pointwise) constant positive sectional curvature (cf. Sect. 12.3.1). It is proved using the maximum principle (which we discuss in Chap. 7) that a bound can be obtained on a function of the form

$$F_\sigma = \frac{|\overset{\circ}{R}|^2}{\mathrm{Scal}^{2-\sigma}},$$

for $\sigma \geq 0$, yielding the following result.[3]

Proposition 4.25 ([Hui85, Sect. 3]). *If the inequality (4.17), that is*

$$|\overset{\circ}{R}|^2 \leq \delta_n \,(1-\varepsilon)^2 \,\frac{2}{n(n-1)}\mathrm{Scal}^2,$$

holds at time $t = 0$, then it remains so on $0 \leq t < T$. Moreover, there are constants $C < \infty$ and $\sigma > 0$, depending on n and the initial metric, such that

$$|\overset{\circ}{R}|^2 \leq C\,\mathrm{Scal}^{2-\sigma}$$

holds for $0 \leq t < T$.

Remark 4.26. From this analysis we see that the main obstacle to obtaining similar sphere-type theorem, under weaker pinching conditions than that of Theorem 4.23, is the Weyl curvature tensor W. Controlling the behaviour of W under the Ricci flow, for dimensions $n \geq 4$, has proved to be a major technical hurdle, which was finally overcome by the efforts of Böhm and Wilking [BW08]. We discuss their method in Chap. 12.

---

[3] The quantity $|\overset{\circ}{R}|^2$ is a higher dimensional analogue of the quantity $|\overset{\circ}{\mathrm{Ric}}|^2 = |\mathrm{Ric}|^2 - \frac{1}{3}\mathrm{Scal}^2$ used by Hamilton [Ham82b] in the $n = 3$ case, which measures how far the eigenvalues of the Ricci tensor diverge from each other.

Chapter 5
Short-Time Existence

An important foundational step in the study of any system of evolutionary partial differential equations is to show short-time existence and uniqueness. For the Ricci flow, unfortunately, short-time existence does not follow from standard parabolic theory, since the flow is only weakly parabolic. To overcome this, Hamilton's seminal paper [Ham82b] employed the deep Nash–Moser implicit function theorem to prove short-time existence and uniqueness. A detailed exposition of this result and its applications can be found in Hamilton's survey [Ham82a]. DeTurck [DeT83] later found a more direct proof by modifying the flow by a time-dependent change of variables to make it parabolic. It is this method that we will follow.

5.1 The Symbol

To investigate short-time existence and uniqueness for the Ricci flow, one naturally looks to the theory of non-linear PDE's on vector bundles. Here we establish the symbol which will be used to determine a PDE's type.

5.1.1 Linear Differential Operators

Let E and F be bundles over a manifold M. We say $L : \Gamma(E) \to \Gamma(F)$ is a *linear* differential operator of order k if it is of the form

$$L(u) = \sum_{|\alpha| \leq k} L_\alpha \partial^\alpha u,$$

where $L_\alpha \in \mathrm{Hom}(E, F)$ is a bundle homomorphism and α is a multi-index.

B. Andrews and C. Hopper, *The Ricci Flow in Riemannian Geometry*, Lecture Notes in Mathematics 2011, DOI 10.1007/978-3-642-16286-2_5,

Example 5.1. Let (e_k) be local frame for E over a neighbourhood of $p \in M$ with local coordinates (x^i). Then a second order linear differential operator $\mathcal{P} : \Gamma(E) \to \Gamma(E)$ has the form

$$\mathcal{P}(u) = \left((\lambda_{ij})^k_\ell \frac{\partial^2 u^\ell}{\partial x^i \partial x^j} + (\mu_i)^k_\ell \frac{\partial u^\ell}{\partial x^i} + \nu^k_\ell u^\ell \right) e_k.$$

Here $\lambda \in \Gamma(\mathrm{Sym}^2(T^*M) \otimes \mathrm{Hom}(E,E))$, while $\mu \in \Gamma(T^*M \otimes \mathrm{Hom}(E,E))$ and $\nu \in \Gamma(\mathrm{Hom}(E,E))$.

The *total symbol* σ of L in direction $\zeta \in \mathscr{X}(M)$ is the bundle homomorphism

$$\sigma[L](\zeta) = \sum_{|\alpha| \leq k} L_\alpha \zeta^\alpha.$$

Thus in Example 5.1, $(\sigma[\mathcal{P}](\zeta))(u) = ((\lambda_{ij})^k_\ell \zeta^i \zeta^j u^\ell + (\mu_i)\ell^k \zeta^i u^\ell + \lambda^k_\ell u^\ell) e_k$. In the familiar case of scalar equations, the bundle is simply $M \times \mathbb{R}$ so that $\mathrm{Hom}(E,E)$ is one-dimensional, and we can think of λ as a section of $\mathrm{Sym}^2(T^*M)$, μ as a vector field and ν as a scalar function.

The *principal symbol* $\widehat{\sigma}$ of L in direction ζ is defined to be the bundle homomorphism of only the highest order terms, that is

$$\widehat{\sigma}[L](\zeta) = \sum_{|\alpha| = k} L_\alpha \zeta^\alpha.$$

The principal symbol captures algebraically the analytic properties of L that depend only on its highest derivatives. In Example 5.1, the principal symbol in direction ζ is $\widehat{\sigma}[\mathcal{P}](\zeta) = (\lambda_{ij})^k_\ell \zeta^i \zeta^j (e^*)^\ell \otimes e_k$. As Hamilton noted in [Ham82b, Sect. 4], computing the symbol is easily obtained (at least heuristically) by replacing the derivative $\frac{\partial}{\partial x^i}$ by the Fourier transformation variable ζ_i.

The principal symbol determines whether an evolution equation is of parabolic type: We say a differential operator L of order $2m$ is *elliptic* if for every direction $\zeta \in T_xM$ the eigenvalues of the principal symbol $\widehat{\sigma}[L](\zeta)$ have strictly positive real part, or equivalently if there exists $c > 0$ such that for all ζ and u we have $\langle \widehat{\sigma}[L](\zeta)u, u \rangle \geq c|\zeta|^{2m}|u|^2$. This implies in particular that the principal symbol in any direction is a linear isomorphism of the fibre. A linear equation of the form $\partial_t u = Lu$ is parabolic if L is elliptic.

An important property of the principal symbol is the following: If G in another vector bundle over M with a operator $\mathcal{O} : \Gamma(F) \to \Gamma(G)$ of order ℓ, then the symbol of $\mathcal{O} \circ L$ in direction ζ is the bundle homomorphism

$$\sigma[\mathcal{O} \circ L](\zeta) = \sigma[\mathcal{O}](\zeta) \circ \sigma[L](\zeta) : E \to G \tag{5.1}$$

It is of degree at most $k + \ell$ in direction ζ.

5.1.2 Nonlinear Differential Operators

When faced with a *nonlinear* partial differential equation, one attempts to *linearise* the equation in such a way that linear theory can be applied. That is, if one has a solution u_0 to a given nonlinear PDE, it is possible to *linearise* the equation by considering a smooth family $u = u(s)$ of solutions with a variation $v = \delta u = \frac{\partial}{\partial s} u\big|_{s=0}$. By differentiating the PDE with respect to s, the result is a linear PDE in terms of v.

For example, if the nonlinear PDE is of the form

$$\frac{\partial u}{\partial t} = F(D^2 u, Du, u, x, t),$$

where $F = F(p, q, r, x, t) : \mathrm{Sym}^2(\mathbb{R}^n) \times \mathbb{R}^n \times \mathbb{R} \times \mathbb{R}^n \times \mathbb{R} \to \mathbb{R}$. The linearisation about a solution u_0 is given by

$$\frac{\partial v}{\partial t} = \frac{\partial F}{\partial p_{ij}}\bigg|_{u_0} D_i D_j v + \frac{\partial F}{\partial q_k}\bigg|_{u_0} D_k v + \frac{\partial F}{\partial r}\bigg|_{u_0} v.$$

The result is a linear PDE with coefficients depending on u_0.

Moreover for a (nonlinear) differential operator $P : \Gamma(E) \to \Gamma(F)$ with a given solution u_0, the *linearisation* DP of P at u_0 (if it exists) is defined to be the linear map $DP : \Gamma(E) \to \Gamma(F)$ given by

$$DP\big|_u(v) = \frac{\partial}{\partial t} P(u(t))\Big|_{t=0},$$

where $u(0) = u_0$ and $u'(0) = v$. From this, we say a nonlinear equation is parabolic if and only if its linearisation about any $u_0 \in \Gamma(E)$ is parabolic.

Of particular interest is the evolution equation of the form

$$\frac{\partial f}{\partial t} = \mathcal{E}(f),$$

where $\mathcal{E}(f)$ is a nonlinear differential operator of degree 2 in f over a bundle $\pi : E \to M$. If $\bar{f}$ is a variation of f then

$$\frac{\partial \bar{f}}{\partial t} = D\mathcal{E}\big|_f(\bar{f})$$

where $D\mathcal{E}\big|_f$ is the linear operator of degree 2. If $D\mathcal{E}\big|_f$ is elliptic, then the evolution equation $\frac{\partial f}{\partial t} = \mathcal{E}(f)$ has a unique smooth solution for the initial value problem $f = f_0$ at $t = 0$ for at least a short time interval $0 \leq t < \varepsilon$ (where ε may depend on the initial data f_0).

5.2 The Linearisation of the Ricci Tensor

When considering the Ricci flow $\frac{\partial}{\partial t} g = -2\mathrm{Ric}(g)$, one would like to regard the Ricci tensor as a nonlinear partial differential operator on the space of metrics g. That is, as an operator $\mathrm{Ric} : \Gamma(\mathrm{Sym}^2_+ T^*M) \to \Gamma(\mathrm{Sym}^2 T^*M)$. By using the variation formula from Proposition 4.9 with $h_{ij} = \frac{\partial}{\partial t} g_{ij}$, the linearisation of Ric is given by

$$\begin{aligned}(D\mathrm{Ric}_g)(h)_{ik} &= \frac{\partial}{\partial t}\mathrm{Ric}_g(g(t))\Big|_{t=0} \\ &= \frac{1}{2} g^{pq}\Big(\nabla_q \nabla_k h_{ip} + \nabla_q \nabla_i h_{kp} - \nabla_i \nabla_k h_{qp} - \nabla_q \nabla_p h_{ik}\Big).\end{aligned}$$

So the principal symbol

$$\widehat{\sigma}[D\mathrm{Ric}_g]\zeta : \mathrm{Sym}^2_+ T^*M \to \mathrm{Sym}^2 T^*M$$

in direction ζ can be obtained by replacing the covariant derivative ∇_i by the covector ζ_i. Hence

$$(\widehat{\sigma}[D\mathrm{Ric}_g]\zeta)(h)_{ik} = \frac{1}{2} g^{pq}\Big(\zeta_q \zeta_k h_{ip} + \zeta_q \zeta_i h_{kp} - \zeta_i \zeta_k h_{qp} - \zeta_q \zeta_p h_{ik}\Big) \quad (5.2)$$

5.3 Ellipticity and the Bianchi Identities

In investigating the principal symbol $\widehat{\sigma}[D\mathrm{Ric}(g)]$, Hamilton [Ham82b, Sect. 4] observed that the failure of ellipticity is principally due to the Bianchi identities. We present this result with a discussion on the link between the Bianchi identities and the diffeomorphism invariance of the curvature tensor. The result of this investigation shows the failure of Ricci flow to be parabolic is a consequence of its geometric nature.

To begin, recall that the divergence operator $\delta_g : \Gamma(\mathrm{Sym}^2 T^*M) \to \Gamma(T^*M)$ is defined by

$$(\delta_g h)_k = -g^{ij}\nabla_i h_{jk}. \quad (5.3)$$

The adjoint of δ_g, denoted by $\delta_g^* : \Gamma(T^*M) \to \Gamma(\mathrm{Sym}^2 T^*M)$, with respect to the L^2-inner product is given by[1]

$$(\delta_g^* \omega)_{jk} = \frac{1}{2}(\nabla_j \omega_k + \nabla_k \omega_j) = \frac{1}{2}(\mathcal{L}_{\omega^\sharp} g)_{jk}. \quad (5.4)$$

[1] Since $\int_M (\delta_g h)^k \omega_k d\mu = \int_M h^{jk} (\delta_g^* \omega)_{jk} d\mu$, for compact M.

Now consider the composition

$$D\text{Ric}(g) \circ \delta_g^* : \Gamma(T^*M) \to \Gamma(\text{Sym}^2 T^*M).$$

By (5.1) this is *a priori* a (2+1)-order differential operator, so its principal symbol $\widehat{\sigma}[D\text{Ric}_g \circ \delta_g^*](\zeta)$ is the degree 3 part of its total symbol. However, by commuting derivatives and the contracted second Bianchi identity one can show that

$$(D\text{Ric}(g) \circ \delta_g^*)(\omega) = \frac{1}{2}(\omega^p \nabla_p R_{ik} + R_k^p \nabla_i \omega_p + R_i^p \nabla_j \omega_p) = \frac{1}{2}\mathcal{L}_{\omega^\sharp}(\text{Ric}(g)).$$

As the right-hand side involves only *one* derivative of ω, its total symbol is of degree at most 1. In other words, the principal (degree 3) symbol $\widehat{\sigma}[D\text{Ric}_g \circ \delta_g^*](\zeta)$ is in fact the zero map. Thus

$$0 = \widehat{\sigma}[D\text{Ric}_g \circ \delta_g^*](\zeta) = \widehat{\sigma}[D\text{Ric}_g](\zeta) \circ \widehat{\sigma}[\delta_g^*](\zeta),$$

and so $\text{Im}\,\widehat{\sigma}[\delta_g^*](\zeta) \subset \ker \widehat{\sigma}[D\text{Ric}_g](\zeta)$. Therefore $\widehat{\sigma}[D\text{Ric}_g](\zeta)$ has *at least* an n-dimensional kernel in each fibre.

In fact one can go further by showing $\dim \ker \widehat{\sigma}[D\text{Ric}_g](\zeta) = n$. Here we briefly sketch the idea as follows: First consider the first order linear operator $B_g : \Gamma(\text{Sym}^2 T^*M) \to \Gamma(T^*M)$ defined by

$$B_g(h)_k = g^{ij}\left(\nabla_i h_{jk} - \frac{1}{2}\nabla_k h_{ij}\right). \tag{5.5}$$

As any metric satisfies the contracted second Bianchi identity,

$$B_g(\text{Ric}(g)) = 0.$$

By linearising this PDE we obtain

$$B_g((D\text{Ric}_g)(h)) + (DB_g)(\text{Ric}(g+h)) = 0.$$

Now $B_g \circ D\text{Ric}_g$ is *a priori* a degree 3 differential operator. However DB_g is of order 1, so its degree 3 symbol is zero. Thus the principal (degree 3) symbol $\widehat{\sigma}[B_g \circ D(\text{Ric}_g)](\zeta)$ must be the zero map, and so

$$\text{Im}\,\widehat{\sigma}[D\text{Ric}_g](\zeta) \subset \ker \widehat{\sigma}[B_g](\zeta) \subset \text{Sym}^2 T^*M.$$

From here one can combine maps $B_g \circ D\text{Ric}_g$ and $D\text{Ric}_g \circ \delta_g^*$ to show

$$0 \longrightarrow T^*M \xrightarrow{\widehat{\sigma}_\zeta[\delta_g^*]} \text{Sym}^2 T^*M \xrightarrow{\widehat{\sigma}_\zeta[D\text{Ric}_g]} \text{Sym}^2 T^*M \xrightarrow{\widehat{\sigma}_\zeta[B_g]} T^*M \longrightarrow 0$$

constitutes a short exact sequence. A discussion on this can be found in [CK04, pp. 77-8]. The desired result follows, however the emphasis lies in the fact there are no degeneracies other than those implied by the contracted second Bianchi identity.

5.3.1 *Diffeomorphism Invariance of Curvature and the Bianchi Identities*

So far we have seen the degeneracy of the Ricci tensor results from the contracted second Bianchi identity. In this section we show the Bianchi identities are a consequence of the invariance of the curvature tensor under the full diffeomorphism group (which is of course infinite dimensional). An upshot of this – as mentioned in [Kaz81] – is a natural and conceptually transparent proof of the Bianchi identities. As a consequence, the failure of $\mathrm{Ric}(g)$ to be (strongly) elliptic is due entirely to this geometric invariance.

The simplest illustration of this involves the scalar curvature Scal. To start, let ϕ_t be the one-parameter group of diffeomorphisms generated by the vector field X with $\phi_0 = \mathrm{id}_M$. By the diffeomorphism invariance of the curvature,

$$\phi_t^*(\mathrm{Scal}(g)) = \mathrm{Scal}(\phi_t^* g).$$

Now the linearisation of Scal is given by

$$D\mathrm{Scal}_g(\mathcal{L}_X g) = \frac{d}{dt}\mathrm{Scal}(\phi_t^* g)\Big|_{t=0} = \frac{d}{dt}\phi_t^*(\mathrm{Scal}(g))\Big|_{t=0} = \mathcal{L}_X \mathrm{Scal} = \nabla_X \mathrm{Scal}.$$

On the other hand, Proposition 4.10 implies that

$$\begin{aligned} D\mathrm{Scal}_g(h) &= -\Delta \mathrm{tr}_g h + \delta^2 h - \langle h, \mathrm{Ric}\rangle \\ &= -g^{ij} g^{k\ell}(\nabla_i \nabla_j h_{k\ell} - \nabla_i \nabla_k h_{j\ell} + R_{ik} h_{j\ell}), \end{aligned}$$

where h is the arbitrary variation of g. By setting this variation $h_{ij} = (\mathcal{L}_X g)_{ij} = \nabla_i X_j + \nabla_j X_i$ we find (by commuting covariant derivatives) that

$$D\mathrm{Scal}_g(\mathcal{L}_X g) = 2 g^{jk} X^i \nabla_k R_{ij} = X^i \nabla_i \mathrm{Scal}.$$

As X is arbitrary, it follows that

$$g^{jk}\nabla_k R_{ij} = \frac{1}{2}\nabla_i \mathrm{Scal}.$$

So the diffeomorphism invariance of Scal implies the contracted second Bianchi identity.

The same method works for the full curvature tensor (see [CLN06, Ex. 1.26]). Here we sketch the main result. The diffeomorphism invariance $R(\phi_t^* g) = \phi_t^*(R(g))$ implies

$$DR_g(\mathcal{L}_X g) = \frac{d}{dt} R(\phi_t^* g)\Big|_{t=0} = \frac{d}{dt}\phi_t^*(R(g))\Big|_{t=0} = \mathcal{L}_X R,$$

where

$$\begin{aligned}(\mathcal{L}_X R)_{ijk}{}^{\ell} &= X^p \nabla_p R_{ijk}{}^{\ell} + R_{pjk}{}^{\ell} \nabla_i X^p + R_{ipk}{}^{\ell} \nabla_j X^p \\ &\quad + R_{ijp}{}^{\ell} \nabla_k X^p - R_{ijk}{}^{p} \nabla_p X^{\ell}.\end{aligned}$$

However by Proposition 4.7 and (4.10) we also have

$$\begin{aligned}2[DR_g(h)]_{ijk}{}^{\ell} = g^{\ell p}\Big(&\nabla_j \nabla_k h_{ip} + \nabla_j \nabla_i h_{kp} + \nabla_j \nabla_p h_{ik} \\ &- \nabla_i \nabla_k h_{jp} - \nabla_i \nabla_j h_{kp} - \nabla_i \nabla_p h_{jk}\Big).\end{aligned}$$

By substituting $h = \mathcal{L}_X g$ into the previous equation and rewriting $\mathcal{L}_X R$ as

$$\begin{aligned}(\mathcal{L}_X R)_{ijk}{}^{\ell} = g^{\ell p}\Big(&-\nabla_i (R_{kpj}{}^{q} X_q) - \nabla_j (R_{pki}{}^{q} X_q) - \nabla_k X_q R_{ijp}{}^{q} \\ &- \nabla_q X_q R_{ijk}{}^{q} + X_q\big(g^{qr}\nabla_r R_{ijkp} + \nabla_i R_{kpj}{}^{q} + \nabla_j R_{pki}{}^{q}\big)\Big),\end{aligned}$$

we find that

$$\begin{aligned}0 &= [DR_g(\mathcal{L}_X g)]_{ijk}{}^{\ell} - (\mathcal{L}_X R)_{ijk}{}^{\ell} \\ &= \frac{1}{2} g^{\ell p}\Big(-\nabla_i\big((R_{jpk}{}^{q} - R_{kpj}{}^{q} - R_{jkp}{}^{q})X^q\big) \\ &\qquad + \nabla_j\big((R_{ipk}{}^{q} - R_{kpi}{}^{q} - R_{ikp}{}^{q})X^q\big) \\ &\qquad - 2X^q(\nabla_q R_{ijkp} + \nabla_i R_{kpjq} + \nabla_j R_{pkiq})\Big).\end{aligned}$$

To get the desired Bianchi identities from this, we evaluate this expression pointwise with an appropriate choice of X. Firstly, by prescribing $X(p) = 0$ and $\nabla_i X_j(p) = g_{ij}(p)$ it follows that

$$0 = -(R_{j\ell ki} - R_{k\ell ji} - R_{jk\ell i}) + (R_{i\ell k} - R_{k\ell ij} - R_{ik\ell j}).$$

In which case symmetries (i)–(iii) in Sect. 2.7.4.1 imply the first Bianchi identity: $0 = R_{ijk\ell} + R_{ik\ell j} + R_{i\ell jk}$. Similarly, if X is chosen to be an element of a local orthonormal frame, we can obtain the second Bianchi identity $0 = \nabla_q R_{ijk\ell} + \nabla_i R_{jqk\ell} + \nabla_j R_{qik\ell}$.

5.4 DeTurck's Trick

Despite the failing of $\mathrm{Ric}(g)$ (as a nonlinear differential operator) to be elliptic, the Ricci flow still enjoys short-time existence and uniqueness:

Theorem 5.2. *If (M, g_0) is a compact Riemannian manifold, there exists a unique solution $g(t)$, defined for time $t \in [0, \varepsilon)$, to the Ricci flow such that $g(0) = g_0$ for some $\varepsilon > 0$.*

In proving this theorem, DeTurck [DeT83] showed it is possible to modify the Ricci flow in such a way that the nonlinear PDE in fact becomes parabolic. As we shall see, this is done by adding an extra term which is a Lie derivative of the metric with respect to a certain time dependent vector field.

5.4.1 Motivation

Here we closely examination the Ricci tensor. To motivate DeTurck's idea, rewrite the linearisation of the Ricci tensor as

$$-2[D\mathrm{Ric}_g(h)]_{ik} = \Delta h_{ik} + g^{pq}(\nabla_i \nabla_k h_{qp} - \nabla_q \nabla_i h_{kp} - \nabla_q \nabla_k h_{ip}).$$

Define the 1-form $V = B_g(h) \in \Gamma(T^*M)$, where B_g is defined by (5.5). Observe that

$$\begin{aligned} V_k &= g^{pq}\left(\nabla_q h_{pk} - \frac{1}{2}\nabla_k h_{pq}\right) \\ \nabla_i V_k &= g^{pq}\left(\nabla_i \nabla_q h_{pk} - \frac{1}{2}\nabla_i \nabla_k h_{pq}\right). \end{aligned}$$

Thus

$$-2[D\mathrm{Ric}_g(h)]_{ik} = \Delta h_{ik} - \nabla_i V_k - \nabla_k V_i + S_{ik}, \tag{5.6}$$

where the lower order tensor

$$\begin{aligned} S_{ik} &= g^{pq}\Big(\frac{1}{2}(\nabla_i \nabla_k h_{qp} - \nabla_k \nabla_i h_{qp}) \\ &\qquad + (\nabla_i \nabla_q h_{kp} - \nabla_q \nabla_i h_{kp}) + (\nabla_k \nabla_q h_{ip} - \nabla_q \nabla_k h_{ip})\Big) \\ &= g^{pq}\Big(\frac{1}{2}(R_{ikq}{}^r h_{rp} + R_{ikp}{}^r h_{rq}) \\ &\qquad + (R_{iqp}{}^r h_{rk} + R_{iqk}{}^r h_{rp}) + (R_{kqi}{}^r h_{rp} + R_{kqp}{}^r h_{ri})\Big) \\ &= g^{pq}(2R_{qik}{}^r h_{rp} - R_{ip} h_{kq} - R_{kp} h_{iq}) \end{aligned}$$

since $\nabla_i \nabla_j h_{pq} - \nabla_j \nabla_i h_{pq} = R_{ijp}{}^r h_{rq} + R_{ijq}{}^r h_{rp}$ by Proposition 2.43. It is clear S_{ik} is symmetric and involves no derivatives of h. Therefore (5.6) implies that the linearisation of the Ricci tensor is equal to a Laplacian term minus a Lie derivative term $\nabla_i V_k + \nabla_k V_i$ with a lower order symmetric term S_{ik}.

Moreover, by Proposition 4.6 we can write V (at least locally) as

$$V_k = \frac{1}{2} g^{pq}(\nabla_p h_{qk} + \nabla_q h_{pk} - \nabla_k h_{pq}) = g^{pq} g_{kr} (D\Gamma_g(h))^r_{pq},$$

where h is the variation of g and

$$D\Gamma_g : \Gamma(\mathrm{Sym}^2 T^*M) \to \Gamma(\mathrm{Sym}^2 T^*M \otimes TM)$$

is the linearisation of the Levi–Civita connection $\Gamma(g)$. We now wish to add an appropriate correction term to the Ricci tensor to make it elliptic.

To do this, fix a background metric $\widetilde{g}$ on M with Levi–Civita connection $\widetilde{\Gamma}$. By our above investigation, define a vector field W by

$$W^k = g^{pq}\left(\Gamma^k_{pq} - \widetilde{\Gamma}^k_{pq}\right). \tag{5.7}$$

As it is the difference of two connections, it is a globally well defined vector field. Since W only involves one derivative of the metric g, the operator

$$P = P(\widetilde{\Gamma}) : \Gamma(\mathrm{Sym}^2 T^*M) \to \Gamma(\mathrm{Sym}^2 T^*M),$$

define by

$$P(g) := \mathcal{L}_W g,$$

is a second order differential operator in g. The linearisation of P is given by

$$\big(DP(h)\big)_{ik} = \nabla_i V_k + \nabla_k V_i + T_{ik}, \tag{5.8}$$

where T_{ik} is a linear first order expression in h. Comparing this with (5.6) leads one to consider the operator

$$Q := -2\mathrm{Ric} + P : \Gamma(\mathrm{Sym}^2 T^*M) \to \Gamma(\mathrm{Sym}^2 T^*M).$$

Therefore by (5.6) and (5.8) we find that

$$DQ(h) = \Delta h + A,$$

where $A_{ik} = T_{ik} - 2S_{ik}$ is a first order linear term in h. Hence the principal symbol of DQ is

$$\widehat{\sigma}[DQ](\zeta)h = |\zeta|^2 h. \tag{5.9}$$

Therefore Q is elliptic, and so by the standard theory of partial differential equations the modified Ricci flow

$$\frac{\partial}{\partial t} g = -2\mathrm{Ric}(g) + P(g) = -2\mathrm{Ric}(g) + \mathcal{L}_W g,$$

also referred to as the Ricci–DeTurck flow, enjoys short-time existence and uniqueness.

5.4.2 Relating Ricci–DeTurck Flow to Ricci Flow

We now follow DeTurck's strategy of proving short time existence and uniqueness for the Ricci flow.

Proof (Theorem 5.2). We proceed in stages, first starting with existence.

Step 1: Fix a background metric $\widetilde{g}$ on M, define the vector field W by (5.7) and let the Ricci–DeTurck flow be given by

$$\begin{aligned} \frac{\partial}{\partial t} g_{ij} &= -2R_{ij} + \nabla_i W_j + \nabla_j W_i \\ g(0) &= g_0 \end{aligned} \tag{5.10}$$

where $W_j = g_{jk} W^k = g_{jk} g^{pq} (\Gamma^k_{pq} - \widetilde{\Gamma}^k_{pq})$. From (5.9), the Ricci–DeTurk flow is strictly parabolic. So for any smooth initial metric g_0 there exists $\varepsilon > 0$ such that a unique smooth solution $g(t)$ to (5.10) flow exists for $0 \leq t < \varepsilon$.

Step 2: As there exists a solution to the Ricci–DeTurck flow, the one-parameter family of vector fields $W(t)$ defined by (5.7) exists for $0 \leq t < \varepsilon$. In which case there is a 1-parameter family of maps $\varphi_t : M \to M$ (i.e. the flow along the vector field $-W$) defined by the ODE

$$\begin{aligned} \frac{\partial}{\partial t} \varphi_t(p) &= -W(\varphi_t(p), t) \\ \varphi_0 &= \mathrm{id}_M . \end{aligned}$$

As M is compact, one can combine the escape lemma (for instance see [Lee02, p. 446]) with the existence and uniqueness of time-dependent flows (for instance see [Lee02, p. 451]) to conclude there exists a unique family of diffeomorphisms $\varphi_t(p)$ which is defined for all times t in the interval of existence $0 \leq t < \varepsilon$.

Step 3: We now show:

Claim 5.3. The family of metrics $\bar{g}(t) := \varphi_t^* g(t)$ defined for $0 \leq t < \varepsilon$ is a unique solution to the Ricci flow

$$\frac{\partial}{\partial t}\bar{g} = -2\mathrm{Ric}(\bar{g})$$
$$\bar{g}(0) = g_0. \tag{5.11}$$

To show that $\bar{g}$ satisfies (5.11), note that $\bar{g}(0) = g(0) = g_0$ (since $\varphi_0 = \mathrm{id}_M$) and that

$$\begin{aligned}
\frac{\partial}{\partial t}\bar{g} &= \frac{\partial}{\partial t}(\varphi_t^* g(t)) \\
&= \frac{\partial}{\partial s}\Big|_{s=0} \varphi_{s+t}^* g(t+s) \\
&= \varphi_t^* \left(\frac{\partial}{\partial t} g(t)\right) + \frac{\partial}{\partial s}\Big|_{s=0} (\varphi_{s+t}^* g(t)) \\
&= \varphi_t^*(-2\mathrm{Ric}(g(t)) + \mathcal{L}_{W(t)} g(t)) + \frac{\partial}{\partial s}\Big|_{s=0} \varphi_{s+t}^* g(t) \\
&= -2\mathrm{Ric}(\varphi_t^* g(t)) + \varphi_t^*(\mathcal{L}_{W(t)} g(t)) + \frac{\partial}{\partial s}\Big|_{s=0} \varphi_{s+t}^* g(t).
\end{aligned}$$

As $\varphi_{t+s}^* = (\varphi_t^{-1} \circ \varphi_{t+s} \circ \varphi_t)^* = (\varphi_t^{-1} \circ \varphi_{t+s})^* \circ \varphi_t^*$ and

$$\frac{\partial}{\partial s}\Big|_{s=0} (\varphi_t^{-1} \circ \varphi_{t+s}) = (\varphi_t^{-1})_* \left(\frac{\partial}{\partial s}\Big|_{s=0} \varphi_{t+s}\right) = (\varphi_t^{-1})_* W(t),$$

we conclude that

$$\begin{aligned}
\frac{\partial}{\partial t}\bar{g} &= -2\mathrm{Ric}(\varphi_t^* g(t)) + \varphi_t^*(\mathcal{L}_{W(t)} g(t)) - \mathcal{L}_{(\varphi_t^{-1})_* W(t)}(\varphi_t^* g(t)) \\
&= -2\mathrm{Ric}(\bar{g}(t)).
\end{aligned}$$

Hence $\bar{g}(t) = \varphi_t^* g(t)$ is indeed a solution of (5.11) for $t \in [0, \varepsilon)$. This completes the existence part of the claim.

Step 4: All we need to show now is the uniqueness for the Ricci flow. It suffices to prove that a solution of the Ricci–DeTurck flow is produced from a solution of Ricci flow after a reparametrisation defined by harmonic map heat flow.

Precisely, let $(M, \bar{g}(t))$ satisfy Ricci flow. Fix N with a background metric $\widetilde{h}$ and an associated Levi–Civita connection $\widetilde{\nabla}$. Let $\varphi_0 : M \to N$ be a diffeomorphism. Define $\varphi : M \times [0, T) \to N$ by the harmonic map heat flow from the (time-dependent) metric $\bar{g}$ to $\widetilde{h}$ so that

$$\varphi_* \partial_t = \Delta_{\bar{g}(t), \widetilde{h}} \varphi = \bar{g}^{ij} \nabla_i \varphi_*(\partial_j),$$

where we take ∇ to be defined by the Levi–Civita connection $\nabla^{\bar{g}}$ (at time t) on T^*M and the pullback connection ${}^{\varphi(t)}\widetilde{\nabla}$ on $\varphi(t)^* TN$.

In which case

$$\begin{aligned}\Delta_{\bar{g},\widetilde{h}}\varphi &= \bar{g}^{ij}\left({}^{\varphi(t)}\nabla_i\left(\varphi(t)_*\partial_j\right) - \varphi(t)_*\left(\nabla_i^{\bar{g}}\partial_j\right)\right) \\ &= \bar{g}^{ij}\left(\partial_i\partial_j\varphi^\gamma + \widetilde{\Gamma}_{\alpha\beta}{}^\gamma\partial_i\varphi^\alpha\partial_j\varphi^\beta - \bar{\Gamma}_{ij}{}^k\partial_i\varphi^\gamma\right)\partial_\gamma. \end{aligned} \quad (5.12)$$

Step 5: Now define $g(t) = (\varphi(t)^{-1})^*\bar{g}(t)$, a time-dependent metric on N. We claim this metric g is a solution of the Ricci–DeTurck flow. To show this, note that a direct computation – similar to that of Step 3 – gives the following:

Lemma 5.4.
$$\frac{\partial g}{\partial t} = (\varphi^{-1})^*\left(\frac{\partial \bar{g}}{\partial t}\right) + \mathcal{L}_V g,$$

where $V_p = -\varphi_*\partial_t\big|_{\varphi^{-1}(p)}$.

The geometric invariance of the curvature implies $(\varphi^{-1})^*(\frac{\partial\bar{g}}{\partial t}) = -2\mathrm{Ric}(g)$, since g is the pullback of $\bar{g}$. So all we need to do is relate the Lie derivative term to that in the Ricci–DeTurck equation, i.e. we want to show that $V = W$.
The key observation here is the geometric invariance of the 'map Laplacian' reflected in the following proposition.

Proposition 5.5. *Suppose K, M and N are smooth manifolds with a diffeomorphism $\psi : K \to M$ and a smooth map $\varphi : M \to N$. Let $\bar{g}$ be a metric on M, $\widetilde{h}$ be a metric on N and $g = \psi^*\bar{g}$. Then*

$$\Delta_{g,\widetilde{h}}\left(\varphi\circ\psi\right) = \left(\Delta_{\bar{g},\widetilde{h}}\varphi\right)_\psi.$$

Note that $\Delta_{\bar{g},\widetilde{h}}\varphi \in \Gamma(\varphi^*TN)$, so by restriction $(\Delta_{\bar{g},\widetilde{h}}\varphi)_\psi \in \Gamma(\psi^*\varphi^*TN) = \Gamma\left((\varphi\circ\psi)^*TN\right)$.

The result is not surprising: It states that the harmonic map Laplacian of a map from M to K is unchanged if we apply an isometry to M.

Proof. For any $p \in K$, we need to show that

$$\Delta_{g,\widetilde{h}}\left(\varphi\circ\psi\right)(p) = \left(\Delta_{\bar{g},\widetilde{h}}\varphi\right)(\psi(p)).$$

To do this for any fixed p, choose local coordinates (x^i) for M near $\psi(p)$ and induce local coordinates on K near p by $y^i = x^i\circ\psi$. Fix local coordinates (z^α) for N near $\varphi\circ\psi(x)$. In these coordinates we have that $z\circ\varphi\circ x^{-1} = z\circ(\varphi\circ\psi)\circ y^{-1}$, $\bar{g}_{ij} = g_{ij}$ and hence

$\bar{\Gamma}_{ij}{}^k = \Gamma_{ij}{}^k$ everywhere on the chart. Therefore in these coordinates $\varphi^\alpha = (\varphi \circ \psi)^\alpha$ and so

$$\begin{aligned}(\Delta_{\bar{g},\tilde{h}}\varphi)(\psi(p)) &= \bar{g}^{ij}\left(\partial_i\partial_j\varphi^\alpha - \bar{\Gamma}_{ij}{}^k\partial_k\varphi^\alpha + \widetilde{\Gamma}_{\beta\gamma}{}^\alpha\partial_i\varphi^\beta\,\partial_j\varphi^\gamma\right)\frac{\partial}{\partial z^\alpha} \\ &= g^{ij}\Big(\partial_i\partial_j(\varphi\circ\psi)^\alpha - \Gamma_{ij}{}^k\partial_k(\varphi\circ\psi)^\alpha \\ &\qquad + \widetilde{\Gamma}_{\beta\gamma}{}^\alpha\partial_i(\varphi\circ\psi)^\beta\,\partial_j(\varphi\circ\psi)^\gamma\Big)\frac{\partial}{\partial z^\alpha} \\ &= \left(\Delta_{g,\tilde{h}}\left(\varphi\circ\psi\right)\right)(p).\end{aligned}$$

□

We can now complete the argument. As $V(p) = -\left(\varphi_*\partial_t\right)\left(\varphi^{-1}(p)\right) = -(\Delta_{\bar{g},\tilde{h}}\varphi)_{\varphi^{-1}}(p)$, Proposition 5.5 (with $\psi = \varphi^{-1}$, $N = M$) and (5.12) gives

$$V = -\Delta_{g,\tilde{h}}\left(\varphi\circ\varphi^{-1}\right) = -g^{ij}\left(\widetilde{\Gamma}_{ij}{}^k - \Gamma_{ij}{}^k\right)\partial_k = W.$$

Step 6: Finally we now prove the uniqueness result for the Ricci flow. Suppose there are solutions $\bar{g}_i(t)$ of the Ricci flow, for $i = 1, 2$, with initial condition $\bar{g}_1(0) = \bar{g}_2(0)$. Taking $N = M$ and $\varphi_0(x) = x$, we produce solutions $g_i(t)$ of the Ricci–DeTurck flow, with $g_2(0) = \bar{g}_2(0) = \bar{g}_1(0) = g_1(0)$. By uniqueness of solutions of the Ricci–DeTurck flow, $g_2(t) = g_1(t)$ for all t in their common interval of existence. Hence $W = g^{ij}(\Gamma_{ij}{}^k - \widetilde{\Gamma}_{ij}{}^k)$ is the same for the two solutions. The diffeomorphisms $\varphi_i(t)$ are given by the harmonic map heat flow, and as before Proposition 5.5 gives

$$\begin{aligned}\partial_t\varphi_i(x,t) &= (\Delta_{\bar{g}_i,\tilde{h}}\varphi_i)(x,t) = -W(\varphi_i(x,t)); \\ \varphi_i(x,0) &= x.\end{aligned}$$

Thus φ_1 and φ_2 are solutions of the same initial value problem for a system of ordinary differential equations, and hence $\varphi_1(x,t) = \varphi_2(x,t)$ and $\bar{g}_2 = \varphi_2^* g_2 = \varphi_1^* g_1 = \bar{g}_1$, proving uniqueness for the Ricci flow. □

Chapter 6
Uhlenbeck's Trick

In Theorem 4.14 we derived an evolution equation for the curvature R under the Ricci flow, which took the form

$$\begin{aligned}\frac{\partial}{\partial t}R_{ijk\ell} &= \Delta R_{ijk\ell} + 2(B_{ijk\ell} - B_{ij\ell k} - B_{i\ell jk} + B_{ikj\ell}) \\ &\quad - (R_i^p R_{pjk\ell} + R_j^p R_{ipk\ell} + R_k^p R_{ijp\ell} + R_\ell^p R_{ijkp})\end{aligned}$$

where the time derivative is interpreted as in Sect. 4.3.1. In this chapter we examine a trick attributed to Karen Uhlenbeck [Ham86, p. 155] that allows one to simplify the above equation by removing the last collection of terms with a 'change of variables'. We pursue this idea via three different methods which correspond to different bundle constructions.

6.1 Abstract Bundle Approach

Let $V \to M$ be an abstract vector bundle isomorphic to the tangent bundle TM over M with fixed (time-independent) metric $h = g(0)$. Let $\iota_0 : V \to TM$ be the identity map, so that $h = \iota_0^*(g(0))$. We aim to extend ι_0 to a family of bundle isometries ι_t, so that $h = \iota_t^*(g(t))$ for every t. To do this, we evolve ι according to the ODE

$$\frac{\partial}{\partial t}\iota = \mathrm{Ric}^\sharp \circ \iota \tag{6.1a}$$

$$\iota(\cdot, 0) = \iota_0(\,\cdot\,) \tag{6.1b}$$

where $\mathrm{Ric}^\sharp \in \mathrm{End}(TM)$ is the endomorphism defined by raising the second index (i.e. $\mathrm{Ric}^\sharp(\partial_i) = R_i^{\ j}\partial_j$). This defines a 1-parameter family of bundle isomorphisms $\iota : V \times [0,T) \to TM$.

Lemma 6.1. *If ι evolves by (6.1), then the bundle map $\iota_t : (V,h) \to (TM, g(t))$ is an isometry for every t.*

B. Andrews and C. Hopper, *The Ricci Flow in Riemannian Geometry*, Lecture Notes in Mathematics 2011, DOI 10.1007/978-3-642-16286-2_6,

To see this, observe that

$$\begin{aligned}\frac{\partial}{\partial t}g(\iota(v),\iota(v)) &= \Big(\frac{\partial}{\partial t}g\Big)(\iota(v),\iota(v)) + g\Big(\frac{\partial}{\partial t}\iota(v),\iota(v)\Big) + g\Big(\iota(v),\frac{\partial}{\partial t}\iota(v)\Big)\\ &= -2\operatorname{Ric}(\iota(v),\iota(v)) + g(\operatorname{Ric}^\sharp(\iota(v)),\iota(v)) + g(\iota(v),\operatorname{Ric}^\sharp(\iota(v)))\\ &= 0.\end{aligned}$$

for any $v \in V$. Therefore $\iota^* g(t)$ is independent of time t, and so continues to equal the fixed metric h.

Now consider the pullback of R by ι: Since $\iota_*\partial_a = \iota_a^i\partial_i$ we have

$$R_{abcd} = (\iota^* R_{g(t)})(\partial_a,\partial_b,\partial_c,\partial_d) = \iota_a^i\iota_b^j\iota_c^k\iota_d^\ell R_{ijk\ell}$$

and

$$B_{abcd} = (\iota^* B)(\partial_a,\partial_b,\partial_c,\partial_d) = \iota_a^i\iota_b^j\iota_c^k\iota_d^\ell B_{ijk\ell}.$$

Thus as $\frac{\partial}{\partial t}\iota_a^i = \iota_a^p R_p{}^i$, we find that

$$\begin{aligned}\frac{\partial}{\partial t}R_{abcd} &= \iota_a^i\iota_b^j\iota_c^k\iota_d^\ell\frac{\partial}{\partial t}R_{ijk\ell} + \Big(\Big(\frac{\partial}{\partial t}\iota_a^i\Big)\iota_b^j\iota_c^k\iota_d^\ell R_{ijk\ell} + \cdots + \iota_a^i\iota_b^j\iota_c^k\Big(\frac{\partial}{\partial t}\iota_d^\ell\Big)R_{ijk\ell}\Big)\\ &= \iota_a^i\iota_b^j\iota_c^k\iota_d^\ell\frac{\partial}{\partial t}R_{ijk\ell} + \Big((\iota_a^p R_p^i)\iota_b^j\iota_c^k\iota_d^\ell R_{ijk\ell} + \cdots + \iota_a^i\iota_b^j\iota_c^k\iota_d^q(R_q^\ell R_{ijk\ell})\Big)\\ &= \iota_a^i\iota_b^j\iota_c^k\iota_d^\ell\frac{\partial}{\partial t}R_{ijk\ell} + \Big(\iota_a^q\iota_b^j\iota_c^k\iota_d^\ell\, R_q^p R_{pjk\ell} + \cdots + \iota_a^i\iota_b^j\iota_c^k\iota_d^q\, R_q^p R_{ijkp}\Big).\end{aligned}$$

So by letting $\Delta_D := \operatorname{tr}_g(\iota^*\nabla\circ\iota^*\nabla)$, where $\iota^*\nabla$ is the pullback connection, Theorem 4.14 implies that

$$\begin{aligned}\frac{\partial}{\partial t}R_{abcd} &= \iota_a^i\iota_b^j\iota_c^k\iota_d^\ell\Delta R_{ijk\ell} + 2(B_{abcd} - B_{abdc} - B_{adbc} + B_{acbd})\\ &= \Delta_D R_{abcd} + 2(B_{abcd} - B_{abdc} - B_{adbc} + B_{acbd}) \qquad (6.2)\end{aligned}$$

where $B_{abcd} = -h^{eg}h^{fi}R_{aebf}R_{cgdi}$. That is, the last collection of terms from the evolution equation for R in Theorem 4.14 have been eliminated.

6.2 Orthonormal Frame Approach

There is an alternative approach to Uhlenbeck's trick put forth by Hamilton [Ham93, Sect. 2] (see also [CCG+08, Appendix F]). In this treatment the idea is to work on the bundle of frames (see Sect. 2.2.2) where we define a natural direction in which to take time derivatives. This is a much more involved construction, but we will see later that this can be derived by a rather standard procedure from the more elementary structure introduced in Sect. 6.3.

First, for completeness, we give an overview of the frame bundle machinery used in this approach.

6.2.1 The Frame Bundle

Let M be a differentiable manifold and $\pi : FM \to M$ the general linear frame bundle (thus the elements of FM_x are nonsingular linear maps from $\mathbb{R}^n$ to T_xM). The group $\mathrm{GL}(n)$ acts by composition on each fibre, so that $M = (M^a_b) \in \mathrm{GL}(n)$ acts on a frame $Y = (Y_a) \in FM_x$ to give the frame $Y^M := Y \circ M = (M^a_b Y_a)$. Suppose (x^i) is a local coordinate system defined on an open set $U \subset M$, so that $(\frac{\partial}{\partial x^i})$ is a local frame on U – which is a local section of the frame bundle FM restricted to U. A frame $Y = (Y_a)$ can be written in local coordinates (x^i) as

$$Y_a = y^i_a(Y)\frac{\partial}{\partial x^i},$$

where $y^i_a : FM|_U \to \mathbb{R}$ assigns to a frame Y the i-th component of the a-th vector. The vector valued function $y = (y^i_a) : FM \to \mathrm{GL}(n, \mathbb{R})$ describes the transition of the frame $(\frac{\partial}{\partial x^i})$ to the frame $Y = (Y_a)$.[1]

Now let $\widetilde{x}^i = x^i \circ \pi : FM|_U \to \mathbb{R}$. The collection $(\widetilde{x}^j, y^i_a)$ is a coordinate system defined on the open set $FM|_U \subset FM$. In particular $(\frac{\partial}{\partial \widetilde{x}^j}, \frac{\partial}{\partial y^i_a})$ is a basis for the tangent space of FM at points in $FM|_U$. Note that $(\frac{\partial}{\partial y^i_a})$ is a basis for the tangent space of the fibres F_xM, for $x \in U$: The vectors $(\frac{\partial}{\partial y^i_a})$ are vertical whereas $(\frac{\partial}{\partial \widetilde{x}^j})$ are transverse to the fibres.

Remarkably, the tangent space of FM is trivial: We can define global independent vertical vector fields (Λ^a_b) and horizontal vector fields (∇_a).In local coordinates $(y^i_a(Y), \widetilde{x}^j(Y))$ for FM, these can be expressed as

$$\Lambda^a_b = y^i_b \frac{\partial}{\partial y^i_a}$$
$$\nabla_a = y^j_a \left(\frac{\partial}{\partial \widetilde{x}^j} - y^k_b \Gamma^i_{kj} \frac{\partial}{\partial y^i_b} \right).$$

---

[1] We follow the convention that indices $i, j, k, \ldots$ are reserved for coordinates on M and indices $a, b, c, \ldots$ are used for components of the frame. For instance in local coordinates $g_{ij} = g\left(\frac{\partial}{\partial x^i}, \frac{\partial}{\partial x^j}\right)$, whereas $g_{ab}(Y) = g(Y_a, Y_b)$ for a given frame Y. If $Y_a = y^i_a \frac{\partial}{\partial x^i}$ then we also have $g_{ab} = g_{ij} y^i_a y^j_b$.

Here moving in the direction (∇_a) in FM at Y corresponds to parallel translating a frame along a path in M with initial velocity Y_a; the vectors ∇_a span a natural 'horizontal subspace' in $T(FM)$ defined by the connection. The vectors (Λ_b^a) are generators of the action of $\mathrm{GL}(n)$ on the fibres of FM.

Each frame $Y \in FM_x$ has a dual co-frame (Y_*^a) for T_x^*M, defined by $Y_*^a(Y_b) = \delta_a^b$. Given a (p,q)-tensor on M we can define a *function* from FM to $\bigotimes^p (\mathbb{R}^n)^* \otimes \bigotimes^q \mathbb{R}^n$ with components given by

$$T_{b_1\ldots b_p}{}^{a_1\ldots a_q}(Y) := T\left(Y_{b_1}, \ldots, Y_{b_p}, Y_*^{a_1}, \ldots, Y_*^{a_q}\right).$$

We can then differentiate this function in the direction of any vector field on FM, such as the vector fields defined above. The notation ∇_a for the horizontal vector fields is justified by the observation that differentiating the above function in direction ∇_a gives the components of the covariant derivative $\nabla_a T$. We also have an expression for derivatives in vertical directions:

Lemma 6.2. *If T is a (p,q)-tensor, then*

$$\begin{aligned}(\Lambda_d^c T_{b_1\cdots b_p}{}^{a_1\cdots a_q})(Y) = &\sum_{\ell=1}^{p} \delta_{b_\ell}^c \, T_{b_1\cdots b_{\ell-1}\, d\, b_{\ell+1}\cdots b_p}{}^{a_1\cdots a_q}(Y) \\ &- \sum_{k=1}^{q} \delta_d^{a_k} \, T_{b_1\cdots b_p}{}^{a_1\cdots b_{k-1}\, c\, a_{k+1}\cdots a_q}(Y).\end{aligned}$$

In particular, if $V_{ab} = y_a^i y_b^j V_{ij}$ is a covariant $(2,0)$-tensor then

$$(\Lambda_b^a V_{cd})(Y) = \delta_c^a V_{bd}(Y) + \delta_d^a V_{cb}(Y).$$

The splitting of $T(FM)$ into horizontal and vertical components also makes it possible to define a metric g^F on FM to make $\pi : (FM, g^F) \to (M, g)$ a Riemannian submersion, so that $(\pi^* g)(V, W) = g^F(V, W)$ for V and W horizontal. We have the vertical and horizontal subspaces orthogonal, and on the vertical parts we take the natural metric defined by $\left\langle \Lambda_a^b, \Lambda_c^d \right\rangle = \delta_{ac}\delta^{bd}$.

The *orthonormal frame bundle* OM is the subbundle of FM defined by

$$OM := \{Y \in FM : g_{ab}(Y) = \delta_{ab}\}.$$

As above, the metric $g = (g_{ab})_{a,b=1}^n$ can be considered as a function

$$g : FM \to \mathrm{Sym}_+(n),$$

where $\mathrm{Sym}_+(n)$ is the space of symmetric positive definite $n \times n$ matrices; in which case $OM = g^{-1}(I_{n\times n})$, where $I_{n\times n}$ is the identity matrix. Also, to be tangent to OM a vector field V on FM must satisfy $V(g_{ab}) = 0$ on OM.

Lemma 6.3. *The globally defined vector fields* (∇_a) *and* (Λ_c^b) *form a basis on* FM *which – when restricted to the subbundle* OM *– is orthonormal with respect to the Riemannian metric* g^F.

Proof. By construction $g^F(\nabla_a, \Lambda_c^b) = 0$, and the vertical vectors (Λ_c^b) are orthonormal. Also, $g^F(\nabla_a, \nabla_c) = g(\pi_* \nabla_a, \pi_* \nabla_b) = g(Y_a, Y_c)$ and as $g(Y_a, Y_c) = \delta_{ac}$ whenever $Y \in OM$, we also have $g^F(\nabla_a, \nabla_c) = \delta_{ac}$. □

Remark 6.4. The global framing of FM by the n^2+n vector fields Λ_a^b and ∇_a was noted by Ambrose–Singer [AS53] and Nomizu [Nom56, p. 49], who called them 'fundamental vector fields' and 'basic vector fields' respectively. The correspondence between tensors on M and functions on the frame bundle was observed by Wong [Won61, Theorem 2.4], who also noted that the basic vector fields ∇_a act to give the components of the covariant derivative [Won61, Theorem 2.15].

6.2.2 Time-Dependent Frame Bundles and the Ricci Flow

When considering a solution to the Ricci flow $(M, g(t))$, for $t \in [0, T)$, it is natural to work on the space-time manifold $M \times [0, T)$. Likewise the product of the frame bundle with the time axis gives a bundle over $M \times [0, T)$ with projection $\widetilde{\pi} : FM \times [0, T) \to M \times [0, T)$ defined by $\widetilde{\pi} = \pi \times \mathrm{id}_{[0,T)}$. The time-dependent metric $g(t)$ can be considered as a function

$$g : FM \times [0, T) \to \mathrm{Sym}(n),$$

where $\mathrm{Sym}(n)$ is the space of symmetric $n \times n$ matrices.

The orthonormal frame bundles at each time combine to give a bundle over space-time also:

$$\widetilde{OM} := \bigcup_{t \in [0,T)} OM_{g(t)} \times \{t\} = g^{-1}(I_{n \times n}) \subset FM \times [0, T).$$

Note that the time-like vector field $\frac{\partial}{\partial t}$ is not always tangent to $\widetilde{OM}$ unless $g(t)$ is independent of time! Since the metric varies according to

$$\frac{\partial}{\partial t} g_{ab} = -2R_{ab},$$

the orthonormal frame bundle, with $g_{ab} = \delta_{ab}$, will now vary with time. Therefore all that is needed is to modify the time-like vector field to make it tangent to the orthonormal frame bundle $\widetilde{OM}$.

To achieve this we consider the time derivative as the directional derivative with respect to the following vector field.

Definition 6.5. The vector field ∇_t on $FM \times [0,T)$ is defined by

$$\nabla_t := \frac{\partial}{\partial t} + R_{ab} g^{bc} \Lambda_c^a. \tag{6.3}$$

This is characterised as follows:

Lemma 6.6. *The vector field ∇_t, restricted to $\widetilde{OM} \subset FM \times [0,T)$, is the unique vector field tangent to the subbundle $\widetilde{OM}$ for which $\nabla_t - \frac{\partial}{\partial t}$ is vertical and perpendicular to each $OM_{g(t)} \subset FM$.*

Proof. On $\widetilde{OM}$ we have $g_{ab} = \delta_{ab}$. Now if $\nabla_t - \frac{\partial}{\partial t}$ is vertical and perpendicular to each $OM_{g(t)} \subset FM$, it follows that:

1. $\nabla_t - \frac{\partial}{\partial t} = \alpha_c^d \Lambda_d^c$ as it is vertical.
2. $0 = \langle \Lambda_b^a - \Lambda_a^b, \alpha_c^d \Lambda_d^c \rangle = \alpha_a^b - \alpha_b^a$, as $\nabla_t - \frac{\partial}{\partial t}$ is perpendicular to the subbundle $OM_{g(t)}$, so that α is symmetric.
3. As $\nabla_t g_{ab} = 0$, we have that

$$\begin{aligned} 0 &= \frac{\partial}{\partial t} g_{ab} + \alpha_a^d g_{bd} + \alpha_b^c g_{ac} \\ &= -2R_{ab} + \alpha_a^b + \alpha_b^a \end{aligned}$$

since ∇_t is tangent to $\widetilde{OM}$.

It follows that $\nabla_t - \frac{\partial}{\partial t} = R_{ab} \Lambda_b^a$. Conversely, if ∇_t is defined in this way then it certainly satisfies the above properties. □

Lemma 6.7. *If T is a time-dependent (p,q)-tensor then*

$$\begin{aligned} \nabla_t T_{b_1 \cdots b_p}{}^{a_1 \cdots a_q} &= \frac{\partial}{\partial t} T_{b_1 \cdots b_p}{}^{a_1 \cdots a_q} \\ &\quad + \sum_{c,d=1}^{n} \sum_{j=1}^{p} R_{b_j}^d T_{b_1 \cdots b_{j-1}\, d\, b_{j+1} \cdots b_p}{}^{a_1 \cdots a_q} \\ &\quad + \sum_{c,d=1}^{n} \sum_{i=1}^{q} R_d^{a_i} T_{b_1 \cdots b_p}{}^{a_1 \cdots a_{i-1}\, d\, a_{i+1} \cdots a_q}. \end{aligned} \tag{6.4}$$

In particular, if T is a $(2,0)$-tensor then

$$\nabla_t T_{ab} = \frac{\partial}{\partial t} T_{ab} + R_{ac} g^{cd} T_{db} + R_{bc} g^{cd} T_{ad}.$$

Proof. When T is a $(2,0)$-tensor, Lemma 6.2 implies that $\Lambda_c^d T_{ab} = \delta_a^d T_{cb} + \delta_b^d T_{ac}$ so we find that

$$\begin{aligned}\nabla_t T_{ab} &= \frac{\partial}{\partial t} T_{ab} + R_{de}\, g^{ec}\, \Lambda_c^d V_{ab} \\ &= \frac{\partial}{\partial t} T_{ab} + R_{ae}\, g^{ec}\, T_{cb} + R_{be}\, g^{ec}\, T_{ac}.\end{aligned}$$

The equation for general tensors follows likewise. □

Finally, by defining the Laplacian $\Delta := \sum_{e=1}^n \nabla_e \nabla_e$ – which acts on vector valued functions on $O\widetilde{M}$ – we see by (6.4) that

$$\nabla_t R_{abcd} = \frac{\partial}{\partial t} R_{abcd} + R_a^p R_{pbcd} + R_b^p R_{apcd} + R_c^p R_{abpd} + R_d^p R_{abcp}.$$

So by Theorem 4.14 it follows that the evolution equation of R_{abcd} takes the desired form

$$(\nabla_t - \Delta) R_{abcd} = 2(B_{abcd} - B_{abdc} + B_{acbd} - B_{adbc}). \tag{6.5}$$

Remark 6.8. This approach is in fact equivalent to the method discussed in Sect. 6.1. To see this, start by evolving a frame Y according to

$$\frac{d}{dt} Y_a - R_{ab} g^{bc} Y_c. \tag{6.6}$$

With this one finds that $\frac{\partial}{\partial t}(g(Y_a, Y_b)) = 0$ under the Ricci flow; in particular, if a frame is initially orthonormal then it remains so. We see that the path $\gamma(t) := (Y(t), t)$ lies in $O\widetilde{M}$, so $\frac{d}{dt}\gamma(t) \in T(O\widetilde{M})$ and

$$\frac{d}{dt}\gamma(t) = \nabla_t.$$

To check this equation, observe that $\frac{d}{dt}\gamma(t) = \mathrm{Ric}(y) + \frac{\partial}{\partial t}$ which in local coordinates takes the form

$$\begin{aligned}\frac{d}{dt}\gamma(t) &= R_{ab} g^{bc} y_c^k(Y) \frac{\partial}{\partial y_a^k}\Big|_{\gamma(t)} + \frac{\partial}{\partial t}\Big|_{\gamma(t)} \\ &= R_{ab} g^{bc} \Lambda_c^a\big|_{\gamma(t)} + \frac{\partial}{\partial t}\Big|_{\gamma(t)} \\ &= \nabla_t.\end{aligned}$$

In which case ∇_t, defined by (6.3), corresponds to taking the time-derivative $\frac{\partial}{\partial t}(T(Y_{a_1}, \ldots, Y_{a_p}))$ of a tensor T where Y_{a_i} satisfy (6.6). Thus the evolution of the curvature $R_{abcd} = R_{ijk\ell} F_a^i F_b^j F_c^k F_d^\ell$ in a frame moving according to (6.6) is equivalent to (6.2) seen in Sect. 6.1.

6.3 Time-Dependent Metrics and Vector Bundles Over $M \times \mathbb{R}$

Now we consider an alternative approach: Instead of using an abstract bundle and constructing an identification with the tangent bundle at each time, we put the tangent bundles at different time together to form a vector bundle over the 'space-time' $M \times \mathbb{R}$ and place a natural connection on this bundle. This reproduces very simply the first method described above, and we will see that the frame bundle machinery also relates closely to this method.

6.3.1 Spatial Tangent Bundle and Time-Dependent Metrics

We begin with a rather general setting: Let $g(t)$ be an arbitrary smooth family of Riemannian metrics on a manifold M, parametrised by 'time' t. That is, we have for each $(p, t) \in M \times \mathbb{R}$ an inner product $g_{(p,t)}$ on T_pM. We interpret this as a metric acting on the *spatial tangent bundle* $\mathfrak{S}$, defined by

$$\mathfrak{S} := \{v \in T(M \times \mathbb{R}) : \ dt(v) = 0\},$$

as a vector bundle over $M \times \mathbb{R}$. With this, the metric g is naturally a metric on $\mathfrak{S}$, since $\mathfrak{S}_{(p,t)}$ is naturally isomorphic to T_pM via the projection $\pi : (p, t) \to p$. In which case a local frame for $\mathfrak{S}$ consists of the coordinate tangent vector fields (∂_i) for TM, and from Example 2.62 we have the decomposition of $T(M \times \mathbb{R})$ as a direct sum

$$T(M \times \mathbb{R}) = \mathfrak{S} \oplus \mathbb{R}\partial_t.$$

Since $\mathfrak{S}$ is a subbundle of $T(M \times \mathbb{R})$, any section of $\mathfrak{S}$ is also a section of $T(M \times \mathbb{R})$; we call these *spatial vector fields.*

6.3.1.1 The Canonical Connection on the Spatial Tangent Bundle

The next step is establish a result analogous to the Levi–Civita theorem – which says that for any Riemannian metric there is a unique compatible connection which is symmetric. We want to construct a canonical connection on $\mathfrak{S}$ for any given metric g on $\mathfrak{S}$. This is provided by the following theorem:

Theorem 6.9. *Let g be a metric on the spatial tangent bundle $\mathfrak{S} \to M \times \mathbb{R}$. Then there exists a unique connection ∇ on $\mathfrak{S}$ satisfying the following three conditions:*

1. *∇ is compatible with g: For any $X \in \Gamma(T(M \times \mathbb{R}))$ and any spatial vector fields $Y, W \in \Gamma(\mathfrak{S})$,*

$$Xg(Y, W) = g(\nabla_X Y, W) + g(Y, \nabla_X W).$$

2. *∇ is spatially symmetric: If $X, Y \in \Gamma(\mathfrak{S})$ are any two spatial vector fields, so they are in particular vector fields on $M \times \mathbb{R}$,*

$$\nabla_X Y - \nabla_Y X = [X, Y].$$

3. *∇ is irrotational: The tensor $\mathcal{S} \in \Gamma(\mathfrak{S}^* \otimes \mathfrak{S})$, defined by*

$$\mathcal{S}(V) = \nabla_{\partial_t} V - [\partial_t, V] \tag{6.7}$$

for any $V \in \Gamma(\mathfrak{S})$, is symmetric with respect to g:

$$g(\mathcal{S}(V), W) = g(V, \mathcal{S}(W)),$$

for any $V, W \in \mathfrak{S}_{(p,t)}$.

Remark 6.10. Observe that $\mathcal{S}$ is $\mathfrak{S}$-valued and tensorial, since

$$\begin{aligned}\mathcal{S}(fV) &= \nabla_{\partial_t}(fV) - [\partial_t, fV] \\ &= f\nabla_{\partial_t} V + (\partial_t f)V - (\partial_t f)V - f[\partial_t, V] \\ &= f\mathcal{S}(V)\end{aligned}$$

and $[\partial_t, V] = [\partial_t, V^i \partial_i] = (\partial_t V^i)\partial_i \in \mathfrak{S}$.

Proof. We show uniqueness. Working in local coordinates, we can write ∇ in terms of its coefficients:

$$\begin{aligned}\nabla_{\partial_i}\partial_j &= {\Gamma_{ij}}^k \partial_k \qquad 1 \le i, j, k \le n \\ \nabla_{\partial_t}\partial_j &= {\Gamma_{0j}}^k \partial_k \qquad 1 \le j, k \le n\end{aligned}$$

Now the first conditions implies that $\partial_t g_{ij} = g(\nabla_t \partial_i, \partial_j) + g(\partial_i, \nabla_t \partial_j) = \Gamma_{0ij} + \Gamma_{0ji}$. Combined with the second condition, we find that the spatial components ${\Gamma_{ij}}^k$ are given by the Christoffel symbols of the metric at fixed time t, that is ${\Gamma_{ij}}^k = \frac{1}{2} g^{kl}(\partial_i g_{jl} + \partial_j g_{il} - \partial_l g_{ij})$. The third is used as follows: As $\mathcal{S}(\partial_i) = \nabla_{\partial_t}\partial_i - [\partial_t, \partial_i] = {\Gamma_{0i}}^k \partial_k$, the symmetry with respect to g amounts to

$$\Gamma_{0ij} = g(\mathcal{S}(\partial_i), \partial_j) = g(\partial_i, \mathcal{S}(\partial_j)) = \Gamma_{0ji}.$$

So with compatibility with g, we have $2\Gamma_{0ij} = \partial_t g_{ij}$. Therefore

$${\Gamma_{0i}}^k = \frac{1}{2} g^{kj} \partial_t g_{ij}. \tag{6.8}$$

It is now a simple matter to check that these formulas define a connection with the required properties. □

Remark 6.11. The connection restricted to each fixed t is simply the Levi–Civita connection of the metric $g(t)$. The importance of the compatibility condition will become apparent when we discuss the maximum principle for vector bundles in Sect. 7.4. The irrotational condition says that parallel transport in the time direction does not have any rotation component (recall that antisymmetric matrices are the generators of rotations).

6.3.1.2 The Time Derivative of the Curvature Tensor

In the special case when the metric evolves by Ricci flow, the above proof – in particular (6.8) – implies in local coordinates that

$$\nabla_{\partial_t}\partial_i = \Gamma_{0i}{}^j\partial_j = \frac{1}{2}g^{jp}\partial_t g_{ip}\partial_j = -\mathrm{Ric}_i^j\partial_j.$$

From this we compute $\nabla_{\partial_t}R$:

$$\begin{aligned}
(\nabla_{\partial_t}R)_{ijkl} &= \partial_t\left(R(\partial_i,\partial_j,\partial_k,\partial_l)\right)\\
&\quad - R(\nabla_{\partial_t}\partial_i,\partial_j,\partial_k,\partial_l) - \cdots - R(\partial_i,\partial_j,\partial_k,\nabla_{\partial_t}\partial_l)\\
&= \partial_t R_{ijkl} - R(-R_i^p\partial_p,\partial_j,\partial_k,\partial_l) - \cdots - R(\partial_i,\partial_j,\partial_k,-R_l^p\partial_p)\\
&= \partial_t R_{ijkl} + R_i^pR_{pjkl} + \cdots + R_l^pR_{ijkp}\\
&= \Delta R_{ijkl} + 2\left(B_{ijkl} - B_{ijlk} - B_{iljk} + B_{ikjl}\right).
\end{aligned}$$

Thus again, the last collection of terms has been eliminated by computing with respect to the connection on $\mathfrak{S}$ rather than the time derivative defined in Sect. 4.3.1.

6.3.1.3 Relation to the Frame Bundle Method

Now we clarify the relationship between our method and that involving computation on the frame bundle: The construction given in Sect. 6.2.1 can be straightforwardly generalised to the following situation: Let E be an arbitrary vector bundle of rank k over M, with a connection ∇. Then TE has a canonical splitting into horizontal and vertical subspaces, where the vertical subspace is tangent to the fibre, and the horizontal consists of the directions corresponding to parallel translation in E.

As in Sect. 2.2.2, the frame bundle FE of E is the bundle with fibre at $x \in M$ given by the space of nonsingular linear maps from $\mathbb{R}^k$ to E_x. As before this bundle has a left action of $\mathrm{GL}(k)$, and each frame (ξ_α) has a dual frame (ϕ^α) for E^*. An arbitrary tensor field T constructed on E (that is, a multilinear function at each $x \in M$ acting on copies of E_x and E_x^*) can be written as a vector-valued function on FE: If $T \in \Gamma(\bigotimes^p E^* \otimes \bigotimes^q E)$, then we associate to T the function from FE to $T_q^p(\mathbb{R}^k) \simeq \mathbb{R}^{k(p+q)}$ given by

$$(T(\xi))_{\alpha_1 \ldots \alpha_p}{}^{\beta_1 \ldots \beta_q} = T\left(\xi_{\alpha_1}, \ldots, \xi_{\alpha_p}, \phi^{\beta_1}, \ldots, \phi^{\beta_q}\right).$$

We are not quite in the situation we had previously where we could define a global frame for $T(FM)$. In general this can be achieved by taking a direct product of FE with the frame bundle FM. Assuming we have a connection on TM, there is again a canonical choice of horizontal subspace, and a canonical framing of $T(FM \oplus FE)$.

However in the present situation a simpler construction suffices: Here we replace M by $M \times [0, T)$, and E is the spatial tangent bundle $\mathfrak{S}$. Then $F(\mathfrak{S})$ is isomorphic to $FM \times [0, T)$. Since $T(M \times [0, T)) = \mathfrak{S} \oplus \mathbb{R}\partial_t$, we have a global framing for $TF(\mathfrak{S})$, given by the fibre directions Λ_a^b, $1 \leq a, b \leq n$, and the horizontal directions, given by ∇_a for $1 \leq a \leq n$ and ∇_t given by the direction corresponding to parallel translation in the ∂_t direction. This is exactly the framing constructed in Sect. 6.2.2.

In most of the computations we will undertake in this book we will find it much more convenient to work directly with the connection on $\mathfrak{S}$ rather than working on the frame bundle.

6.3.2 Alternative Derivation of the Evolution of Curvature Equation

We will now discuss an alternative derivation of the evolution equation for the curvature tensor, making use of the canonical connection on the spatial tangent bundle defined above. We will first carry this out in the setting of a general time-dependent metric, then specialise to the Ricci flow.

6.3.2.1 A Canonical Connection on the Space-Time Tangent Bundle

We have constructed a connection on the spatial subbundle $\mathfrak{S}$ of $T(M \times \mathbb{R})$. It will be convenient to extend this to a connection on all of $T(M \times \mathbb{R})$ (mostly for the application of the Bianchi identity). There are several ways to do this, but we will choose the following obvious construction:

Theorem 6.12. *There exists a unique connection ∇ on $T(M \times \mathbb{R})$ for which $\nabla \partial_t = 0$ and ∇X is as constructed in Theorem 6.9 for all $X \in \Gamma(\mathfrak{S})$.*

Remark 6.13. This choice of connection is not symmetric: We have in general that $\nabla_{\partial_t} \partial_i \neq 0$, while $\nabla_i \partial_t = 0$ always. However, this choice of connection has several good points: Each of the submanifolds $M \times \{t\}$ is totally geodesic (and so importantly for us computing derivatives of spatial tangent vector fields gives the same answer whether computed as section of $\mathfrak{S}$ or of $T(M \times \mathbb{R})$. Also, choosing ∂_t to be parallel has some benefits: In particular it ensures that the projection from $T(M \times \mathbb{R})$ to $\mathfrak{S}$ is a parallel tensor.

We compute the torsion tensor: This clearly vanishes if both arguments are spatial, so the only non-zero components (up to symmetry) are

$$\tau(\partial_t, \partial_i) = \nabla_{\partial_t}\partial_i - \nabla_{\partial_i}\partial_t = \mathcal{S}(\partial_i), \tag{6.9}$$

where $\mathcal{S}$ is defined by (6.7).

6.3.2.2 The Spatial and Temporal Curvature Tensors

We wish to relate the curvatures of the connections we have constructed to the curvature of the Levi–Civita connections of each of the metrics $g(t)$.

To do this we define the *spatial curvature* $R \in \Gamma(\otimes^4 \mathfrak{S}^*)$ by

$$R(X, Y, W, Z) = R_\nabla(X, Y, W, Z) = g(R_\nabla(X, Y)W, Z)$$

for any $X, Y, W, Z \in \Gamma(\mathfrak{S})$. Moreover, the *temporal curvature tensor* $\mathcal{P} \in \Gamma(\otimes^3 \mathfrak{S}^*)$ is defined by

$$\mathcal{P}(X, Y, Z) = g(R_\nabla(\partial_t, X)Y, Z) \tag{6.10}$$

for any $X, Y, Z \in \Gamma(\mathfrak{S})$. Note that our choice of connections ensures that these are the same whether we interpret the connection as acting on $\mathfrak{S}$ or on $T(M \times \mathbb{R})$.

Furthermore, if $X, Y, W, Z \in T_{(p,t)}M$, then $R(X, Y, W, Z)$ is the Riemannian curvature of the metric $g(t)$ at the point p, acting on $\pi_* X$, $\pi_* Y$, $\pi_* W$ and $\pi_* Z$ where $\pi : M \times \mathbb{R} \to M; (p, t) \mapsto p$ is the projection map. The other components (up to symmetry) of the curvature of the connection ∇ on $T(M \times \mathbb{R})$ all vanish.

Our next step will be to compute the temporal curvature tensor $\mathcal{P}$ in terms of the tensor $\mathcal{S}$, but to do this we must first make a digression to discuss Bianchi identities.

6.3.2.3 Generalised Bianchi Identities

Let ∇ be a connection over a vector bundle E over a manifold M, and suppose there is also a symmetric connection $\widehat{\nabla}$ defined on TM. As $R_\nabla \in \Gamma(\otimes^2 T^*M \otimes E^* \otimes E)$ we can – by the discussion in Sect. 2.5.3 – make sense of the covariant derivative $\nabla R_\nabla \in \Gamma(\otimes^3 T^*M \otimes E^* \otimes E)$.

Theorem 6.14. *For any vector fields $X, Y, Z \in \mathscr{X}(M)$ and $\xi \in \Gamma(E)$,*

$$(\nabla_X R_\nabla)(Y, Z, \xi) + (\nabla_Y R_\nabla)(Z, X, \xi) + (\nabla_Z R_\nabla)(X, Y, \xi) = 0.$$

Proof. Since the equation is tensorial, it is enough to check the identity with $X = \partial_i$, $Y = \partial_j$ and $Z = \partial_k$ about a point $p \in M$. We work in local coordinates defined by exponential coordinates about p, so that $\nabla_i \partial_j|_p = 0$ for all i and j. Extend $\xi \in E_p$ to a section of E such that $\nabla \xi|_p = 0$ (for example, construct ξ on a neighbourhood of p by parallel transport along geodesics from p). Then as $[\partial_i, \partial_j] = 0$ everywhere, we find at the point p that

$$\begin{aligned}(\nabla_i R_\nabla)(\partial_j, \partial_k, \xi) &= \nabla_i \left(R_\nabla(\partial_j, \partial_k, \xi)\right) \\ &\quad - R_\nabla(\widehat{\nabla}_i \partial_j, \partial_k, \xi) - R_\nabla(\partial_j, \widehat{\nabla}_i \partial_k, \xi) - R_\nabla(\partial_j, \partial_j, \nabla_i \xi) \\ &= \nabla_i \left(\nabla_j \left(\nabla_k \xi\right) - \nabla_k \left(\nabla_j \xi\right)\right). \end{aligned} \tag{6.11}$$

Using this, we find a the point p that

$$\begin{aligned}&(\nabla_i R_\nabla)(\partial_j, \partial_k, \xi) + (\nabla_j R_\nabla)(\partial_k, \partial_i, \xi) + (\nabla_k R_\nabla)(\partial_i, \partial_j, \xi) \\ &\quad = \nabla_i \left(\nabla_j \left(\nabla_k \xi\right)\right) - \nabla_i \left(\nabla_k \left(\nabla_j \xi\right)\right) \\ &\qquad + \nabla_j \left(\nabla_k \left(\nabla_i \xi\right)\right) - \nabla_j \left(\nabla_i \left(\nabla_k \xi\right)\right) \\ &\qquad + \nabla_k \left(\nabla_i \left(\nabla_j \xi\right)\right) - \nabla_k \left(\nabla_j \left(\nabla_i \xi\right)\right) \\ &\quad = R_\nabla(\partial_j, \partial_i, \nabla_k \xi) + R_\nabla(\partial_k, \partial_j, \nabla_i \xi) + R_\nabla(\partial_i, \partial_k, \nabla_j \xi) \\ &\quad = 0 \end{aligned}$$

where we grouped the first term on the first line with the second on the third, the first term on the second line with the second on the first, and the first term on the third line with the second on the second. □

Unfortunately we will not always be in this convenient situation: It will turn out to be natural to work with a *non*-symmetric connection when working with time-dependent metrics, so we will need the following variation on the Bianchi identity in the non-symmetric case. In this case we define the torsion tensor $\tau \in \Gamma(\otimes^2 T^*M \otimes TM)$ by

$$\tau(X, Y) = \widehat{\nabla}_X Y - \widehat{\nabla}_Y X - [X, Y]. \tag{6.12}$$

Theorem 6.15. *For any vector fields $X, Y, Z \in \mathscr{X}(M)$ and $\xi \in \Gamma(E)$,*

$$\begin{aligned}&(\nabla_X R_\nabla)(Y, Z, \xi) + (\nabla_Y R_\nabla)(Z, X, \xi) + (\nabla_Z R_\nabla)(X, Y, \xi) \\ &\quad + R_\nabla(\tau(X, Y), Z, \xi) + R_\nabla(\tau(Y, Z), X, \xi) + R_\nabla(\tau(Z, X), Y, \xi) = 0.\end{aligned}$$

Proof. As before, choose $X = \partial_i$, $Y = \partial_j$, and $Z = \partial_k$ about a point $p \in M$. Extend $\xi \in E_p$ to a smooth section with $\nabla\xi|_p = 0$. The difference arises because the connection in geodesic coordinates from p is no longer symmetric: Instead we have $\widehat{\nabla}_{v^i\partial_i}(v^j\partial_j)|_p = 0$ for every vector $v \in T_pM$ (this is the geodesic equation along the geodesic with direction $v^i\partial_i$ at p). It follows that $\widehat{\Gamma}_{ij}{}^k$ is antisymmetric in the first two arguments at p, so we have

$$2\widehat{\Gamma}_{ij}{}^k = \widehat{\Gamma}_{ij}{}^k - \widehat{\Gamma}_{ji}{}^k = \tau_{ij}{}^k.$$

So rather than (6.11), we find that

$$\begin{aligned}(\nabla_i R_\nabla)(\partial_j, \partial_k, \xi) &= \nabla_i\left(R_\nabla(\partial_j, \partial_k, \xi)\right)\\ &\quad - R_\nabla(\widehat{\nabla}_i\partial_j, \partial_k, \xi) - R_\nabla(\partial_j, \widehat{\nabla}_i\partial_k, \xi) - R_\nabla(\partial_j, \partial_j, \nabla_i\xi)\\ &= \nabla_i\left(\nabla_j\left(\nabla_k\xi\right) - \nabla_k\left(\nabla_j\xi\right)\right)\\ &\quad - \frac{1}{2}R_\nabla(\tau_{ij}, \partial_k, \xi) - \frac{1}{2}R_\nabla(\partial_j, \tau_{ik}, \xi).\end{aligned}$$

The result now follows as before by observing the antisymmetry of τ and of the first two argument of R_∇. □

In the special case where we are working with a connection on TM, we also have a version of the first Bianchi identity, which we state in the case of connections which may be non-symmetric:

Theorem 6.16. *For any vector fields $X, Y, Z \in \mathscr{X}(M)$,*

$$\begin{aligned}&R(X,Y)Z + R(Y,Z)X + R(Z,X)Y\\ &\quad = \nabla_Y\tau(X,Z) + \nabla_Z\tau(Y,X) + \nabla_X\tau(Z,Y)\\ &\qquad + \tau(\tau(Y,X),Z) + \tau(\tau(Z,Y),X) + \tau(\tau(X,Z),Y).\end{aligned}$$

Proof. As before we work in exponential coordinates, so that at a point p we have $\nabla_i\partial_j = \frac{1}{2}\tau(\partial_i, \partial_j)$. Then we compute

$$\begin{aligned}&R(\partial_i,\partial_j)\partial_k + R(\partial_j,\partial_k)\partial_i + R(\partial_k,\partial_i)\partial_j\\ &\quad = \nabla_j(\nabla_i\partial_k) - \nabla_i(\nabla_j\partial_k) + \nabla_k(\nabla_j\partial_i)\\ &\qquad - \nabla_j(\nabla_k\partial_i) + \nabla_i(\nabla_k\partial_j) - \nabla_k(\nabla_i\partial_j)\\ &\quad = \nabla_j\left(\tau(\partial_i,\partial_k)\right) + \nabla_k\left(\tau(\partial_j,\partial_i)\right) + \nabla_i\left(\tau(\partial_k,\partial_j)\right)\\ &\quad = \nabla_j\tau_{ik} + \nabla_k\tau_{ji} + \nabla_i\tau_{kj} + \tau(\tau_{ji},\partial_k) + \tau(\tau_{kj},\partial_i) + \tau(\tau_{ik},\partial_j)\end{aligned}$$

where we used the antisymmetry of τ in the last line. □

Finally, a result which applies when the connection is compatible with a metric g (or, more generally, a parallel symmetric bilinear form).

Theorem 6.17. *Let E be a vector bundle over M, ∇ a connection on E compatible with the form g. For all $X, Y \in \mathscr{X}(M)$ and $\xi, \eta \in \Gamma(E)$,*

$$R(X,Y,\xi,\eta) + R(X,Y,\eta,\xi) = 0$$

where we define $R(X,Y,\xi,\eta) = g(R(X,Y)\xi,\eta)$.

Proof. By compatibility and the definition of the Lie bracket,

$$\begin{aligned}
0 &= Y(X\,g(\xi,\eta)) - X(Y\,g(\xi,\eta)) - [Y,X]\,g(\xi,\eta)\\
&= Y\big(g(\nabla_X\xi,\eta) + g(\xi,\nabla_X\eta)\big) - X\big(g(\nabla_Y\xi,\eta) + g(\xi,\nabla_Y\eta)\big)\\
&\quad - g(\nabla_{[Y,X]}\xi,\eta) - g(\xi,\nabla_{[Y,X]}\eta)\\
&= \quad g(\nabla_Y(\nabla_X\xi),\eta) + g(\nabla_X\xi,\nabla_Y\eta) + g(\nabla_Y\xi,\nabla_X\eta) + g(\xi,\nabla_Y(\nabla_X\eta))\\
&\quad - g(\nabla_X(\nabla_Y\xi),\eta) - g(\nabla_Y\xi,\nabla_X\eta) - g(\nabla_X\xi,\nabla_Y\eta) - g(\xi,\nabla_X(\nabla_Y\eta))\\
&\quad - g(\nabla_{[Y,X]}\xi,\eta) - g(\xi,\nabla_{[Y,X]}\eta)\\
&= R(X,Y,\xi,\eta) + R(X,Y,\eta,\xi).
\end{aligned}$$

□

6.3.2.4 Expression for the Temporal Curvature

We see that the temporal curvature tensor $\mathcal{P}$ can be computed in terms of the tensor $\mathcal{S}$:

Theorem 6.18. *For $X, Y, Z \in \Gamma(\mathfrak{S})$,*

$$\mathcal{P}(X,Y,Z) = \nabla_Z\mathcal{S}(Y,X) - \nabla_Y\mathcal{S}(Z,X).$$

Proof. Observing that $R(\cdot,\cdot,\partial_t,\cdot) = 0$ and the expression for the torsion given by (6.9) with the Bianchi identity proved in Theorem 6.16 gives

$$\begin{aligned}
R(\partial_t,X,Y,Z) - R(\partial_t,Y,X,Z) &= g(\nabla_X\mathcal{S}(Y),Z) - g(\nabla_Y\mathcal{S}(X),Z)\\
&= \nabla_X\mathcal{S}(Y,Z) - \nabla_Y\mathcal{S}(X,Z)
\end{aligned}$$

since the connection is spatially symmetric. From this we obtain, using Theorem 6.17, the following expression:

$$\begin{aligned}
2R(\partial_t,X,Y,Z) &= R(\partial_t,X,Y,Z) - R(\partial_t,X,Z,Y)\\
&= \quad R(\partial_t,Y,X,Z) + \nabla_X\mathcal{S}(Y,Z) - \nabla_Y\mathcal{S}(X,Z)\\
&\quad - R(\partial_t,Z,X,Y) - \nabla_X\mathcal{S}(Z,Y) + \nabla_Z\mathcal{S}(X,Y)\\
&= R(\partial_t,Z,Y,X) - R(\partial_t,Y,Z,X) - \nabla_Y\mathcal{S}(X,Z) + \nabla_Z\mathcal{S}(X,Y)\\
&= -2\nabla_Y\mathcal{S}(Z,X) + 2\nabla_Z\mathcal{S}(Y,X)
\end{aligned}$$

where we used the symmetry of $\mathcal{S}$ from Theorem 6.9. □

6.3.2.5 The Temporal Bianchi Identity and Evolution of Curvature

The Bianchi identity from Theorem 6.15 gives the usual second Bianchi identity in spatial directions. The remaining identities are those where one of the vector fields is ∂_t, which provide the following evolution equation for the spatial curvature tensor:

Theorem 6.19. *For $X, Y, Z \in \Gamma(\mathfrak{S})$,*

$$\begin{aligned}\nabla_{\partial_t} R(X,Y,W,Z) &= \nabla_X \nabla_Z \mathcal{S}(Y,W) - \nabla_X \nabla_W \mathcal{S}(Y,Z)\\ &\quad - \nabla_Y \nabla_Z \mathcal{S}(X,W) + \nabla_Y \nabla_W \mathcal{S}(X,Z)\\ &\quad - R(\mathcal{S}(X),Y,W,Z) - R(X,\mathcal{S}(Y),W,Z).\end{aligned}$$

Proof. From the Bianchi identity in Theorem 6.15 and the torsion identity (6.9) we have:

$$\begin{aligned}\nabla_{\partial_t} R(X,Y,W,Z) &= -\nabla_X R(Y,\partial_t,W,Z) - \nabla_Y R(\partial_t,X,W,Z)\\ &\quad - R(\tau(\partial_t,X),Y,W,Z) - R(\tau(X,Y),\partial_t,W,Z)\\ &\quad - R(\tau(Y,\partial_t),X,W,Z)\\ &= \nabla_X \mathcal{P}(Y,W,Z) - \nabla_Y \mathcal{P}(X,W,Z)\\ &\quad - R(\mathcal{S}(X),Y,W,Z) - R(X,\mathcal{S}(Y),W,Z)\\ &= \nabla_X \nabla_Z \mathcal{S}(Y,W) - \nabla_X \nabla_W \mathcal{S}(Y,Z)\\ &\quad - \nabla_Y \nabla_Z \mathcal{S}(X,W) + \nabla_Y \nabla_W \mathcal{S}(X,Z)\\ &\quad - R(\mathcal{S}(X),Y,W,Z) - R(X,\mathcal{S}(Y),W,Z) \qquad (6.13)\end{aligned}$$

where we used the result of Theorem 6.18. Note the difference in the last two terms compared to the expression in Proposition 4.8. □

6.3.2.6 The Evolution of Curvature in the Ricci Flow Case

Suppose now that the metric g evolves by Ricci flow. With the vector bundle machinery presented so far, we can now derive the evolution equation for the spatial curvature. Working with the connection ∇ on $M \times \mathbb{R}$ constructed from Theorems 6.9 and 6.12, we first observe directly from Proposition 4.2 – as ∇ is spatially symmetric – that:

Theorem 6.20. *For $X, Y, W, Z \in \Gamma(\mathfrak{S})$, the Laplacian of the spatial curvature tensor R satisfies*

$$\begin{aligned}(\Delta R)(X,Y,W,Z) = {} & \nabla_X \nabla_W \mathrm{Ric}(Y,Z) - \nabla_Y \nabla_W \mathrm{Ric}(X,Z) \\ & + \nabla_Y \nabla_Z \mathrm{Ric}(X,W) - \nabla_X \nabla_Z \mathrm{Ric}(Y,W) \\ & - 2\big(B(X,Y,W,Z) - B(X,Y,Z,W) \\ & + B(X,W,Y,Z) - B(X,Z,Y,W)\big) \\ & + \mathrm{Ric}(R(W,Z)X,Y) - \mathrm{Ric}(R(W,Z)Y,X).\end{aligned}$$

Now since the time-dependent metric $g = g(t)$ evolves according to $\partial_t g_{ij} = -2\mathrm{Ric}_{ij}$, (6.8) implies that

$$\begin{aligned}\mathcal{S}(\partial_i) &= \nabla_{\partial_t} \partial_i - [\partial_t, \partial_i] \\ &= {\Gamma_{0i}}^k \partial_k \\ &= \frac{1}{2}(g^{kj} \partial_t g_{ij}) \partial_k \\ &= -{\mathrm{Ric}_i}^k \partial_k.\end{aligned}$$

So we conclude for $X, Y \in \Gamma(\mathfrak{S})$ that

$$\mathcal{S}(X) = -\mathrm{Ric}^\sharp(X) \qquad \text{and} \qquad \mathcal{S}(X,Y) = -\mathrm{Ric}(X,Y).$$

As ∇ is compatible with g, we also see that

$$\nabla_X \nabla_Y \mathcal{S}(Z,W) = -\nabla_X \nabla_Y \mathrm{Ric}(Z,W) = -g(\nabla_X \nabla_Y \mathrm{Ric}^\sharp(Z), W).$$

Thus, by combining Theorems 6.19 and 6.20, we obtain the following desired reaction-diffusion type equation for the curvature.

Theorem 6.21. *If the metric g evolves by Ricci flow, then the curvature R evolves according to*

$$\begin{aligned}\nabla_{\partial_t} R(X,Y,Z,W) = {} & (\Delta R)(X,Y,Z,W) \\ & + 2\big(B(X,Y,Z,W) - B(X,Y,W,Z) \\ & + B(X,W,Y,Z) - B(X,Z,Y,W)\big),\end{aligned}$$

where X, Y, Z and W are spatial vector fields.

Chapter 7
The Weak Maximum Principle

The maximum principle is the main tool we will use to understand the behaviourof solutions to the Ricci flow. While other problems arising in geometric analysis and calculus of variations make strong use of techniques from functional analysis, here – due to the fact that the metric is changing – most of these techniques are not available; although methods in this direction are developed in the work of Perelman [Per02]. The maximum principle, though very simple, is also a very powerful tool which can be used to show that pointwise inequalities on the initial data of parabolic PDE are preserved by the evolution. As we have already seen, when the metric evolves by Ricci flow the various curvature tensors R, Ric, and Scal do indeed satisfy systems of parabolic PDE. Our main applications of the maximum principle will be to prove that certain inequalities on these tensors are preserved by the Ricci flow, so that the geometry of the evolving metrics is controlled.

7.1 Elementary Analysis

Suppose $U \subset \mathbb{R}^n$ is open and let $f : U \subset \mathbb{R}^n \to \mathbb{R}$ be a smooth function. If f has a local minimum at some $p \in U$, then it follows that $\nabla f(p) = 0$ and $\Delta f(p) \geq 0$. This follows from the second derivative test for functions of one variable: Given any direction $v \in \mathbb{R}^n$, the function $t \mapsto f_v(t) := f(p+tv)$ has a local minimum at $t = 0$, so $f_v'(0) = \nabla_v f(p) = 0$ and $f_v''(0) = \nabla_v \nabla_v f(p) \geq 0$. The Laplacian is proportional to the average of $f_v''(0)$ over all unit vectors v, and so is also non-negative.

On a Riemannian manifold the same argument applies if we define $f_v(t) := f\left(\exp_p(tv)\right)$, so that we have the following result:

Lemma 7.1 (Second Derivative Test). *Let M be a n-dimensional Riemannian manifold, and $u : M \to \mathbb{R}$ a C^2-function. If u has a local minimum at a point $p \in M$, then*

$$\nabla u(p) = 0 \qquad \text{and} \qquad \Delta u(p) \geq 0.$$

B. Andrews and C. Hopper, *The Ricci Flow in Riemannian Geometry*, Lecture Notes in Mathematics 2011, DOI 10.1007/978-3-642-16286-2_7,

As we shall see, Lemma 7.1 is the main ingredient in the proof of the maximum principle.

7.2 Scalar Maximum Principle

The simplest manifestation of the weak maximum principle is the following scalar maximum principle for time dependent metrics.

Proposition 7.2. *Suppose $g(t)$, $t \in [0,T)$, is a smooth family of metrics on a compact manifold M such that $u : M \times [0,T) \to \mathbb{R}$ satisfies*

$$\frac{\partial}{\partial t}u - \Delta_{g(t)}u \geq 0.$$

If $u \geq c$ at $t = 0$, for some $c \in \mathbb{R}$, then $u \geq c$ for all $t \geq 0$.[1]

Proof. Fix $\varepsilon > 0$ and define $u_\varepsilon = u + \varepsilon(1+t)$, so by hypothesis $u_\varepsilon > c$ at $t = 0$. We claim $u_\varepsilon > c$ for all $t > 0$.

To prove this, suppose the result is false. That is, there exists $\varepsilon > 0$ such that $u_\varepsilon \leq c$ somewhere in $M \times [0,T)$. As M is compact, there exists $(x_1, t_1) \in M \times (0,T)$ such that $u_\varepsilon(x_1,t_1) = c$ and $u_\varepsilon(x,t) \geq c$ for all $x \in M$ and $t \in [0, t_1]$. From this it follows that at (x_1, t_1) we have $\frac{\partial u_\varepsilon}{\partial t} \leq 0$ and $\Delta u_\varepsilon \geq 0$, so that

$$0 \geq \frac{\partial u_\varepsilon}{\partial t} \geq \Delta_{g(t)} u_\varepsilon + \varepsilon > 0,$$

which is a contradiction. Hence $u_\varepsilon > c$ on $M \times [0,T)$; and since $\varepsilon > 0$ is arbitrary, $u \geq c$ on $M \times [0,T)$. □

This proposition can be generalised by considering the semi-linear second-order parabolic operator:

$$Lu := \frac{\partial u}{\partial t} - \Delta_{g(t)}u - \langle X(t), \nabla u\rangle - F(u,t),$$

where $X(t)$ is a time-dependent vector field and $F = F(x,t) : \mathbb{R} \times [0,T) \to \mathbb{R}$ is continuous in t and locally Lipschitz in x. We say u is a *supersolution* if $Lu \geq 0$, and a *subsolution* if $Lu \leq 0$.

Proposition 7.3 (Comparison Principle). *Suppose that u and v are C^2 and satisfy $Lv \leq Lu$ on $M \times [0,T)$, and $v(x,0) \leq u(x,0)$ for all $x \in M$. Then*

$$v(x,t) \leq u(x,t)$$

holds on $M \times [0,T)$.

---

[1] Note that the Laplacian is defined by $\Delta_{g(t)} := g(t)^{ij}\nabla_i^{(t)}\nabla_j^{(t)}$, where $\nabla^{(t)}$ is the covariant derivative associated to $g(t)$.

Proof. We apply an argument to $w = u - v$ similar to that of the previous proposition. Firstly, compute

$$0 \leq Lu - Lv = \frac{\partial w}{\partial t} - \Delta w - \langle X, \nabla w \rangle - F(u,t) + F(v,t),$$

and note that the main difficulty is in controlling the last two terms. To do this, let $\tau \in (0,T)$, so that u and v are C^2 on $M \times [0,\tau]$. In particular, since $M \times [0,\tau]$ is compact and F is locally Lipschitz in the first argument, there exists a constant C such that $|F(u(x,t),t) - F(v(x,t),t)| \leq C|u(x,t) - v(x,t)|$, for all $(x,t) \in M \times [0,\tau]$. Now let $\varepsilon > 0$, and define $w_\varepsilon(x,t) = w(x,t) + \varepsilon e^{2Ct}$. Then $w_\varepsilon(x,0) \geq \varepsilon > 0$ for all $x \in M$, while

$$\frac{\partial w_\varepsilon}{\partial t} \geq \Delta w + \langle X, \nabla w \rangle - C|w| + 2C\varepsilon e^{2Ct}.$$

At a first point and time (x_0, t_0) where $w_\varepsilon(x_0,t_0) = 0$, we have $w = -\varepsilon e^{2Ct_0}$, $\nabla w = 0$, and $\Delta w \geq 0$, while $\frac{\partial w_\varepsilon}{\partial t} \leq 0$. So at this point

$$0 \geq \frac{\partial w_\varepsilon}{\partial t} \geq \Delta w + \langle X, \nabla w \rangle - C\varepsilon e^{2Ct_0} + 2C\varepsilon e^{2Ct_0} \geq C\varepsilon e^{2Ct_0} > 0,$$

which is a contradiction. Therefore $w_\varepsilon > 0$ for all $\varepsilon > 0$, and hence $w \geq 0$ on $M \times [0,\tau]$. Since $\tau \in (0,T)$ is arbitrary, $w \geq 0$ on $M \times [0,T)$. □

From the above result we can conclude that super/sub-solutions of heat type equations can be bounded by solutions to associated ODE:

Theorem 7.4 (The Scalar Maximum Principle). *Suppose u is C^2 and satisfies $Lu \geq 0$ on $M \times [0,T)$, and $u(x,0) \geq c$ for all $x \in M$. Let $\phi(t)$ be the solution to the associated* ODE*:*

$$\frac{d\phi}{dt} = F(\phi, t),$$
$$\phi(0) = c.$$

Then

$$u(x,t) \geq \phi(t)$$

for all $x \in M$ and all $t \in [0,T)$ in the interval of existence of ϕ.

Proof. Apply Proposition 7.3 with $v(x,t) = \phi(t)$: This can be done since $Lu \geq 0 = L\phi$, and $u(x,0) \geq c = \phi(0)$. □

Remark 7.5. Of course a similar result holds if both inequalities are reversed, and in particular if u satisfies $Lu = 0$ then both upper and lower bounds for u can be obtained from solutions of the associated ODE.

7.2.1 Lower Bounds on the Scalar Curvature

As a simple illustration of how the scalar maximum principle can be applied to the Ricci flow, consider the evolution equation for the scalar curvature given in Corollary 4.20. Since the reaction term is always non-negative, we have

$$\frac{\partial}{\partial t}\mathrm{Scal} \geq \Delta \mathrm{Scal}.$$

So any initial lower bound for scalar curvature is preserved (while upper bounds are in general not preserved). A stronger result can be obtained by estimating the reaction term more carefully: Since

$$|\mathrm{Ric}|^2 \geq \frac{1}{n}\mathrm{Scal}^2,$$

we have

$$\frac{\partial}{\partial t}\mathrm{Scal} \geq \Delta \mathrm{Scal} + \frac{2}{n}\mathrm{Scal}^2.$$

Writing $\mathscr{S}_0 = \inf\{\mathrm{Scal}(p,0) : p \in M\}$, we can apply Theorem 7.4 to prove $\mathrm{Scal}(p,t) \geq \phi(t)$, where

$$\phi(t) = \begin{cases} -\frac{n|\mathscr{S}_0|}{n+2|\mathscr{S}_0|t} & \text{if } \mathscr{S}_0 < 0, \\ 0 & \text{if } \mathscr{S}_0 = 0, \\ \frac{n\mathscr{S}_0}{n-2\mathscr{S}_0 t} & \text{if } \mathscr{S}_0 > 0. \end{cases}$$

In particular, if $\mathscr{S}_0 > 0$ then the lower bound $\phi(t)$ approaches infinity as $t \to \frac{n}{2\mathscr{S}_0}$, thus giving an upper bound on the maximal interval existence for the solution of Ricci flow.

7.2.2 Doubling-Time Estimates

Another useful application of the scalar maximum principle is the so-called doubling time estimate, which gives a lower bound on the time taken for the curvature to become large. From the evolution equation for curvature (as in the proof of Lemma 8.3) we find that

$$\frac{\partial}{\partial t}|R|^2 \leq \Delta |R|^2 + C(n)|R|^3.$$

So if $|R| \leq K$ at $t = 0$, then

$$|R|^2(x,t) \leq \left(\frac{1}{K} - \frac{C(n)}{2}t\right)^{-2} =: \rho(t) \tag{7.1}$$

for all $x \in M$ and $t \geq 0$ where $\rho(t)$ is the solution of the ODE: $\frac{d\rho}{dt} = C(n)\rho^{3/2}$ with $\rho(0) = K^2$. In particular $|R|(t) \leq 2K$, whenever $t \in [0, 1/C(n)K]$. This proves the following result which states that the maximum of the curvature cannot grow too fast:

Lemma 7.6 (Doubling Time Estimate). *Let $g(t)$ be a solution of the Ricci flow on a compact manifold M, with $|R(g(0))| \leq K$. Then $|R(g(t))| \leq 2K$, for all $t \in [0, 1/C(n)K]$.*

7.3 Maximum Principle for Symmetric 2-Tensors

To go beyond controlling the scalar curvature to controlling the Ricci curvature, it is natural to consider a generalisation of the maximum principle which applies to symmetric 2-tensors. This was done by Hamilton in his paper on three-manifolds [Ham82b]. In order to state his result, we need the following definition:

Definition 7.7 (Null-eigenvector Assumption). We say $\beta : \mathrm{Sym}^2 T^*M \times [0,T) \to \mathrm{Sym}^2 T^*M$ satisfies the null-eigenvector assumption if whenever ω_{ij} is a nonnegative symmetric 2-tensor at a point x, and if $V \in T_xM$ is such that $\omega_{ij}V^j = 0$, then

$$\beta_{ij}(\omega, t)V^iV^j \geq 0$$

for any $t \in [0, T)$.

Note that a symmetric tensor ω_{ij} is defined to be non-negative if and only if $\omega_{ij}v^iv^j \geq 0$ for all vectors v^i (i.e. if the quadratic form induced by ω_{ij} is positive semi-definite). In this situation we write $\omega_{ij} \geq 0$.

Theorem 7.8. *Suppose that $g(t)$, $t \in [0,T)$, is a smooth family of metrics on a compact manifold M. Let $\alpha(t) \in \Gamma(\mathrm{Sym}^2 T^*M)$ be a symmetric 2-tensor satisfying*

$$\frac{\partial \alpha}{\partial t} \geq \Delta_{g(t)}\alpha + \langle X, \nabla\alpha\rangle + \beta,$$

where X is a (time-dependent) vector field and $\beta = \beta(\alpha, t)$ is a symmetric 2-tensor locally Lipschitz in α and continuous in t.

If $\alpha(p, 0) \geq 0$ for all $p \in M$ and β satisfies the null-eigenvector assumption, then $\alpha(p,t) \geq 0$ for all $p \in M$ and $t \in [0, T)$.

We naturally want to apply this theorem to the evolution of Ric_{ij} under the Ricci flow, which by Corollary 4.18 takes the form

$$\frac{\partial}{\partial t}\mathrm{Ric}_{ij} = \Delta \mathrm{Ric}_{ij} + 2(\mathrm{Ric}^{pq}R_{pijq} - \mathrm{Ric}_{ip}\mathrm{Ric}_j^p).$$

The problem here is trying to control the curvature term R_{pijq} which potentially could be quite complicated. Fortunately Hamilton [Ham82b] noted that the Weyl curvature tensor, discussed in Sect. 4.5.2, vanishes when the dimension $n = 3$.[2] The significance of this is that it allows one to write the full curvature tensor:

$$R_{ijk\ell} = g_{ik}\mathrm{Ric}_{j\ell} + g_{j\ell}\mathrm{Ric}_{ik} - g_{i\ell}\mathrm{Ric}_{jk} - g_{jk}\mathrm{Ric}_{i\ell} - \frac{1}{2}\mathrm{Scal}(g_{ik}g_{j\ell} - g_{i\ell}g_{jk}).$$

It follows that the evolution of the Ricci tensor, when $n = 3$, is given by

$$\frac{\partial}{\partial t}\mathrm{Ric}_{ij} = \Delta \mathrm{Ric}_{ij} - Q_{ij},$$

where $Q_{ij} = 6\mathrm{Ric}_{ip}\mathrm{Ric}^p{}_j - 3\mathrm{Scal}\,\mathrm{Ric}_{ij} - (\mathrm{Scal}^2 - 2|\mathrm{Ric}|^2)g_{ij}$ is completely expressed in terms of the Ricci tensor Ric_{ij} and its contractions with the metric g_{ij}. With the equation now in this form, one can simply check the null-eigenvector condition in Theorem 7.8, with $X = 0$, $\alpha_{ij} = \mathrm{Ric}_{ij}$ and $\beta_{ij} = -Q_{ij}$, to prove the following:

Theorem 7.9 ([Ham82b, Sect. 9]). *Suppose $g(t)$, $t \in [0, T)$, is a solution of the Ricci flow on a closed 3-manifold M. If $\mathrm{Ric}_{ij} \geq 0$ at $t = 0$, then $\mathrm{Ric}_{ij} \geq 0$ on $0 \leq t < T$.*

Remark 7.10. However in general for $n \geq 3$ neither the condition of nonnegative Ricci curvature nor the condition of nonnegative sectional curvature is preserved under the Ricci flow on closed manifolds (see [CCG+08, Chap. 13]). It is precisely the problem of controlling the Weyl part in the evolutionary equation that has remained a major obstruction to pinching results in higher dimensions. We will return to this problem in Chaps. 12 and 13.

7.4 Vector Bundle Maximum Principle

Consider a compact Riemannian manifold (M, g) with a (possibly time-varying) metric g with Levi–Civita connection $\widehat{\nabla}$, and a rank k-vector bundle $\pi : E \to M \times \mathbb{R}$ with a connection ∇. Let h be a metric on the bundle E which is compatible with the connection ∇.

Now consider a heat-type PDE in which sections $u \in \Gamma(E)$ evolve by

$$\nabla_{\partial_t} u = \Delta u + \nabla_V u + F(u), \tag{7.2}$$

[2] When $n = 3$ there are only two possible types of nonzero components of W. Either there are three distinct indices, such as W_{1231}, or there are two distinct indices such as W_{1221}. Using the trace-free property, $W_{1231} = -W_{2232} - W_{3233} = 0$. Also, $W_{1221} = -W_{2222} - W_{3223} = -W_{3223} = W_{3113} = -W_{2112} = -W_{1221}$ so $W_{1221} = 0$.

where the Laplacian Δ acting on sections of E is defined by

$$\Delta u = g^{ij}(\nabla_i(\nabla_j u) - \nabla_{\hat{\nabla}_i \partial_j} u);$$

V is a smooth section of the spatial tangent bundle $\mathfrak{S}$, i.e. a smooth time-dependent vector field on M; and F is a time-dependent vertical vector field on each fibre of E, i.e. $F \in \Gamma(\pi^* E \to E)$ where $\pi^* E$ is the pullback bundle. Thus we have $F(x,t,u) \in (\pi^* E)_{(x,t,u)} = E_{(x,t)}$ for any $(x,t,u) = (x,t,u(x,t)) \in E$. In particular $F_{(x,t)} := F|_{\pi^{-1}(x,t)} : u \to v = F(x,t,u)$ is a map from $E_{(x,t)}$ to itself; in which case (7.2) takes the form $\nabla_{\partial_t} u(x,t) = \Delta u(x,t) + F(x,t,u(x,t))$.

7.4.1 Statement of Maximum Principle

In order to state an analogous maximum principle for vector bundles, we need the following three conditions.

Definition 7.11 (Convex in the Fibre). A subset $\Omega \subset E$ is said to be *convex in the fibre* if for each $(x,t) \in M \times \mathbb{R}$, the set $\Omega_{(x,t)} = \Omega \cap E_{(x,t)}$ is a convex subset of the vector space $E_{(x,t)}$.

With reference to Appendix B, we define the *support function* $s : E^* \to \mathbb{R}$ of Ω by

$$s(x,t,\ell) = \sup\{\ell(v) : v \in \Omega_{(x,t)} \subset E_{(x,t)}\}.$$

The *normal cone* $\mathcal{N}_v \Omega_{(x,t)}$ of $\Omega_{(x,t)}$ at a point $v \in \partial \Omega_{(x,t)}$ is then defined by

$$\mathcal{N}_v \Omega_{(x,t)} = \{\ell \in E^*_{(x,t)} : \ell(v) = s(x,t,\ell)\},$$

and the *tangent cone* is defined by

$$\mathcal{T}_v \Omega_{(x,t)} = \bigcap_{\ell \in \mathcal{N}_v \Omega_{(x,t)}} \left\{z \in E_{(x,t)} : \ell(z) \le 0\right\}.$$

Definition 7.12 (Vector Field Points into the Set). Let Ω be a subset of E which is convex in the fibre. The vector field $F \in \Gamma(\pi^* E)$ is said to *point into* Ω if $F(x,t,v) \in \mathcal{T}_v \Omega_{(x,t)}$ for every $(x,t,v) \in E$ with $v \in \partial \Omega_{(x,t)}$. Furthermore, F is said to *strictly point into* Ω if $F(x,t,v)$ is in the interior of $\mathcal{T}_v \Omega_{(x,t)}$ for every $(x,t,v) \in E$ with $v \in \partial \Omega_{(x,t)}$.

Remark 7.13. We will prove in Corollary 7.17 that F points into Ω precisely when the flow of the vector field F on each fibre $E_{(x,t)}$ takes $\Omega_{(x,t)}$ into itself.

Definition 7.14 (Invariance Under Parallel Transport). Let $\Omega \subset E$ be a subset. We say Ω is *invariant under parallel transport* by the connection ∇ if for every curve γ in $M \times \mathbb{R}$ and vector $V_0 \in \Omega_{\gamma(0)}$, the unique parallel section V along γ with $V(0) = V_0$ is contained in Ω.

We now state the maximum principle for vector bundles as follows.

Theorem 7.15 (Maximum Principle for Vector Bundle). *Let $\Omega \subset E$ be closed, convex in the fibre, and invariant under parallel transport with respect to ∇. Let F be a vector field which points into Ω. Then any solution u to the* PDE (7.2) *which starts in Ω will remain in Ω: If $u(x, t_0) \in \Omega$ for all $x \in M$, then $u(x,t) \in \Omega$ for all $x \in M$ and $t \geq t_0$ in the interval of existence.*

Proof. We apply a maximum principle argument to a function f on the 'sphere bundle' S^* over $M \times \mathbb{R}$ (with fibre at (x,t) given by $S^*_{(x,t)} = \{\ell \in E^*_{(x,t)} : \|\ell\| = 1\}$) defined by

$$f(x,t,\ell) = \ell(u(x,t)) - s(x,t,\ell)$$

where s is the support function defined above. By Theorem B.2 in Appendix B we note that for each (x,t), the supremum of $f(x,t,\ell)$ over $\ell \in S^*_{(x,t)}$ gives the distance of $u(x,t)$ from $\Omega_{(x,t)}$. Thus the condition: $u(x,t) \in \Omega$, for all (x,t), is equivalent to the inequality $f \leq 0$ on S^*. In particular the initial condition $u(x,t_0) \in \Omega_{(x,t_0)}$, for all $x \in M$ is equivalent to the condition:

$$f(x,t_0,\ell) \leq 0, \quad \forall\, (x,t_0,\ell) \in S^*.$$

We seek to prove the same is true for positive times t.

Our strategy will be to show that $f - \varepsilon e^{Ct}$ remains negative for all small $\varepsilon > 0$, for a suitable constant C. The choice of C will depend on Lipschitz bounds for the vector field F, so we must first restrict to a suitable compact region: Let $t_1 > t_0$ be any time less than the maximal time of existence of the solution, so that u is a smooth section on E over $M \times [t_0, t_1]$. Since $M \times [t_0, t_1]$ is compact, there exists $K > 0$ such that $\|u(x,t)\| \leq K$. Since F is smooth, there exists a constant L such that $\|F(x,t,v_2) - F(x,t,v_1)\| \leq L\|v_2 - v_1\|$ for all $(x,t) \in M \times [t_0,t_1]$ and all $v_2, v_1 \in E_{(x,t)}$ with $\|v_i\| \leq 2K$, $i = 1, 2$. Now choose $\varepsilon_0 > 0$ such that $\varepsilon_0 e^{(L+1)(t_1 - t_0)} \leq K$. We will prove that:

$$f - \varepsilon e^{(L+1)(t-t_0)} \leq 0 \quad \text{on } M \times [t_0, t_1]$$

for any $\varepsilon \leq \varepsilon_0$.

First note that the inequality $f - \varepsilon e^{(L+1)(t-t_0)}$ holds strictly for $t = t_0$. If the inequality does not hold on the entire domain $M \times [t_0, t_1]$, then there exists $(\tilde{x},\tilde{t},\tilde{\ell})$ with $t_0 < \tilde{t} \le t_1$ and $\ell \in S^*_{(\tilde{x},\tilde{t})}$ such that $f(\tilde{x},\tilde{t},\ell_0) = \varepsilon e^{(L+1)(\tilde{t}-t_0)}$ and $f(x,t,\ell) \le \varepsilon e^{(L+1)(t-t_0)}$ for all $(x,t,\ell) \in S^*$ with $t_0 \le t \le \tilde{t}$.

Let v be the closest point in $\Omega_{(\tilde{x},\tilde{t})}$ to $u(\tilde{x},\tilde{t})$. By Theorem B.2 we have $\|v - u(\tilde{x},\tilde{t})\| = \varepsilon e^{(L+1)(\tilde{t}-t_0)} \le K$ since $\varepsilon \le \varepsilon_0$, and $\|u(\tilde{x},\tilde{t})\| \le K$, so $\|v\| \le 2K$. Thus by the definition of L we have

$$\begin{aligned}\big\|F(\tilde{x},\tilde{t},v) - F(\tilde{x},\tilde{t},u(\tilde{x},\tilde{t}))\big\| &\le L\|v - u(\tilde{x},\tilde{t})\| \\ &= \varepsilon L e^{(L+1)(\tilde{t}-t_0)}.\end{aligned}$$

We claim that $\ell \in \mathcal{N}_v\Omega_{(\tilde{x},\tilde{t})}$: From the proof of Theorem B.2, for ℓ to attain the maximum of f we must have $\ell(\,\cdot\,) = \big\langle u(\tilde{x},\tilde{t}) - v, \cdot\big\rangle/\|u(\tilde{x},\tilde{t}) - v\|$, and then the proof of Theorem B.1 shows that $\ell(v) = s(\tilde{x},\tilde{t},\ell)$, so $\ell \in \mathcal{N}_v\Omega_{(\tilde{x},\tilde{t})}$ as required. The assumption that F points into Ω then implies that $\ell(F(\tilde{x},\tilde{t},v)) \le 0$.

In a neighbourhood U of $(\tilde{x},\tilde{t})$, extend $\tilde{\ell}$ to a smooth section ℓ of E^* by parallel translation along spatial geodesics from $\tilde{x}$, then along lines of fixed x in the t direction. Since the metric is compatible with the connection on E, we have $\ell(x,t) \in S^*$ for $(x,t) \in U$.

Claim 7.16. For this ℓ, the function $(x,t) \mapsto s(x,t,\ell_{(x,t)})$ is constant on U.

Proof of Claim. Parallel transporting along the curves indicated above defines the parallel transport maps

$$P\colon\ U \times E_{(\tilde{x},\tilde{t})} \to \pi_E^{-1}U \subset E \quad \text{and} \quad P^*:\ U \times E^*_{(\tilde{x},\tilde{t})} \to \pi_{E^*}^{-1}(U) \subset E^*.$$

Thus $\ell(x,t) = P^*(x,t,\tilde{\ell})$ by definition. The invariance of Ω under parallel transport implies that $P(x,t,\Omega_{(\tilde{x},\tilde{t})}) = \Omega_{(x,t)}$ for every $(x,t) \in U$. The definition of the dual connection implies that $(P^*(x,t,\tilde{\ell}))(P(x,t,v))$ is parallel, hence constant, along each of the curves. In particular this implies

$$\begin{aligned}s(x,t,P^*(x,t,\tilde{\ell})) &= \sup\big\{(P^*(x,t,\tilde{\ell}))(v) :\ v \in \Omega_{(x,t)}\big\} \\ &= \sup\big\{(P^*(x,t,\tilde{\ell}))(P(x,t,v)) :\ v \in \Omega_{(\tilde{x},\tilde{t})}\big\} \\ &= \sup\big\{\tilde{\ell}(v) :\ v \in \Omega_{(\tilde{x},\tilde{t})}\big\} \\ &= s(\tilde{x},\tilde{t},\tilde{\ell})\end{aligned}$$

which proves the claim. □

Define a function $\bar{f}$ on U by setting $\bar{f}(x,t) = f(x,t,\ell(x,t))$. By the construction of ℓ, $\bar{f}$ is smooth on U. Furthermore, $\bar{f}(\tilde{x},\tilde{t}) = \varepsilon e^{(L+1)(\tilde{t}-t_0)}$ and $\bar{f}(x,t) \le \varepsilon e^{(L+1)(t-t_0)}$ for all (x,t) with $t_0 \le t \le \tilde{t}$. It follows from this that spatial derivatives of $\bar{f}$ vanish at $(\tilde{x},\tilde{t})$, $\Delta\bar{f}(\tilde{x},\tilde{t}) \le 0$, and $\frac{\partial}{\partial t}\bar{f}(\tilde{x},\tilde{t}) - \varepsilon(L+1)e^{(L+1)(\tilde{t}-t_0)} \ge 0$.

Since ℓ is parallel along spatial geodesics through $(\tilde{x},\tilde{t})$, we have that $(\nabla^2\ell(\tilde{x},\tilde{t}))(v,v) = 0$ for every v, and hence $\nabla^2\ell(\tilde{x},\tilde{t}) = 0$ since $\hat{\nabla}$ is symmetric. In particular, $\Delta\ell(\tilde{x},\tilde{t}) = 0$.

Now we compute

$$\begin{aligned}
\frac{\partial}{\partial t}\bar{f}\Big|_{(\tilde{x},\tilde{t})} &= \frac{\partial}{\partial t}\big(\ell(u) - s(\ell)\big) \\
&= \cancel{\nabla_{\partial_t}\ell}(u) + \ell(\nabla_{\partial_t}u) - \cancel{\frac{\partial}{\partial t}s}(\ell) \\
&= \ell\big(\Delta u + \nabla_V u + F(u)\big) \\
&= \Delta\bar{f}(\tilde{x},\tilde{t}) + \nabla_V\bar{f}(\tilde{x},\tilde{t}) \\
&\quad + \ell\big(F(\tilde{x},\tilde{t},u(\tilde{x},\tilde{t})) - F(\tilde{x},\tilde{t},v)\big) + \ell(F(\tilde{x},\tilde{t},v)),
\end{aligned}$$

since $\nabla\bar{f} = \ell(\nabla u) + (\nabla\ell)(u) - \nabla(s(\ell)) = \ell(\nabla u)$ and $\Delta\bar{f} = (\Delta\ell)(u) + 2\nabla_i\ell(\nabla_i u) + \ell(\Delta u) - \Delta(s(\ell)) = \ell(\Delta u)$.

Combining the inequalities $\Delta\bar{f} \le 0$, $\nabla\bar{f} = 0$, $\frac{\partial}{\partial t}(\bar{f} - \varepsilon e^{(L+1)(\tilde{t}-t_0)}) \ge 0$, $\|F(u) - F(v)\| \le L\|u - v\|$ and $\ell(F(v)) \le 0$ noted previously, we see that

$$\begin{aligned}
0 &\le \frac{\partial}{\partial t}\bar{f}(\tilde{x},\tilde{t}) - \varepsilon(L+1)e^{(L+1)(\tilde{t}-t_0)} \\
&= \Delta\bar{f}(\tilde{x},\tilde{t}) + \nabla_V\bar{f}(\tilde{x},\tilde{t}) + \ell(F(\tilde{x},\tilde{t},v)) \\
&\quad + \ell(F(\tilde{x},\tilde{t},u(\tilde{x},\tilde{t})) - F(\tilde{x},\tilde{t},v)) - \varepsilon(L+1)e^{(L+1)(\tilde{t}-t_0)} \\
&\le L\|u(\tilde{x},\tilde{t}) - v\| - \varepsilon(L+1)e^{(L+1)(\tilde{t}-t_0)} \\
&= -\varepsilon e^{(L+1)(\tilde{t}-t_0)} < 0
\end{aligned}$$

which is a contradiction, since $\|u(\tilde{x},\tilde{t}) - v\| = \varepsilon e^{(L+1)(\tilde{t}-t_0)}$.

Therefore the inequality

$$f - \varepsilon e^{(L+1)(t-t_0)} \le 0$$

holds true on $M \times [t_0,t_1]$ for any $0 < \varepsilon < \varepsilon_0$, and so $f \le 0$ on $M \times [t_0,t_1]$. Since t_1 is an arbitrary time before the maximal time, we have $f \le 0$ for all $t \ge t_0$ in the interval of existence, and hence $u(x,t) \in \Omega$ for all $x \in M$ and $t \ge t_0$ as desired. □

As a first application of the maximum principle for vector bundles, we apply Theorem 7.15 in the simplest possible case to prove the following characterisation of the flow when a vector field preserves a convex set:

Corollary 7.17. *Let Ω be a closed convex subset of a finite-dimensional vector space V, and let $F : [0,T] \times V \to V$ be a smooth time-dependent vector field which points into Ω, so that $F(t,v) \in \mathcal{T}_v\Omega$ for all $v \in \partial\Omega$ and $t \in [0,T]$. Then the flow of F preserves Ω, in the sense that for any $u_0 \in \Omega$ the solution of the* ODE

$$\frac{d}{dt}u(t) = F(t, u(t))$$
$$u(0) = u_0$$

has $u(t) \in \Omega$ *for all* $t \geq 0$ *in its interval of existence.*

Proof. Let M to be the zero-dimensional manifold consisting of a single point $\{x\}$. Let E the trivial bundle $\{x\} \times V$ with the trivial connection given by differentiation in the t direction, and metric given by a time-independent inner product on V. Also take Ω constant. Then $\Delta u = 0$ since there are no spatial directions. Equation (7.2) reduces to the ordinary differential equation above, and the maximum principle applies. □

Remark 7.18. The converse also holds: If F does not point into Ω, then there exists some $v \in \partial\Omega$ and $\ell \in \mathcal{N}_v\Omega$ such that $\ell(F(v)) > 0$. But then the solution of the ODE with initial value v has $\frac{d}{dt}\ell(u(t)) > 0$ at $t = 0$. Therefore $\ell(u(t)) > \ell(v) = s(\ell) = \sup\{\ell(x) : x \in \Omega\}$ for small $t > 0$, so $u(t) \notin \Omega$.

7.5 Applications of the Vector Bundle Maximum Principle

7.5.1 Maximum Principle for Symmetric 2-Tensors Revisited

To illustrate the application of the vector bundle maximum principle, we show how Theorem 7.15 implies the maximum principle for symmetric 2-tensors given in Theorem 7.8. Here we take $\mathfrak{S}$ to be the spatial tangent bundle defined in Sect. 6.3.1, with a time-dependent metric g and connection given by Theorem 6.9. We take E to be the bundle of symmetric 2-tensors over $\mathfrak{S}$, i.e. $E = \mathrm{Sym}^2\mathfrak{S}^*$, with the metric and connection induced from $\mathfrak{S}$, and consider evolution equation for $A \in \Gamma(E)$ given in local coordinates by

$$\frac{\partial}{\partial t}A_{ij} = \Delta A_{ij} + \nabla_V A_{ij} + F_{ij}(A),$$

where V is a smooth section of $\mathfrak{S}$ and $F : E \to E$ is a smooth section of $\pi_E^* E$, i.e. for each $\alpha \in E_{(x,t)}$, $F(\alpha) \in E_{(x,t)}$. We note that since $\nabla_t \partial_i = \mathcal{S}(\partial_i)$, the equation can be rewritten as follows:

$$\begin{aligned}
\nabla_{\partial_t} A_{ij} &= \frac{\partial}{\partial t}A_{ij} - A(\mathcal{S}(\partial_i), \partial_j) - A(\partial_i, \mathcal{S}(\partial_j)) \\
&= \Delta A_{ij} + \nabla_V A_{ij} + F_{ij}(A) - A(\mathcal{S}(\partial_i), \partial_j) - A(\partial_i, \mathcal{S}(\partial_j)) \\
&= \Delta A_{ij} + \nabla_V A_{ij} + \bar{F}_{ij}(A).
\end{aligned}$$

Now let $\Omega \subset E$ be the set of positive definite symmetric 2-tensors acting on $\mathfrak{S}$. The set Ω is certainly convex in the fibres of E, since it is given as an intersection of half-spaces by

$$\Omega_{(x,t)} = \bigcap_{v \in \mathfrak{S}_{(x,t)} :\, \|v\|=1} \{\alpha \in E_{(x,t)} :\, \ell_v(\alpha) \leq 0\} \tag{7.3}$$

where $\ell_v \in E^*_{(x,t)}$ is defined by $\ell_v(\alpha) = -\alpha(v,v)$. This has the form of (B.1), where $B = \{\ell_v :\, v \in \mathfrak{S}, \|v\| = 1\}$ and $\phi(\ell_v) = 0$ for each v.

Note that Ω is invariant under parallel transport: Let γ be a smooth curve in M, and let P_s be the parallel transport operator along γ from $\gamma(0)$ to $\gamma(s)$. Suppose $\alpha \in \Omega_{\gamma(0)}$. Then $P_s\alpha \in E_{\gamma(s)}$ is also in Ω, since $\frac{d}{ds}(P_s(\alpha))(P_sv, P_sv) = (\nabla_s P_s\alpha)(P_sv, P_sv) + 2(P_s\alpha)(\nabla_s P_sv, P_sv) = 0$, and so

$$(P_s(\alpha))(v,v) = \alpha(P_s^{-1}v, P_s^{-1}v) \geq 0$$

for any $v \in \mathfrak{S}_{\gamma(s)}$.

For the maximum principle to apply, we need the vector field $\bar{F}$ to point into Ω. That is, we require that for any $\alpha \in \partial\Omega$, $\bar{F}(\alpha) \in \mathcal{T}_\alpha\Omega$. Since Ω is given in the form (7.3), Theorem B.7 implies that $\bar{F}(\alpha) \in \mathcal{T}_\alpha\Omega$ if and only if $\ell_v(\bar{F}(\alpha)) \leq 0$ for any $v \in \mathfrak{S}$ with $\|v\| = 1$ and $\ell_v(\alpha) = 0$. That is, if $\alpha(v,v) = 0$ and α is non-negative definite (so that v is a null eigenvector of α) then we need $\bar{F}(v,v) \geq 0$. This reduces to the same condition for F, since

$$\bar{F}(v,v) = F(v,v) - 2\alpha(\mathcal{S}(v),v) = F(v,v)$$

because v is a null-eigenvector of α. Thus the null-eigenvector condition of Definition 7.7 is precisely the condition that F points into the set Ω of weakly positive definite symmetric bilinear forms. The tensor maximum principle now follows.

7.5.2 Reaction-Diffusion Equation Applications

Reaction-diffusion systems describe interaction and diffusion of a density or concentration – often typified by a chemical process. As we have see, the systems is characterised by the PDE:

$$\partial_t q = \kappa \Delta q + F(q)$$

where $q = q(x,t)$ represents the concentration of a substance, κ is the diffusion coefficient and F is the reaction term.

Example 7.19. One of the simplest examples is the chemical reaction:

$$X + Y \rightleftharpoons 2Y$$

This reaction is one in which a molecule of species X interacts with a molecule of species Y. The X molecule is converted into a Y molecule. The final product consists of the original Y molecule plus the Y molecule created in the reaction. It is know as an autocatalytic reaction as at least one of the products (in this case species Y) is also a reactant.

Let x denote the concentration of X and y denote the concentration of Y. In a well stirred reaction the concentration can be modeled as satisfying the rate reaction ODE:

$$\begin{aligned}\frac{dx}{dt} &= -k_1xy + k_2y^2 \\ \frac{dy}{dt} &= k_1xy - k_2y^2\end{aligned}$$

where k_1 and k_2 are the respective forward and backwards reaction rates. The system is clearly in equilibrium when $k_1xy = k_2y^2$.

If both chemicals are able to diffuse within the underlying medium the (unstirred) reaction satisfies the reaction-diffusion PDE:

$$\begin{aligned}\frac{\partial x}{\partial t} &= \kappa_1 \Delta x - k_1xy + k_2y^2 \\ \frac{\partial y}{\partial t} &= \kappa_2 \Delta y + k_1xy - k_2y^2.\end{aligned}$$

Example 7.20. The Lotka–Volterra equations (also known as the predator-prey equations) are used to describe the dynamics of biological systems in which two species interact, one a predator and one its prey. They were proposed independently by Alfred J. Lotka in 1925 and Vito Volterra in 1926. The ODE describing the interaction between x number of prey (i.e. rabbits) and y number of some predator (i.e. foxes) is given by:

$$\begin{aligned}\frac{dx}{dt} &= x(\alpha - \beta y) \\ \frac{dy}{dt} &= -y(\gamma - \delta x)\end{aligned}$$

where α represents the exponential growth of the prey, γ represents the decay of the predator and β, δ quantifies the interaction of the species (all are positive). The fixed points of this system are $(0, 0)$ and $(\gamma/\delta, \alpha/\beta)$; the first of which is a saddle point.

Our interest here is to show the corresponding reaction-diffusion equations obeys the weak maximum principle discussed in the above. Crucially, this system has an integral of motion of the form

$$K = y^{\alpha} e^{-\beta y} x^{\gamma} e^{-\delta x}.$$

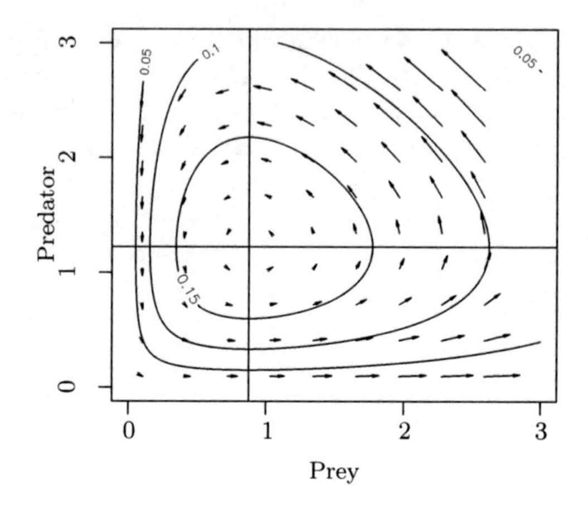

Fig. 7.1 Contour plots for Lotka–Volterra model with fixed point $(\gamma/\delta, \alpha/\beta)$. Note the level sets are convex in the plane and the vector field (7.4) is tangent to them

We observe K is log-concave, so the super-level sets are convex by the following lemma:

Lemma 7.21. *If $f : C \to \mathbb{R}$ is a convex function on a convex domain $C \subset \mathbb{R}^n$, then the super-level sets $L_\lambda = \{x : f(x) \leq \lambda\}$ are convex.*

Proof. For any points $x_1, x_2 \in L_\lambda$, one has $f(x_1) \leq \lambda$, $f(x_2) \leq \lambda$ and $x_1, x_2 \in C$. Consider the point $x = \theta x_1 + (1-\theta)x_2$, for any $0 < \theta < 1$. Clearly $x \in C$ by the convexity of C and since f is a convex function, $f(x) \leq \theta f(x_1) + (1-\theta) f(x_2) \leq \theta\lambda + (1-\theta)\lambda = \lambda$. Thus $x \in L_\lambda$ and so L_λ is a convex set. □

This is easily seen Fig. 7.1. The orbits are cyclic with four phases in the positive quadrant. Since K is a constant of the motion, the flow of the vector field

$$F(x,y) = (\alpha x - \beta xy, -\gamma y + \delta xy). \tag{7.4}$$

preserves the super-level sets of K. By Remark 7.18 the vector field F points into Ω. Thus by the maximum principle, solutions $u : M \times [0,T) \to \mathbb{R}^2$ of the reaction-diffusion equation $\frac{\partial}{\partial t} u = \Delta u + F(u)$ which start in a super-level set of K will remain so.

Remark 7.22. In the last example one could deduce the same result by applying the scalar maximum principle to $\psi \circ u$, where ψ is the convex function $\psi = -\log K$: A direct computation gives

$$\frac{\partial}{\partial t}(\psi \circ u) = \Delta(\psi \circ u) - D^2\psi(\nabla u, \nabla u) + D_{F(u)}\psi(u).$$

The fact that K is a constant of the motion implies that the last term vanishes. The convexity of ψ makes the second term non-positive, so we can apply the maximum principle to show that the maximum of $\psi \circ u$ does not increase.

This gives some intuition for why the convexity condition is important. The power of the vector bundle maximum principle, Theorem 7.15, is that it can be applied to cases where the set Ω is not smooth or is not naturally described as the sub-level set of a smooth convex function.

7.5.3 Applications to the Ricci Flow When $n = 3$

In three dimensions we can define an isomorphism from $\mathfrak{S}$ to $\Lambda^2(\mathfrak{S})$ (up to sign in the non-oriented case) by $\iota : \ \partial_i \mapsto \frac{1}{2}\varepsilon_{ijk}g^{jp}g^{kq}\partial_p \wedge \partial_q$, where ε_{ijk} is the Levi–Civita symbol, or elementary alternating 3-tensor (i.e. the volume form), so that in an oriented orthonormal basis $\{e_1, e_2, e_3\}$ we have $\varepsilon_{ijk} = 1$ if $(i\,j\,k)$ is a positive permutation of $(1\,2\,3)$, -1 for a negative permutation, and zero otherwise. Using this we can rewrite the Riemann tensor as a symmetric bilinear form $\Lambda \in \Gamma(\mathfrak{S}^* \otimes \mathfrak{S}^*)$ by $\Lambda(u, v) = R(\iota u, \iota v)$, yielding

$$\Lambda_{ij} = \frac{1}{4}\varepsilon_{iab}\varepsilon_{jcd}R_{abcd}. \tag{7.5}$$

Thus we find that

$$\Lambda = \begin{pmatrix} R_{2323} & R_{1332} & R_{1223} \\ R_{2331} & R_{1313} & R_{2113} \\ R_{3221} & R_{3112} & R_{1212} \end{pmatrix}.$$

The geometric interpretation of Λ is that $\Lambda(v, v)$ is $\|v\|^2$ times the sectional curvature of the 2-plane normal to v. We will now consider applications of the maximum principle to prove that the Ricci flow keeps Λ inside suitable subsets of the bundle $\mathrm{Sym}^2(\mathfrak{S})$ of symmetric 2-tensors acting on the spatial tangent bundle $\mathfrak{S}$, if this is true initially.

7.5.3.1 Subsets of the Bundle of Symmetric Tensors

Let E be a vector bundle of rank k over a manifold M, equipped with a metric g and a compatible connection ∇. Then the constructions in Sects. 2.4 and 2.5 provide metrics and compatible connections on each of the tensor bundles $T^q_p(E) = \bigotimes^p E \otimes \bigotimes^q E^*$ constructed from E, so in particular on the bundle of $(2, 0)$-tensors. By Example 2.64 the bundle $\mathrm{Sym}^2(E)$ of symmetric 2-tensors acting on E is a parallel subbundle of $T^2_0(E)$, and so inherits a metric and compatible connection. We will discuss a natural construction which produces subsets of $\mathrm{Sym}^2(E)$ which are automatically convex and invariant under parallel transport.

The idea is as follows: Let K be a closed convex subset in the space $\mathrm{Sym}^2(\mathbb{R}^k)$ of symmetric $k \times k$ matrices, which is invariant under the action of $O(k)$ (or, if E is oriented, of $SO(k)$): That is, for $\mathbb{O} \in O(k)$ and $T \in \mathrm{Sym}^2(\mathbb{R}^k)$, define $T^{\mathbb{O}}$ by

$$T^{\mathbb{O}}(u,v) = T(\mathbb{O}u, \mathbb{O}v)$$

for all $u, v \in \mathbb{R}^k$. Then our requirement is that $T \in K$ implies that $T^{\mathbb{O}} \in K$.

Now define $\Omega \subset \mathrm{Sym}^2(E)$ as follows: $T \in \mathrm{Sym}^2(E_x) \in \Omega$ if and only if for any orthonormal frame Y for E_x (that is, $Y : (\mathbb{R}^k, \delta) \to (E_x, g_x)$ is a linear isometry), the element T_Y of $\mathrm{Sym}^2(\mathbb{R}^k)$ defined by $\tilde{T}_Y(u,v) = T(Y(u), Y(v))$ lies in K.[3] We note that this is independent of the choice of Y, since any other frame is of the form $Y \circ \mathbb{O}$ for some $\mathbb{O} \in O(k)$, and

$$T_{Y\circ\mathbb{O}}(u,v) = T_Y(\mathbb{O}u, \mathbb{O}v) = T_Y^{\mathbb{O}}(u,v),$$

and by assumption $T_Y^{\mathbb{O}} \in K$ if $T_Y \in K$.

We can check that Ω is convex in the fibre: If T and S are in Ω_x and $a \in (0,1)$, then for any frame Y, T_Y and S_Y are in K, so $(aT + (1-a)S)_Y = aT_Y + (1-a)S_Y \in K$ since K is convex, and so $aT + (1-a)S \in \Omega_x$.

Finally, we can check that Ω is invariant under parallel transport: Suppose σ is a smooth curve in M, and T is a parallel section of $\sigma^*(\mathrm{Sym}^2(E))$ with $T(0) \in \Omega_{\sigma(0)}$. Let Y be a parallel frame along σ, defined by ${}^{\sigma}\nabla_{\partial_s}(Y(u)) = 0$ for all $u \in \mathbb{R}^k$ (see Sect. 2.8.5). Then $Y(s)$ is an orthonormal frame for $E_{\sigma(s)}$ for each s, since

$$\frac{\partial}{\partial s} g(Y(u), Y(v)) = g({}^{\sigma}\nabla_{\partial_s}(Y(u)), Y(v)) + g(Y(u), {}^{\sigma}\nabla_{\partial_s} Y(v)) = 0$$

for every $u, v \in \mathbb{R}^k$, where we used the result from Proposition 2.58 that shows the compatibility of the pullback connection and metric. But since T, $Y(u)$ and $Y(v)$ are parallel, we have

$$\frac{\partial}{\partial s}(T_{Y(s)}(u,v)) = \frac{\partial}{\partial s}(T(Y(u), Y(v))) = 0,$$

and so $(T(s))_{Y(s)} = (T(0))_{Y(0)} \in K$, so $T(s) \in \Omega_{\sigma(s)}$ for every s, as required.

Now let us look a little closer at the convex set K: A special property of the space of symmetric $k \times k$ matrices is that they can always be diagonalised by an orthogonal transformation. That is, for each $T \in \mathrm{Sym}^2(\mathbb{R}^k)$ there

[3] This can be conveniently expressed in terms of the frame bundle machinery discussed in Sect. 6.2: As noted there, a symmetric 2-tensor gives rise to a *function* from the orthonormal frame bundle to $\mathrm{Sym}^2(\mathbb{R}^k)$. Our condition is simply that this function takes values in K.

exists $\mathbb{O} \in O(k)$ such that $T^{\mathbb{O}}$ is diagonal. Then by the invariance of K, T is in K if and only if the diagonal matrix $T^{\mathbb{O}}$ is in K. Thus it is sufficient for us to consider the set K_D given by the intersection of K with the space of diagonal matrices, which we identify with $\mathbb{R}^k$. Since the diagonal matrices are a vector subspace of $\mathrm{Sym}^2(\mathbb{R}^k)$, K_D is convex. Furthermore, K_D is invariant under those orthogonal matrices which send diagonal matrices to diagonal matrices, which are exactly the permutation matrices which act by interchanging basis elements of $\mathbb{R}^k$. That is, K determines a convex set $K_D \subset \mathbb{R}^k$ which in symmetric, in the sense that it is invariant under permutation of the coordinates. In fact the converse is also true: If K_D is any symmetric convex set in $\mathbb{R}^k$, then the space K of symmetric matrices having eigenvalues in K_D is convex and $O(k)$-invariant, and so our construction above gives a subset of $\mathrm{Sym}^2(E)$ which is convex in the fibre and invariant under parallel transport.

7.5.3.2 Checking that the Vector Field Points into the Set

We now apply the above construction taking E to be the spatial tangent bundle $\mathfrak{S}$ over the manifold $M \times I$, for an interval $I \subset \mathbb{R}$. This is a vector bundle of rank 3, so for any symmetric convex set K_D in $\mathbb{R}^3$ we produce a suitable parallel convex set $\Omega \subset \mathrm{Sym}^2(\mathfrak{S})$. We want to use the maximum principle to show that if $\Lambda(x,0) \in \Omega_{(x,0)}$ for all $x \in M$, then $\Lambda_{(x,t)} \in \Omega_{(x,t)}$ for all $x \in M$ and all $t \geq 0$ in I. There is one more condition we need to check: The vector field arising in the reaction-diffusion equation for Λ must point into Ω.

By Theorem 6.21 the PDE system for R is of the form

$$\nabla_{\partial_t} R_{ijk\ell} = \Delta R_{ijk\ell} + 2Q(R)_{ijkl},$$

where $Q(R)_{ijkl} = B_{ijk\ell} - B_{ij\ell k} + B_{ikj\ell} - B_{i\ell jk}$. The fact that the metric is parallel implies also that the alternating tensor ε is parallel, so Λ satisfies the reaction-diffusion equation

$$\nabla_{\partial_t} \Lambda_{ij} = \Delta \Lambda_{ij} + \tilde{Q}(\Lambda)_{ij}.$$

A direct computation shows that in a basis $\{Y_i\}$ for which Λ_Y is diagonal, the reaction term $(\tilde{Q}(\Lambda))_Y$ is also diagonal, and is given as follows:[4]

$$\tilde{Q}\left(\begin{bmatrix} x & 0 & 0 \\ 0 & y & 0 \\ 0 & 0 & z \end{bmatrix}\right) = 2\begin{bmatrix} x^2+yz & 0 & 0 \\ 0 & y^2+xz & 0 \\ 0 & 0 & z^2+xy \end{bmatrix}.$$

[4] See Example 12.14 in Chap. 12 for the derivation of this.

In particular, the flow of this vector field keeps stays inside the subspace of diagonal matrices (with respect to the given basis), and so the flow stays inside Ω provided the flow of Λ_Y by the vector field $\tilde{Q}_Y$ stays inside K. This happens precisely when the vector field

$$V(x,y,z) = (x^2+yz, y^2+xz, z^2+xy)$$

on $\mathbb{R}^3$ points into the symmetric convex set $K_D \subset \mathbb{R}^3$.

We now check this condition for several concrete examples:

1. *The cone of positive sectional curvature operators:* Here

$$K_D = \{(x,y,z) \in \mathbb{R}^3 : \ x,y,z \geq 0\}.$$

This is convex since it is an intersection of three half-spaces: We can write it in the form

$$K_D = \bigcap_{i=1,2,3} \{(x,y,z) : \ \ell_i(x,y,z) \leq 0\}, \tag{7.6}$$

where $\ell_1 = (-1,0,0)$, $\ell_2 = (0,-1,0)$ and $\ell_3 = (0,0,-1)$. We must check that V is in the tangent cone at any boundary point, which by Theorem B.7 in Appendix B amounts to showing that if $\ell_i(x,y,z) = 0$ for some $(x,y,z) \in K_D$, then $\ell_i(V(x,y,z)) \leq 0$. By symmetry we need only prove this for $i=1$, in which case we have $x=0$, y, $z \geq 0$, and

$$\ell_1(V(x,y,z)) = -x^2 - yz = -yz \leq 0$$

as required. (Fig. 7.2)

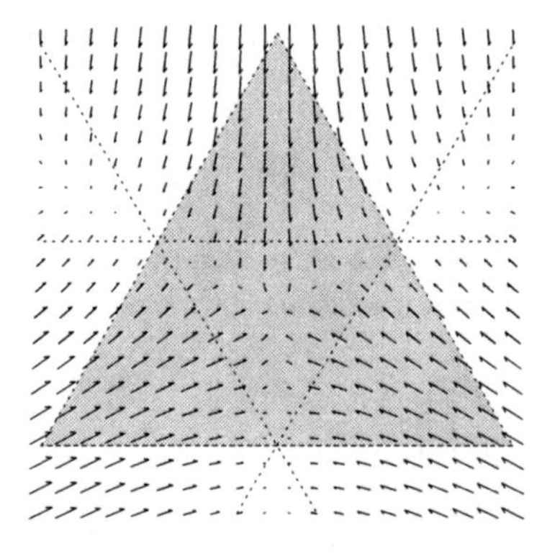

Fig. 7.2 Radial projection of the vector field V onto the plane $x+y+z=1$. The lines shown are lines of vanishing sectional curvatures, and the shaded region corresponds to the cone of positive Ricci curvature. The inner triangle corresponds to the cone of positive sectional curvature

2. *The cone of positive Ricci curvature operators:* Here

$$K_D = \bigcap_{P \in S_3} \{(x, y, z) : \ \ell(P(x, y, z)) \le 0\},$$

where S_3 is the set of permutations of three objects, and $\ell(x, y, z) = -X \cdot (1, 1, 0)$. By symmetry it suffices to check $\ell(V) \le 0$ at a point with $\ell(x, y, z) = x + y = 0$ and $\ell(P(x, y, z)) \le 0$ for all P: This is true, since

$$\ell(V) = -x^2 - yz - y^2 - xz = -x^2 - y^2 - z(x + y) = -x^2 - y^2 \le 0.$$

3. *Cones of pinched Ricci curvatures:* Here we fix $\varepsilon \in (0, 2)$ and define

$$K_D = \bigcap_{P \in S_3} \{(x, y, z) : \ \ell(P(x, y, z)) \le 0\}$$

where $\ell(x, y, z) = \varepsilon z - x - y$. Again by symmetry we need only prove $\ell(V(x, y, z)) \le 0$ at any point with $\varepsilon z - x - y = \ell(x, y, z) = 0$ and $\ell(P(x, y, z)) \le 0$ for all $P \in S_3$:

$$\begin{aligned} \ell(V) &= -x^2 - yz - y^2 - xz + \varepsilon z^2 + \varepsilon xy \\ &= -x^2 - y^2 + \varepsilon xy - z(x + y - \varepsilon z) \\ &= -x^2 - y^2 + \varepsilon xy \\ &\le 0 \end{aligned}$$

since $\varepsilon \le 2$. Notice that the inequality is strict for $\varepsilon < 2$, unless $(x, y, z) = (0, 0, 0)$. (Fig. 7.3)

4. *Pinching sets:* The fact that the vector field points strictly into the cones of pinched Ricci curvature means that we can find a preserved set which gets near the constant curvature line when the curvature is large. Let us

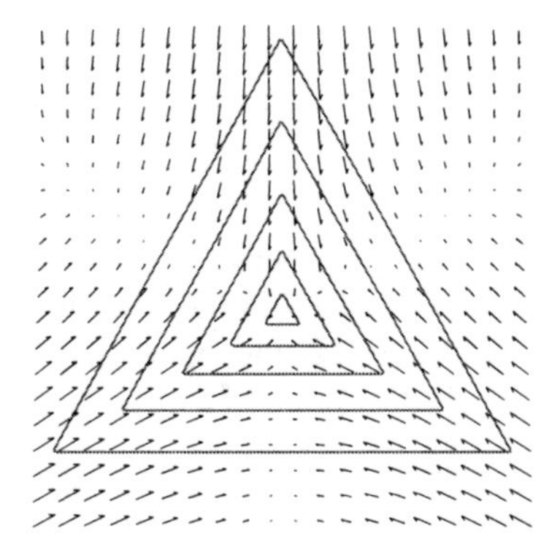

Fig. 7.3 Radial projection of the vector field V onto the plane $x + y + z = 1$, showing the regions corresponding to pinched Ricci curvature cones. Note that the vector field points into each of these sets

consider sets defined by

$$K_D = \bigcap_{P \in S_3} \{(x, y, z) : \varphi(P(x, y, z)) \le 0\},$$

where $\varphi(x, y, z) = z - f\left(\frac{x+y}{2}\right)$ and f is an increasing concave differentiable function on $[0, \infty)$ with $f(0) = 0$. Since f is concave, φ is convex, so the sub-level set $\{\varphi \le 0\}$ is convex, and hence K_D is an intersection of convex sets, hence convex. We can write this as an intersection of half-spaces as follows: Since a concave function always lies below its tangent line, we can write

$$f(a) = \inf\{f(b) + f'(b)(a - b) : b \ge 0\},$$

and hence

$$K_D = \bigcap_{P \in S^3, b \ge 0} \{(x, y, z) : \ell_b(P(x, y, z)) \le f(b)\},$$

where $\ell_b(x, y, z) = z - f'(b)(\frac{x+y}{2} - b)$. This gives K_D as an intersection of half-spaces in the form used in Sect. B.5 of Appendix B. We need to prove that if $\ell_b(x, y, z) = f(b)$ for some $b \ge 0$, and $\ell_{b'}(P(x, y, z)) \le f(b')$ for all b' and P, then $\ell_b(V(x, y, z)) \le 0$. The conditions imply that $\ell_b = \ell_{\frac{x+y}{2}}$ and $z = f\left(\frac{x+y}{2}\right)$, so we have

$$\begin{aligned}
\ell_b(V(x, y, z)) &= z^2 + xy - \frac{f'}{2}(x^2 + yz + y^2 + xz) \\
&= f^2 + \left(\frac{x+y}{2}\right)^2 - \left(\frac{x-y}{2}\right)^2 \\
&\quad - f'\left(\left(\frac{x+y}{2}\right)^2 + \left(\frac{x-y}{2}\right)^2 + \left(\frac{x+y}{2}\right)f\right).
\end{aligned}$$

Thus we require that the right-hand side be non-positive for all such points. Since we are assuming that $f' \ge 0$, we observe that for a fixed value of $\xi = \frac{x+y}{2}$, the right-hand side is maximised by choosing $x = y$, so for the inequality to hold it is necessary and sufficient that

$$f(\xi)^2 + \xi^2 - f'(\xi)\left(\xi^2 + \xi f(\xi)\right) \le 0.$$

This differential equation has the solutions $(f - \xi)^2 = C\xi e^{-f/\xi}$ for arbitrary $C > 0$. One can verify directly that the solutions are concave and increasing and have $f(0) = 0$, and satisfy $f(\xi) \sim \xi - \tilde{C}\sqrt{\xi}$ as $\xi \to \infty$. (Fig. 7.4)

Any metric with strictly positive Ricci curvature on a compact 3-manifold has principal sectional curvatures lying in one of the pinching sets of the last

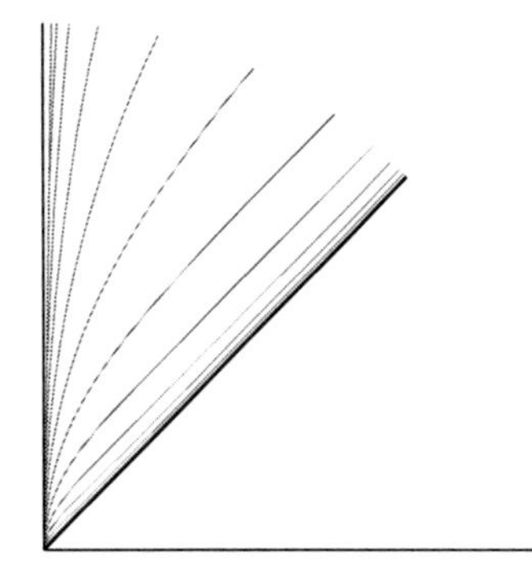

Fig. 7.4 Graphs of f for a family of pinching sets defined by $z \leq f\left(\frac{x+y}{2}\right)$. The diagonal corresponds to the constant curvature line $z = (x + y)/2$, and the entire family of pinching sets fills out the cone of positive Ricci curvature

example for some value of C. In particular this implies that the sectional curvatures have ratios approaching 1 wherever the curvature becomes large. A manifold which has sectional curvatures equal at each point is necessarily a constant curvature space (see Schur's theorem in Sect. 12.3.1). We would like to conclude that the metric on the manifold approaches a constant curvature metric. However to conclude this we must first establish two things: We need to know that the solution of Ricci flow continues to exist until the curvature 'blows up' somewhere, and we need to develop some machinery of compactness for Riemannian metrics in order to produce a limiting metric to which Schur's theorem can be applied. We will address these points in the next two chapters.

Chapter 8
Regularity and Long-Time Existence

In Chaps. 4 and 6 we saw that the curvature under Ricci flow obeys a parabolic equation with quadratic nonlinearity. By appealing to this view, we would expect the same kind of regularity that is seen in parabolic equations to apply to the curvature. In particular we want to show that bounds on curvature automatically induce *a priori* bounds on all derivatives of the curvature for positive times. In the literature these are known as Bernstein–Bando–Shi derivative estimates as they follow the strategy and techniques introduced by Bernstein (done in the early twentieth century) for proving gradient bounds via the maximum principle and were derived for the Ricci flow in [Ban87] and comprehensively by Shi in [Shi89]. Here we will only need the global derivative of curvature estimates (for various local estimates see [CCG+08, Chap. 14]). In the second section we use these bounds to prove long-time existence.

8.1 Regularity: The Global Shi Estimates

We seek to prove the following global derivative estimates – which says that for short times there is an estimate which depends on the initial bound, and for large times there is a bound independent of initial bounds. As we shall see, the short-time estimate follows by applying the maximum principle directly to (8.3).

Theorem 8.1 (Bernstein–Bando–Shi Estimate). *Let $g(t)$, $t \in [0,T)$, be a smooth solution of the Ricci flow on a compact manifold M^n. Then for each $p \in \mathbb{N}$ there exists a constant C_p depending only on n such that if $|R(x,t)|_{g(t)} \leq K$ for all $x \in M$ and $t \in (0,T)$ then*

$$|\nabla^p R(x,t)|_{g(t)} \leq \min\left\{ \sup_{M\times\{0\}} |\nabla^p R(x,0)|_{g(0)}\, e^{C_p K t}, C_p \max\left\{K^{\frac{p+2}{2}}, \frac{K}{t^{\frac{p}{2}}}\right\}\right\}$$

for all $x \in M$ and $t \in (0,T)$.

B. Andrews and C. Hopper, *The Ricci Flow in Riemannian Geometry*, Lecture Notes in Mathematics 2011, DOI 10.1007/978-3-642-16286-2_8,

Remark 8.2. To avoid a notational quagmire in the proof, we adopt the following convention: If A and B are two tensorial quantities on a Riemannian manifold, we let $A * B$ be any linear combination of tensors obtained from the tensor product $A \otimes B$ by one or more of these operations:

1. Summation over pairs of matching upper and lower indices.
2. Contraction on upper indices with respect to the metric.
3. Contraction on lower indices with respect to the dual metric.

Lemma 8.3. *Suppose, under the Ricci flow, that a tensor A satisfies*

$$\nabla_{\partial_t} A = \Delta A + B,$$

where B is a tensor of the same type as A, and ∇ is the connection on the spatial tangent bundle $\mathfrak{S}$ defined in Theorem 6.9. Then the square of its norm satisfies the heat-type equation

$$\frac{\partial}{\partial t}|A|^2 = \Delta|A|^2 - 2|\nabla A|^2 + 2\langle B, A\rangle .$$

Proof. Using the identity $\Delta|A|^2 = 2\langle \Delta A, A\rangle + 2|\nabla A|^2$, and noting that ∇ is compatible with the metric, we find that

$$\begin{aligned}\frac{\partial}{\partial t}\langle A, A\rangle &= 2\Big\langle \frac{\partial}{\partial t}A, A\Big\rangle \\ &= \Delta|A|^2 - 2|\nabla A|^2 + 2\langle B, A\rangle.\end{aligned}$$

□

Moreover, using the formula

$$[\nabla_t, \nabla_i]A = R(\nabla_i, \nabla_t)A = \nabla R * A$$

from Theorem 6.18, and

$$[\nabla, \Delta]A = \nabla\Delta A - \Delta\nabla A = R * \nabla A + \nabla R * A,$$

we see that the derivative ∇A also satisfies a heat-type equation, of the form

$$\nabla_t\nabla A = \Delta(\nabla A) + \nabla B + R * \nabla A + \nabla R * A. \tag{8.1}$$

Proof (Theorem 8.1, sketch only). The proof follows by induction on p, first starting with the case $p = 1$. Using our $*$-convention, the evolution equation for R, derived in Theorem 4.14, takes the form

$$\frac{\partial}{\partial t}R = \Delta R + R * R. \tag{8.2}$$

So by Lemma 8.3, with $A = R$ and $B = R * R$, we find that

$$\frac{\partial}{\partial t}|R|^2 = \Delta|R|^2 - 2|\nabla R|^2 + R * R * R.$$

Moreover, by using Lemma 8.3 again, now with $A = \nabla R$ and $B = R * \nabla R$, it is also possible to show (in conjunction with (8.1)) that

$$\frac{\partial}{\partial t}|\nabla R|^2 = \Delta|\nabla R|^2 - 2|\nabla^2 R|^2 + R * \nabla R * \nabla R. \tag{8.3}$$

We would now like to estimate $|\nabla R|^2$ from these equations. However, there are one or two difficulties. The first is the potentially bad term $R*(\nabla R)^{*2}$ on the right-hand side of (8.3), the second is that we have no control on $|\nabla R|^2$ at $t = 0$. To clear these obstacles, we define

$$F := t|\nabla R|^2 + |R|^2. \tag{8.4}$$

With this choice we find that the uncontrollable $|\nabla R|^2$ vanishes at $t = 0$, so an initial upper bound $F|_{t=0} \leq K^2$ is obtained by hypothesis, and when t is small the bad term from differentiating $|\nabla R|^2$ can be controlled by the good term $-2|\nabla R|^2$ from differentiating $|R|^2$: We have

$$\frac{\partial F}{\partial t} \leq \Delta F + (t\,C_1 K - 1)|\nabla R|^2 + C_2 K^3.$$

Now by applying the maximum principle on $0 < t \leq \frac{1}{C_1 K}$ we get

$$t|\nabla R|^2 \leq K^2 + t\,C_2 K^3 \leq \left(1 + \frac{C_2}{C_1}\right) K^2.$$

If $t > \frac{1}{C_1 K}$, the same argument on the interval $[t - \frac{1}{C_1 K}, t]$ gives

$$\frac{1}{C_1 K}|\nabla R|^2 \leq \left(1 + \frac{C_2}{C_1}\right) K^2.$$

So by combining these two results we find that

$$|\nabla R|^2 \leq \max\left\{\frac{(1 + \frac{C_2}{C_1})K^2}{t}, (C_1 + C_2)K^3\right\}.$$

Thus proving the case $p = 1$.

To get the necessary results for higher derivatives, we bootstrap the $p = 1$ case. Using (8.1) one can prove by induction that

$$\frac{\partial}{\partial t}\nabla^k R = \Delta\nabla^k R + \sum_{j=0}^{k} \nabla^j R * \nabla^{k-j} R.$$

It follows from Lemma 8.3 that

$$\frac{\partial}{\partial t}|\nabla^k R|^2 = \Delta|\nabla^k R|^2 - 2|\nabla^{k+1}R|^2 + \sum_{j=0}^{k} \nabla^j R * \nabla^{k-j} R * \nabla^k R. \tag{8.5}$$

Now by taking

$$G = t^p|\nabla^p R|^2 + \beta_p \sum_{k=1}^{p} \alpha_p t^{p-k}|\nabla^{p-k}R|^2,$$

we look to apply the same PDE techniques as before with appropriately chosen constants β_p. Indeed, on $[0, \frac{1}{CK}]$ one can show that

$$\frac{\partial}{\partial t}G \leq \Delta G + B_p K^3.$$

By the maximum principle we find $\widehat{C}_p$ depending on p and n such that

$$\sup_x G(x,t) \leq \widehat{C}_p^2 K^2.$$

If $t > \frac{1}{CK}$, the same argument on $[t - \frac{1}{CK}, t]$ gives

$$|\nabla^p R|(t) \leq (\widetilde{C}(p)K)K^{p/2} = \widetilde{C}(p)K^{1+p/2}. \qquad \square$$

8.2 Long-Time Existence

Suppose $(M, g(t))$, $t \in [0, T)$, is a solution of the Ricci flow. We say that $[0, T)$ is the *maximum time interval of existence* if either $T = \infty$, or that $T < \infty$ and there does not exist $\varepsilon > 0$ and a smooth solution $\widetilde{g}(t)$, $t \in [0, T + \varepsilon)$, of the Ricci flow such that $\widetilde{g}(t) = g(t)$ for $t \in [0, T)$. In the latter case, when T is finite, we say $g(t)$ forms a *singularity* at time T or simply $g(t)$, $t \in [0, T)$, is a *singular solution*. Henceforth, with this terminology, we will talk about the Ricci flow with initial metric g_0 on a maximal time interval $[0, T)$.

In this section we are interested in proving long-time existence for the Ricci flow with a particular interest in the limiting behaviour of singular solutions.

Theorem 8.4 (Long-time Existence). *On a compact manifold M with smooth initial metric g_0, the unique solution $g(t)$ of Ricci flow with $g(0) = g_0$ exists on a maximal time interval $0 \leq t < T \leq \infty$. Moreover, $T < \infty$ only if*

$$\lim_{t \nearrow T} \sup_{x \in M} |R(x,t)| = \infty.$$

In order to prove this, we need bounds on the metric and its derivatives.

Lemma 8.5. [Ham82b, Lemma 14.2] *Let $g(t)$, $t \in [0,\tau]$ be a solution to the Ricci flow for $\tau \leq \infty$. If $|\mathrm{Ric}| \leq K$ on $M \times [0,\tau]$ then*

$$e^{-2Kt} g(0) \leq g(t) \leq e^{2Kt} g(0)$$

for all $t \in [0,\tau]$.[1]

Proof. For any nonzero $v \in T_x M$, $\frac{\partial}{\partial t} g_{(x,t)}(v,v) = -2\mathrm{Ric}_{(x,t)}(v,v)$ so that

$$\left| \frac{\partial}{\partial t} g_{(x,t)}(v,v) \right|_{g(t)} \leq 2|\mathrm{Ric}|\, g_{(x,t)}(v,v).$$

Hence for times $0 \leq t_1 \leq t_2 \leq \tau$, we find that

$$\begin{aligned} \left| \log \frac{g_{(x,t_2)}(v,v)}{g_{(x,t_1)}(v,v)} \right| &= \left| \int_{t_1}^{t_2} \frac{\frac{\partial}{\partial t} g_{(x,t)}(v,v)}{g_{(x,t)}(v,v)} dt \right| \\ &\leq \int_{t_1}^{t_2} \left| \frac{\frac{\partial}{\partial t} g_{(x,t)}(v,v)}{g_{(x,t)}(v,v)} \right|_{g(t)} dt \\ &\leq 2|\mathrm{Ric}|\, (t_2 - t_1). \end{aligned} \tag{8.6}$$

Therefore by letting $t_1 = 0$ and $t_2 = t$ we have

$$\left| \log \frac{g_{(x,t)}(v,v)}{g_{(x,0)}(v,v)} \right| \leq 2Kt.$$

So the results follows by exponentiation. □

Lemma 8.6. *Fix a metric $\bar{g}$ and connection $\bar{\nabla}$ on M. For any smooth solution $g(t)$ of the Ricci flow on $M \times [0,T)$, for which $\sup_{M\times[0,T)} |R| \leq K$, there exist constants C_q for each $q \in \mathbb{N}$ such that*

$$\sup_{M\times[0,T)} \left| \bar{\nabla}^{(q)} g(x,t) \right|_{\bar{g}} \leq C_q.$$

Proof. It suffices to prove by induction that:

$$\frac{\partial}{\partial t} \bar{\nabla}^q g = \sum_{j_0+j_1+\cdots+j_m=q} \nabla^{j_0} R * \bar{\nabla}^{j_1} g * \cdots * \bar{\nabla}^{j_m} g. \tag{8.7}$$

[1] We say $g_2 \geq g_1$ if $g_2 - g_1$ is weakly positive definite.

The case $q = 1$ is proved as follows: Since $\bar{\nabla}$ is independent of time,

$$\begin{aligned}\frac{\partial}{\partial t}\bar{\nabla}_i g_{kl} &= \bar{\nabla}_i \partial_t g_{kl} \\ &= \bar{\nabla}_i \left(-2\mathrm{Ric}_{kl}\right) \\ &= -2\nabla_i \mathrm{Ric}_{kl} - 2\mathrm{Ric}_{pl}\left(\bar{\Gamma} - \Gamma\right)_{ik}{}^{p} - 2\mathrm{Ric}_{kp}\left(\bar{\Gamma} - \Gamma\right)_{il}{}^{p} \\ &= \nabla R + R * \bar{\nabla} g,\end{aligned}$$

since $(\bar{\Gamma} - \Gamma)_{ik}{}^{p} = \frac{1}{2} g^{pr}\left(\bar{\nabla}_i g_{kr} + \bar{\nabla}_k g_{ir} - \bar{\nabla}_r g_{ik}\right)$.

The induction step is now straightforward, for if (8.7) holds for some q, then

$$\begin{aligned}\frac{\partial}{\partial t}\bar{\nabla}^{q+1} g &= \bar{\nabla}\left(\frac{\partial}{\partial t}\bar{\nabla}^{q} g\right) \\ &= \bar{\nabla}\left(\sum_{j_0 + j_1 + \cdots + j_m = q} \nabla^{j_0} R * \bar{\nabla}^{j_1} g * \cdots * \bar{\nabla}^{j_m} g\right) \\ &= \sum_{j_0 + j_1 + \cdots + j_m = q} \left(\nabla^{j_0+1} R + \bar{\nabla}^{j_0} R * \bar{\nabla} g\right) * \bar{\nabla}^{j_1} g * \cdots * \bar{\nabla}^{j_m} g \\ &\quad + \sum_{j_0 + j_1 + \cdots + j_m = q} \nabla^{j_0} R * \left(\bar{\nabla}^{j_1+1} g * \cdots * \nabla^{j_m} g \right. \\ &\qquad\qquad \left. + \cdots + \bar{\nabla}^{j_1} g * \cdots * \bar{\nabla}^{j_m+1} g\right) \\ &= \sum_{j_0 + j_1 + \cdots + j_m = q+1} \nabla^{j_0} R * \bar{\nabla}^{j_1} g * \cdots * \bar{\nabla}^{j_m} g.\end{aligned}$$

We now can use (8.7) to deduce the bounds needed for the Lemma by induction on q. That is, if we have controlled $\bar{\nabla}$-derivatives of g up to order q, then we have

$$\frac{\partial}{\partial t}\left|\bar{\nabla}^{(q+1)} g\right|_{\bar{g}}^{2} \leq C\left(1 + \left|\bar{\nabla}^{(q+1)} g\right|_{\bar{g}}\right),$$

which implies a bound on the order $q + 1$ derivatives since the time interval is finite. □

Remark 8.7. The derivatives in the t direction are also controlled, since they are related to the spatial derivatives by the Ricci flow equation.

Proof (Theorem 8.4). By taking the contrapositive, suppose there exists a sequence $t_i \nearrow T$ and a constant $K < \infty$ independent of i such that

$$\sup_M |R(\cdot, t_i)| \leq K.$$

In particular, by the doubling time estimate (7.1) we have that $|R(x,t)| \leq K/(1 - CK(t - t_i))$, for $t \geq t_i$ and all $x \in M$. So for large i (i.e. when $t_i = T - \varepsilon$ for some $0 < \varepsilon \ll 1$) we have

$$\sup_M |R(\cdot, t)| \leq \frac{K}{1 - CK(t - (T - \varepsilon))}$$

for $t \in [T - \varepsilon, T]$. Hence $|R| \leq \widetilde{K}$ on $[0, T)$. We claim that:

Claim 8.8. The metric $g(t)$ may be extended smoothly from $[0, T)$ to $[0, T]$.

Proof of Claim. By Lemma 8.5 and (8.6) we find that:

$$\left| \log \frac{g_{(x,t_2)}(v, v)}{g_{(x,t_1)}(v, v)} \right| \leq 2\widetilde{K} |t_2 - t_1|.$$

So $g_{(x,t)}(v, v)$ is Cauchy at $t \to T$. By Lemma 8.6 it follows that $g(x, t)$ is Cauchy in C^k, for every k, as $t \to T$. □

So from this result, we take $g(T)$ as the 'initial' metric. By Theorem 5.2, the short-time existence implies that we can extend the flow for times $t \in [0, T + \epsilon)$, thus contradicting the maximality of the finite final time T. □

Chapter 9
The Compactness Theorem for Riemannian Manifolds

The compactness theorem for the Ricci flow tells us that any sequence of complete solutions to the Ricci flow, having uniformly bounded curvature and injectivity radii uniformly bounded from below, contains a convergent subsequence. This result has its roots in the convergence theory developed by Cheeger and Gromov. In many contexts where the latter theory is applied, the regularity is a crucial issue. By contrast, the proof of the compactness theorem for the Ricci flow is greatly aided by the fact that a sequence of solutions to the Ricci flow enjoy excellent regularity properties (which were discussed in the previous chapter). Indeed, it is precisely because bounds on the curvature of a solution to the Ricci flow imply bounds on all derivatives of the curvature that the compactness theorem produces C^∞-convergence on compact sets.

The compactness result has natural applications in the analysis of singularities of the Ricci flow by 'blow-up', discussed here in Sect. 9.5: The idea is to consider shorter and shorter time intervals leading up to a singularity of the Ricci flow, and to rescale the solution on each of these time intervals to obtain solutions on long time intervals with uniformly bounded curvature. The limiting solution obtained from these gives information about the structure of the singularity.

As a remark concerning notation in this chapter, quantities depending on the metric g_k or $g_k(t)$ will have a subscript k. For instance ∇_k and R_k denote the Riemannian connection and Riemannian curvature tensor of g_k. Quantities without a subscript will depend on the background metric g.

9.1 Different Notions of Convergence

In order to establish the C^∞-convergence of a sequence of metrics (g_k) uniformly on compact sets, we need to recall some of the different ways a sequence of functions can converge.

B. Andrews and C. Hopper, *The Ricci Flow in Riemannian Geometry*, Lecture Notes in Mathematics 2011, DOI 10.1007/978-3-642-16286-2_9,

9.1.1 Convergence of Continuous Functions

Consider the Banach Space $C(X)$ where X is a topological space (typically taken to be compact Hausdorff). Recall that sequence $(f_n) \subset C(X)$ converges *uniformly* to a limiting function f if in heuristic terms 'the rate of convergence of $f_n(x)$ to $f(x)$ is independent of x'. Formally we say $f_n \to f$ uniformly if for every $\varepsilon > 0$ there exists $N = N(\varepsilon) > 0$ such that for all $x \in X$, $|f_n(x) - f(x)| < \varepsilon$ whenever $n \geq N$. A generalisation of this is the notion of *uniform convergence on compact sets.*[1] In this case, a sequence (f_n) converges to f *uniformly on compact subsets of* X (in short we just say $f_n \to f$ compactly) if for every compact $K \subset X$ and for every $\varepsilon > 0$ there exists $N = N(K, \varepsilon) > 0$ such that for all $x \in K$, $|f_n(x) - f(x)| < \varepsilon$ whenever $n > N$. Intuitively, uniform convergence on compact sets says that every point has a neighbourhood in which the convergence is uniform.

To illustrate the different types of convergence, consider $C(0,1)$ with $f_n(x) = x^n$. In this situation f_n converges compactly, but not uniformly, to the zero function. However if we change the space to $C(0,1]$, then f_n converges pointwise, but not uniformly on compact subsets, to the function that is zero on $(0,1)$ and one at $x = 1$.

In general it is easy to see that if $(f_n) \subset C(X)$ and $f_n \to f$ uniformly then $f_n \to f$ compactly. The above example shows the converse is false; however we do have the following partial results:

- If X is compact then $f_n \to f$ compactly implies $f_n \to f$ uniformly.
- If X is locally compact[2] and $f_n \to f$ compactly then the limit f is continuous.

As our metrics g_k are smooth, we need to take into consideration the differentiability aspect of our convergence.

9.1.2 The Space of C^∞-Functions and the C^p-Norm

Let $\Omega \subset \mathbb{R}^n$ be an open set with compact closure, and let $C^p(\Omega)$ be the set of functions with continuous derivatives up to order p. Moreover, let the space $C^p(\bar{\Omega})$ be the set of functions on $\bar{\Omega}$ which extend to a C^p function on some open set containing $\bar{\Omega}$. It is equipped with the following norm:

$$\|u\|_{C^p} = \sup_{0 \leq |\alpha| \leq p} \sup_{x \in \bar{\Omega}} |D^\alpha u(x)|,$$

[1] Also known as *compact convergence* or *the topology of compact convergence.*

[2] That is, if every point has a neighbourhood whose closure is compact.

where D^α is the derivative corresponding to the multi-index α. It can be shown that the norm is complete, thus making the space $(C^p(\bar{\Omega}), \|\cdot\|_{C^p})$ a Banach space. However, if we consider the space of smooth functions

$$C^\infty(\bar{\Omega}) = \bigcap_{p=0}^{\infty} C^p(\bar{\Omega}),$$

one notes that it is only a metric space, not a Banach space, with metric

$$d(f,g) = \sum_{p=0}^{\infty} \frac{1}{2^p} \frac{\|f-g\|_{C^p}}{1+\|f-g\|_{C^p}}.$$

A sequence converges in C^∞ if and only if it converges in C^p for every p.

9.1.3 Convergence of a Sequence of Sections of a Bundle

We will be interested in convergence of a sequence of metrics, which are of course sections of a certain vector bundle. To make sense of this we define what is meant by C^p convergence or C^∞ convergence for sequences of sections of vector bundles.

Definition 9.1 (C^p-Convergence). Let E be a vector bundle over a manifold M, and let metrics g and connections ∇ be given on E and on TM. Let $\Omega \subset M$ be an open set with compact closure $\bar{\Omega}$ in M, and let (ξ_k) be a sequence of sections of E. For any $p \geq 0$ we say that ξ_k *converges in* C^p *to* $\xi_\infty \in \Gamma(E|_{\bar{\Omega}})$ if for every $\varepsilon > 0$ there exists $k_0 = k_0(\varepsilon)$ such that

$$\sup_{0 \leq \alpha \leq p} \sup_{x \in \bar{\Omega}} |\nabla^\alpha(\xi_k - \xi_\infty)|_g < \varepsilon$$

whenever $k > k_0$. We say ξ_k converges in C^∞ to ξ_∞ on $\bar{\Omega}$ if ξ_k converges in C^p to ξ_∞ on $\bar{\Omega}$ for every $p \in \mathbb{N}$.

Note that since we are working on a compact set, the choice of metric and connection on E and TM have no affect on the convergence.

Next we define smooth convergence on compact subsets for a sequence of sections: To do this we require an *exhaustion of* M – that is, a sequence of open sets (U_k) in M such that $\bar{U}_k$ is compact and $\bar{U}_k \subset U_{k+1}$ for all k, and $\bigcup_{k\geq 1} U_k = M$. Note that if $K \subset M$ is compact, then there exists k_0 such that $K \subset U_k$ for all $k \geq k_0$ (in particular if M is compact then $U_k = M$ for all large k).

Definition 9.2 (C^∞-Convergence on Compact Sets). Let (U_k) be an exhaustion of a smooth manifold M, and E a vector bundle over M. Fix metrics g and connections ∇ on E and TM. Let (ξ_i) be a sequence of sections of E defined on open sets $A_i \subset M$, and let $\xi_\infty \in \Gamma(E)$. We say ξ_i converges smoothly on compact sets to ξ_∞ if for every $k \in \mathbb{N}$ there exists i_0 such that $\bar{U}_k \subset A_i$ for all $i \geq i_0$, and the sequence $(\xi_i\big|_{\bar{U}_k})_{i \geq i_0}$ converges in C^∞ to ξ_∞ on $\bar{U}_k$.

Again, we remark that this notion of convergence does not depend on the choice of the metric and connection on E and TM.

9.2 Cheeger–Gromov Convergence

We are interested in analysing the convergence of a sequence of Riemannian manifolds arising from dilations about particular singularities. In order to do so, we need to build into our definitions which part of the manifold we are interested in. This is done by including a base point into the definition. We will see later some examples where different choices for a sequence of base points on the same sequence of Riemannian manifolds can give quite different limits.[3]

Definition 9.3 (Pointed Manifolds). A *pointed Riemannian manifold* is a Riemannian manifold and $O \in M$ together with a choice of origin or basepoint $O \in M$. If the metric g is complete, we say that the tuple is a *complete pointed Riemannian manifold.* If $(M, g(t))$ is a solution to the Ricci flow, we say $(M, g(t), O)$, for $t \in (a, b)$, is a *pointed solution to the Ricci flow.*

Moreover, we would like a notion of convergence for a sequence (M_k, g_k, O_k) of pointed Riemannian manifolds that takes into account the action of basepoint-preserving diffeomorphisms on the space of metrics $\mathfrak{Met}$.

Definition 9.4 (Cheeger–Gromov Convergence in C^∞). A sequence $\{(M_k, g_k, O_k)\}$ of complete pointed Riemannian manifolds converges to a complete pointed Riemannian manifold $(M_\infty, g_\infty, O_\infty)$ if there exists:

1. An exhaustion (U_k) of M_∞ with $O_\infty \in U_k$;
2. A sequence of diffeomorphisms $\Phi_k : U_k \to V_k \subset M_k$ with $\Phi(O_\infty) = O_k$

such that $(\Phi_k^* g_k)$ converges in C^∞ to g_∞ on compact sets in M_∞.

[3] The definitions here are essentially due to Hamilton [Ham95a], but following [CCG+07] we do not include in the definition a choice of orthonormal frame at the basepoint.

This notion of convergence is often referred to as *smooth Cheeger–Gromov convergence*. The corresponding convergence for a sequence of pointed solutions to the Ricci flow is as follows.

Definition 9.5. A sequence $\{(M_k^n, g_k(t), O_k)\}$, for $t \in (a,b)$, of complete pointed solutions to the Ricci flow converges to a complete pointed solution to the Ricci flow $(M_\infty, g_\infty(t), O_\infty)$, for $t \in (a,b)$, if there exists:

1. An exhaustion (U_k) of M_∞ with $O_\infty \in U_k$;
2. A sequence of diffeomorphisms $\Phi_k : U_k \to V_k \subset M_k$ with $\Phi(O_\infty) = O_k$

such that $\big(\Phi_k^* g_k(t)\big)$ converges in C^∞ to $g_\infty(t)$ on compact sets in $M_\infty \times (a,b)$.

9.2.1 Expanding Sphere Example

An explicit illustration of the Cheeger–Gromov convergence can be seen in the following example which shows: A sequence of pointed manifolds $(S_k^n, g_{\text{can}}, N)$, where S_k^n is the standard n-sphere of radius k (taken with the usual canonical metric g_{can}) and N is the basepoint corresponding to the sphere's 'north pole', has a limiting pointed manifold $(\mathbb{R}^n, \delta_{ij}, 0)$ centred at the origin 0. (Fig. 9.1)

To see this, first take an exhaustion $U_k = B_k(0) \nearrow \mathbb{R}^n$ of balls centered at the origin with a sequence of diffeomorphisms $\Phi_k : B_k(0) \to S_k^n \subset \mathbb{R}^{n+1}$, defined by

$$\Phi_k : \; x \mapsto (x, \sqrt{k^2 - |x|^2}),$$

with $V_k = \Phi_k(B_k(0))$. For such Φ_k, we are able to explicitly compute the pullback metric:

$$(\Phi_k^* g_k)_{ij} = \left\langle \frac{\partial \Phi_k}{\partial x^i}, \frac{\partial \Phi_k}{\partial x^j} \right\rangle_{\mathbb{R}^{n+1}} = \delta_{ij} + \frac{x^i x^j}{k^2 - |x|^2},$$

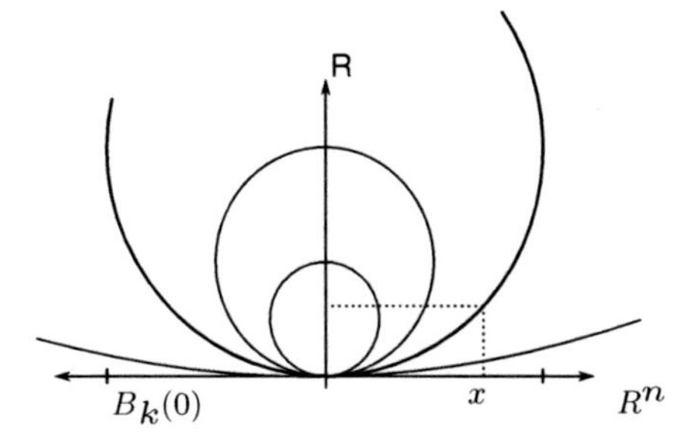

Fig. 9.1 Cheeger–Gromov convergence of expanding spheres to a flat metric

since $\frac{\partial \Phi_k}{\partial x^i} = (e^i, x^i/\sqrt{k^2 - |x|^2})$. Thus, it suffices to show that

$$\left(B_k(0), \delta_{ij} + \frac{x^i x^j}{k^2 - |x|^2}\right) \xrightarrow{C^\infty} (\mathbb{R}^n, \delta_{ij})$$

uniformly on compact sets, which can be achieved simply by showing that $1/(1 - |x/k|^2)$ converges to 1 in C^∞ on compact subsets of $\mathbb{R}^n$.

9.2.2 The Rosenau Metrics

We will illustrate various aspects of Cheeger–Gromov convergence by investigating the following specific family:[4] For each $\alpha \in [0, 1)$, define a metric g_α on the two-dimensional sphere $S^2 \subset \mathbb{R}^3$ by

$$g_\alpha = \frac{1}{1 - \alpha^2 x^2} \bar{g},$$

where $\bar{g}$ is the standard metric on S^2 coming from its inclusion in $\mathbb{R}^3$, and x is one of the coordinate functions on $\mathbb{R}^3$. When $\alpha = 0$ this metric is just the standard metric on S^2, but as $\alpha \to 1$ the manifold becomes longer in the x direction, and becomes close to a long cylinder with capped-off ends. We will make this precise below, by producing the cylinder as a limit in the Cheeger–Gromov sense when we keep the base-points away from the ends $x = \pm 1$, (Fig. 9.2) and produce a limit which looks like a capped-off cylinder when the base-points are at the ends (Fig. 9.3).

Let us consider the convergence of a sequence of metrics $g_k = g_{\alpha(k)}$ with $\lim_{k\to\infty} \alpha(k) = 1$, with a choice of base points away from the poles, such as $O_k = (0, 1, 0)$ for every k. We will prove that the sequence (S^2, g_k, O_k) converges in the Cheeger–Gromov sense to a flat cylinder $(\mathbb{R} \times (\mathbb{R}/(2\pi)), du^2 + dv^2, (0, 0))$. Choose

$$\Phi_k(u, v) = \Phi(u, v) = \frac{(\sinh u, \cos v, \sin v)}{\cosh u},$$

so that $\Phi_k(0, 0) = (0, 1, 0)$ for every k.[5] Then we compute

$$\Phi_*(\partial_u) = \frac{(1, -\cos v \sinh u, -\sin v \sinh u)}{\cosh^2 u}; \quad \Phi_*(\partial_y) = \frac{(0, -\sin v, \cos v)}{\cosh u}.$$

---

[4] Up to rescaling by a suitable factor depending on k, these are exactly the metrics which appear in an important explicit solution of the Ricci flow on S^2 known as the Rosenau solution.

[5] This is a conformal map from the cylinder to the sphere without its poles. It can be produced by composing the complex exponential map (which maps $\mathbb{R} \times (\mathbb{R}/2\pi)$ to $\mathbb{C} \setminus \{0\}$) with the stereographic projection from $\mathbb{C}$ to S^2.

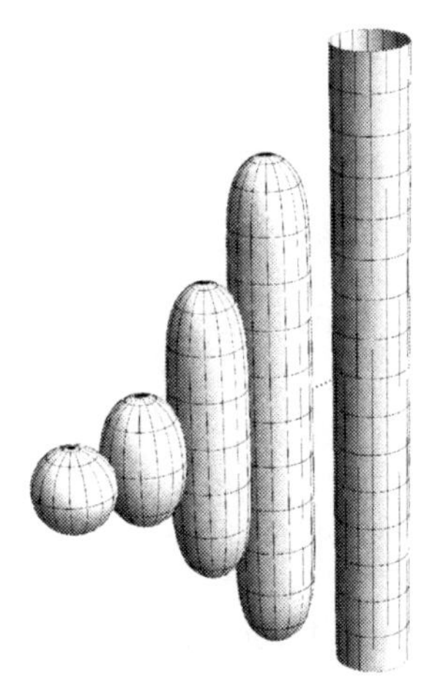

Fig. 9.2 The Rosenau metrics $g_\alpha = \frac{1}{1-\alpha^2 x^2}\bar{g}$ with base points away from the ends, converging to a cylinder as α increases from 0 towards 1 (shown embedded in $\mathbb{R}^3$)

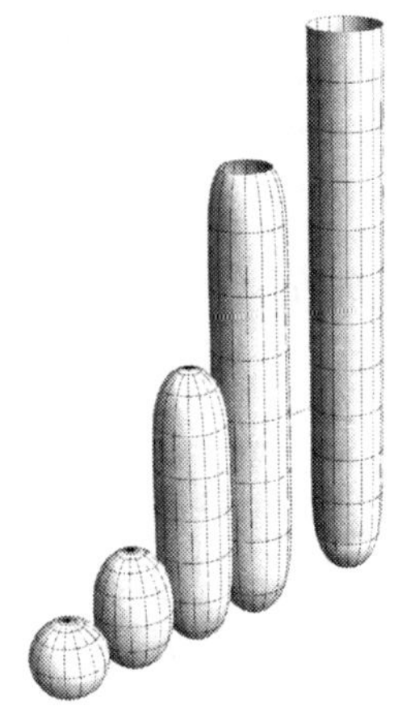

Fig. 9.3 The Rosenau metrics $g_\alpha = \frac{1}{1-\alpha^2 x^2}\bar{g}$ with base points near an end, converging to the cigar metric $\frac{du^2+dv^2}{1+u^2+v^2}$ as α increases from 0 towards 1 (shown embedded in $\mathbb{R}^3$)

So

$$\begin{aligned}
\Phi^* g_k(\partial_u, \partial_u) &= \frac{1}{1-\alpha^2 \tanh^2 u}\bar{g}(\Phi_* \partial_u, \Phi_* \partial_u) \\
&= \frac{1}{1-\alpha^2 \tanh^2 u}\,\frac{1}{\cosh^2 u} \\
&= 1 - \frac{(1-\alpha^2)\tanh^2 u}{1-\alpha^2 \tanh^2 u};
\end{aligned}$$

$$
\begin{aligned}
\Phi^* g_k(\partial_v, \partial_v) &= \frac{1}{1-\alpha^2\tanh^2 u}\bar{g}(\Phi_*\partial_v, \Phi_*\partial_v) \\
&= 1 - \frac{(1-\alpha^2)\tanh^2 u}{1-\alpha^2\tanh^2 u};
\end{aligned}
$$

and

$$
\begin{aligned}
\Phi^* g_k(\partial_u, \partial_v) &= \frac{1}{1-\alpha^2\tanh^2 u}\bar{g}(\Phi_*\partial_u, \Phi_*\partial_v) \\
&= 0.
\end{aligned}
$$

This shows that

$$
\Phi^* g_k = \left(1 - \frac{(1-\alpha^2)\tanh^2 u}{1-\alpha^2\tanh^2 u}\right)(du^2 + dv^2),
$$

and the quantity in the brackets converges in C^∞ to 1 on any compact subset of $\mathbb{R} \times (\mathbb{R}/(2\pi))$. Thus $\Phi^* g_k \to du^2 + dv^2$, so we have proved the Cheeger–Gromov convergence to a the cylinder metric.

Now let us see what happens when we take the base points to be at a pole, say $O_k = (1,0,0)$ for every k. We will prove that the sequence of pointed metrics

$$
(S^2, g_k, O_k) \longrightarrow \left(\mathbb{R}^2, \frac{du^2+dv^2}{1+u^2+v^2}, (0,0)\right)
$$

converge in the Cheeger–Gromov sense. To do this choose Φ_k to be a composition of stereographic projection with a suitable dilation depending on k:

$$
\Phi_k(u,v) = \frac{1}{1+\frac{1-\alpha^2}{2}(u^2+v^2)}\left(1 - \frac{1-\alpha^2}{2}(u^2+v^2), \sqrt{1-\alpha^2}u, \sqrt{1-\alpha^2}v\right).
$$

One can then compute directly

$$
\begin{aligned}
(\Phi_k)^* g_k &= \frac{du^2+dv^2}{1+u^2+v^2-\frac{1-\alpha^2}{2}(u^2+v^2)+\frac{(1-\alpha^2)^2}{16}(u^2+v^2)^2} \\
&\to \frac{du^2+dv^2}{1+u^2+v^2}
\end{aligned}
$$

as $\alpha \to 1$ smoothly on compact subsets.

Remark 9.6. This example illustrates several features of Cheeger–Gromov convergence: First, the need for specifying the base points is clear here, since we produce very different limits for different choices of base-point. Second, the limit of a sequence of compact spaces can be non-compact. Note that this possibility is made possible since we use an exhaustion by open sets on the limit manifold, and say nothing about surjectivity of the diffeomorphisms Φ_k onto the spaces M_k. On the other hand one cannot produce a compact limit from a sequence of complete non-compact spaces.

Remark 9.7. The limiting metric above is also important in Ricci flow: It is an example of a 'soliton' metric, which evolves without changing shape – more precisely, the solution of Ricci flow starting from the above metric is given at any positive time by the same metric pulled back by a suitable diffeomorphism. This metric is called the 'cigar', and it is asymptotic to a cylinder at infinity.

9.3 Statement of the Compactness Theorem

Having established the convergence criterion for a sequence of pointed Riemannian manifolds, we seek to find sufficient conditions under which a given sequence $\{(M_k, g_k, O_k)\}$ has a convergent subsequence. As we shall see, it turns out that there are two such conditions needed for this to occur: The first is that of (globally) uniform bounds on the curvature and its higher derivatives; the second is a lower bound on the injectivity radius at the basepoint:

(a) $|\nabla^p R(g_k)| \leq C_p$ on M_k for each $p \geq 0$; and
(b) $\mathrm{inj}_{g_k}(O_k) \geq \kappa_0$ for some $\kappa_0 > 0$.

We say that the sequence of Riemannian manifolds has (uniformly) *bounded geometry* whenever the curvature and its derivatives of all orders have uniform bounds (i.e. condition (a) holds).

The resulting compactness theorem for metrics is stated as follows. It is a fundamental result in Riemannian geometry independent of the Ricci flow.

Theorem 9.8 (Compactness for Metrics). *Let $\{(M_k, g_k, O_k)\}$ be a sequence of complete pointed Riemannian manifolds satisfying (a) and (ii) above. Then there exists a subsequence (j_k) such that $\{(M_{j_k}, g_{j_k}, O_{j_k})\}$ converges in the Cheeger–Gromov sense to a complete pointed Riemannian manifold $(M_\infty, g_\infty, O_\infty)$ as $k \to \infty$.*

Remark 9.9. Due to a estimate from [CGT82] we only need injectivity radius lower bounds at the basepoints, thereby avoiding any assumption of a uniform lower bounds for the injectivity radius over the whole manifold.

Remark 9.10. Our discussion here will take this theorem for granted as its proof is non-trivial and does not directly relate to the Ricci flow (see [Ham95a, Sect. 3] or [CCG+07, Chap. 4] for further details).

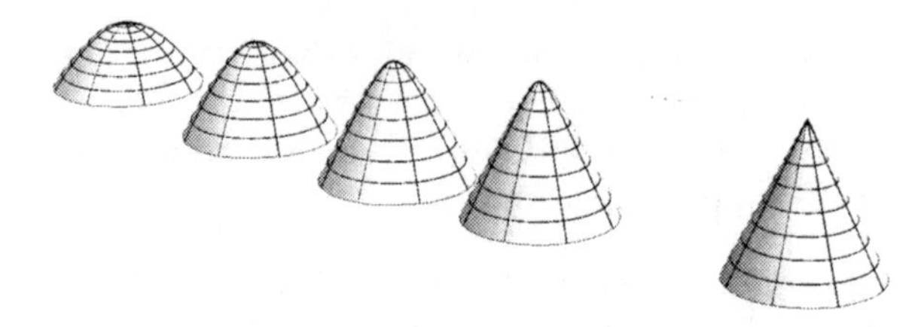

Fig. 9.4 A sequence of manifolds with infinite injectivity radius at the central point but curvature unbounded, and which do not converge to a manifold (cf. [Pet87, Example 1.1])

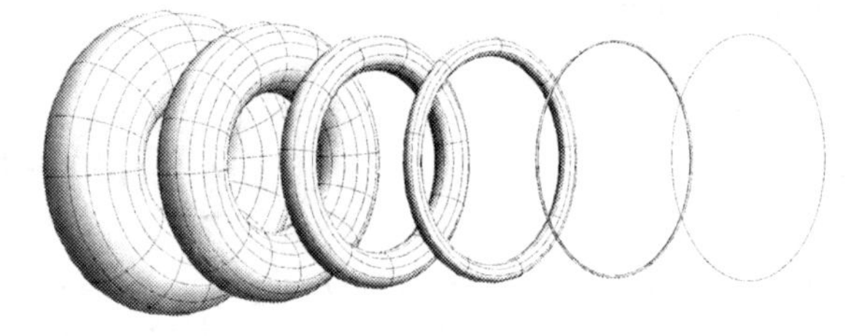

Fig. 9.5 A sequence of tori with injectivity radius approaching zero (cf. [Pet87, Example 1.2])

It is clear that all of the conditions are needed:[6] For example, if there is no bound on the curvatures then one could have a sequence like the one shown in Fig. 9.4 (explicitly, take the manifolds M_k to be hypersurfaces in $\mathbb{R}^{n+1}$ given by the graphs of $y = \sqrt{1/k^2 + |x|^2}$, with the induced metric). Similarly, one can construct examples where smooth convergence fails if the p-th derivative is not uniformly bounded, for any $p > 0$. As we shall see, the Ricci flow's excellent regularity properties mean that this condition does not present too great a challenge in the analysis – at least, once we have a bound on curvature (corresponding to $p = 0$) then the bounds for higher p are guaranteed).

A more subtle matter is that of a positive lower bound on the injectivity radius (see Fig. 9.5). For an explicit example of what can go wrong, consider the sequence given by $M_k = S^1 \times S^1 = \mathbb{R}^2/\mathbb{Z}^2$, with metrics given by $g_k(x, y) = dx^2 + k^{-1}dy^2$, and any choice of base points O_k. These metrics all have zero curvature, so the first condition is fulfilled with $C_p = 0$ for every p. However (M_k, g_k) has a closed geodesic (a circle in the y direction) of length $1/\sqrt{k}$, so the injectivity radius can be no greater than $1/(2\sqrt{k})$. This sequence 'collapses', in a certain sense converging to the lower-dimensional limit given by the circle S^1. This condition has proved to be been a major obstacle in the analysis of Ricci flow. Establishing a suitable bound is highly non-trivial, and was one of the fundamental breakthroughs in the work of Perel'man.

[6] Condition (a) can be weakened in the following way: Rather than requiring uniform bounds on all of M_k, it suffices to have *local* uniform bounds, i.e. for each $p \geq 0$ and each $R > 0$ there exists $C(p, R)$ such that $|\nabla_k^p R_k|_k \leq C(p, R)$ on $B_R(O_k)$.

9.3.1 Statement of the Compactness Theorem for Flows

When considering a sequence of complete pointed solutions to the Ricci flow, the corresponding compactness theorem takes the following form:

Theorem 9.11 (Compactness for Flows). *Let $\{(M_k, g_k(t), O_k)\}$, for $t \in (a,b)$, where $-\infty \leq a < 0 < b \leq \infty$, be a sequence of complete pointed solutions to the Ricci flow such that:*

1. Uniformly bounded curvature:

$$|R_k|_k \leq C_0 \quad \text{on } M_k \times (a,b)$$

for some constant $C_0 < \infty$ independent of k; and

2. Injectivity radius estimate at $t = 0$:

$$\text{inj}_{g_k(0)}(O_k) \geq \kappa_0$$

for some constant $\kappa_0 > 0$.

Then there exists a subsequence (j_k) such that $\{(M_{j_k}, g_{j_k}(t), O_{j_k})\}$ converges to a complete pointed solution to the Ricci flow $(M_\infty, g_\infty(t), O_\infty)$ as $k \to \infty$, for $t \in (a,b)$.

Remark 9.12. As remarked previously, there is no requirement for bounds on derivatives of curvature, since these are deduced from Theorem 8.1. We also remark that the curvature bound can be replaced by *local* uniform bounds, in the sense that we need only require bounds independent of k on each set $B_r(O_k) \times I$, where $r \geq 0$, $B_r(O_k)$ is the ball of radius r with respect to the metric g_k at $t = 0$, and I is a compact sub-interval of (a,b).

9.4 Proof of the Compactness Theorem for Flows

In this section we prove the compactness theorem for flows, Theorem 9.11, given the compactness theorem for metrics, Theorem 9.8. We will give the proof only for the case where the manifolds M_k are compact (though there is no such assumption on the limit M_∞). The proof for the more general case of complete solutions is not significantly more difficult, but requires a stronger version of the regularity result of Theorem 8.1, which we do not wish to prove. The main tools of the proof are the theorem of Arzelà–Ascoli and the regularity results presented in the previous chapter.

9.4.1 The Arzelà–Ascoli Theorem

Let X and Y be metric spaces, and fix $y_0 \in Y$. We say $\mathcal{F} \subset C(X,Y)$ is *equicontinuous* if for every $\varepsilon > 0$ and $x \in X$ there is a $\delta > 0$ such that $d_Y(f(x), f(y)) < \varepsilon$ for every $f \in \mathcal{F}$ and for all $y \in X$ with $d(x,y) < \delta$. Also, we say $\mathcal{F}$ is *pointwise bounded* if for every $x \in X$ there exists $C = C(x) < \infty$ such that $d_Y(f(x), y_0) \leq C(x)$ for all $f \in \mathcal{F}$.

Theorem 9.13 (Arzelà–Ascoli). *Let X and Y be locally compact metric spaces. Then a subset $\mathcal{F}$ of $C(X,Y)$ is compact in the compact-open topology if and only if it is equicontinuous, pointwise bounded and closed.*[7]

Note that if $\mathcal{F} \subset C(X,Y)$ is compact, then any sequence $(f_n) \subset \mathcal{F}$ has a subsequence that *converges uniformly on every compact subset of* X. We can apply this result to extract a convergent subsequence from a sequence of sections of a bundle with bounded derivatives.

Corollary 9.14. *Let (M,g) be a manifold, and E a vector bundle over M, and fix metrics g and connections ∇ on E and TM. Let $\Omega \subset M$ an open set with compact closure $\bar{\Omega}$, and let $p \in \mathbb{N} \cup \{0\}$. Let (ξ_k) be a sequence of sections of E over $\bar{\Omega}$ such that*

$$\sup_{0 \leq \alpha \leq p+1} \sup_{x \in \bar{\Omega}} |\nabla^\alpha \xi_k| \leq C < \infty.$$

Then there exists a section ξ_∞ of $E\big|_{\bar{\Omega}}$ and a subsequence of (ξ_k) which converges to ξ_∞ in C^p on $\bar{\Omega}$.

Remark 9.15. To reduce the proof to the Arzelà–Ascoli theorem, we can fix a finite collection of charts defined on compact domains covering $\bar{\Omega}$, with corresponding trivialisations of E, and apply the Arzelà–Ascoli theorem to the components of $\nabla^p \xi_k$ in each chart. We observe that the components of $\nabla^p \xi_k$ are equicontinuous, since their derivatives are controlled by $|\nabla^{p+1} \xi_k|$.

There are two situations in which Corollary 9.14 is directly applicable to the compactness theorems we are discussing: The first is where E is the bundle of symmetric bilinear forms on TM, and the sections ξ_k are Riemannian metrics on M. The second is where E is the bundle of symmetric bilinear forms on the space-like tangent bundle $\mathfrak{S}$ over $M \times I$, where I is a time interval. Then the sections ξ_k correspond to a sequence of *time-dependent* metrics on M, such as a sequence of solutions to the Ricci flow.

---

[7] Note that the collection of all such $\mathcal{U}_{K,U} = \{f \in C(X,Y) : f(K) \subset U\}$, where $K \subset X$ is compact and $U \subset Y$ open, defines a subbase for the compact-open topology on $C(X,Y)$.

9.4.2 The Proof

With the Arzelà–Ascoli theorem and the regularity for the Ricci flow discussed in Chap. 8, we are now in a position to prove the compactness theorem for flows from the general compactness theorem for metrics.

Proof (Theorem 9.11, given Theorem 9.8). We prove only the case where each M_k is compact. Consider a sequence of pointed solutions $(M_k, g_k(t), O_k)$, for $t \in (a, b)$, to the Ricci flow where $\sup_{M_k \times (a,b)} |R(g_k)| \leq K$. The Bernstein–Bando–Shi estimates, Theorem 8.1, give bounds of the form

$$|\nabla^p R(x,t)| \leq C(p, \varepsilon, K)$$

for all $x \in M$ and $t \in [a + \varepsilon, b)$, for each small $\varepsilon > 0$. The assumption on the injectivity radius at O_k, when $t = 0$, fulfils the conditions of the Cheeger–Gromov compactness theorem for metrics applied to the sequence $(M_k, g_k(0), O_k)$. Thus there exists a subsequence which converges in the Cheeger–Gromov sense to a complete limit $(\bar{M}, \bar{g}, \bar{O})$. That is, (passing to a subsequence if necessary) there exists an exhaustion $\{U_k\}$ of $\bar{M}$, and smooth injective maps $\Phi_k : U_k \to M_k$ taking $\bar{O}$ to O_k such that $\Phi_k^*(g_k(0))$ converges in C^∞ on compact sets of $\bar{M}$ to $\bar{g}$.

The idea now is to obtain uniform C^∞ control on $\tilde{g}_k(t) = \Phi_k^*(g_k(t))$ on compact subsets of $\bar{M} \times (a, b)$ (note these are also solutions of Ricci flow). To do this, fix a compact set Z in $\bar{M}$ and consider only k sufficiently large so that $Z \subset U_k$. The metrics $\tilde{g}_k(t)$ are uniformly comparable to $\bar{g}$, since by the convergence statement they are comparable to $\bar{g}$ at $t = 0$ and by Lemma 8.5 they remain comparable for other $t \in (a, b)$. To bound higher derivatives we apply, for each k, the evolution equation (8.7) derived in the proof of the long-time existence theorem:

$$\frac{\partial}{\partial t} \bar{\nabla}^q \tilde{g}_k = \sum_{j_0 + j_1 + \cdots + j_m = q} \nabla^{j_0} R(\tilde{g}_k) * \bar{\nabla}^{j_1} \tilde{g}_k * \cdots * \bar{\nabla}^{j_m} \tilde{g}_k. \tag{9.1}$$

In the case $q = 1$ this gives

$$\left| \frac{\partial}{\partial t} \bar{\nabla} \tilde{g}_k \right| \leq C \left(1 + |\bar{\nabla} \tilde{g}_k|\right),$$

where C depends on K and ε but not k. Since the time interval is finite and $\bar{\nabla} \tilde{g}_k(0) \to 0$ as $k \to \infty$, this implies $|\bar{\nabla} \tilde{g}_k(t)| \leq C$, independent of k. Proceeding by induction on q: Suppose $|\bar{\nabla}^j \tilde{g}_k(t)|$ is bounded independent of k, for $j = 1, \ldots, q$, so that (9.1) again gives

$$\left| \frac{\partial}{\partial t} \bar{\nabla}^{q+1} \tilde{g}_k \right| \leq C \left(1 + |\bar{\nabla}^{q+1} \tilde{g}_k|\right),$$

which implies a bound independent of k on $Z \times [a+\varepsilon, b)$. Note that derivatives in time directions are also bounded, since they can be written in terms of spatial derivatives via the Ricci flow equation. It follows by the Arzela–Ascoli theorem (in the form of Corollary 9.14) that there is a subsequence which converges in C^∞ on $Z \times [a+\varepsilon, b-\varepsilon]$. A diagonal subsequence argument then produces a subsequence which converges in C^∞ on compact sets of $\bar{M} \times (a,b)$ to a complete solution $\bar{g}(t)$ of Ricci flow, thus proving that the corresponding subsequence of $(M_k, g_k(t), O_k)$ converges to $(\bar{M}, \bar{g}(t), \bar{O})$ in the sense of the theorem. □

9.5 Blowing Up of Singularities

We will apply the compactness results of this chapter, in particular Theorem 9.11, to a solution of the Ricci flow with a finite maximal time of existence T. As the curvature explodes in this situation, we need to choose a sequence of times $t_i \nearrow T$ and rescale of the metric to make the curvatures bounded. By doing this we hope that the limiting manifold will tell us about the nature of the singularity, and hopefully some desirable topological information. We pursue this idea in detail.

Suppose that a solution of the Ricci flow $(M, g(t))$ exists on a maximal time interval $t \in [0,T)$ with finite final time $T < \infty$. By Theorem 8.4,

$$\limsup_{t \nearrow T} |R|(\,\cdot\,, t) = \infty$$

so the maximum value of $|R|$ on M explodes to $+\infty$ as $t \nearrow T$. In which case, choose points $O_i \in M$ and times $t_i \nearrow T$ such that

$$|R|(O_i, t_i) = \sup_{(x,t) \in M \times [0,t_i]} |R|(x,t),$$

and also set $Q_i := |R|(O_i, t_i)$. Applying the parabolic rescaling of the Ricci flow discussed in Sect. 4.1.3, we define

$$g_i(t) := Q_i \, g(t_i + Q_i^{-1} t)$$

so that $(M, g_i(t))$ satisfies Ricci flow on the time interval $[-t_i \, Q_i, (T - t_i) Q_i]$. For each i and times $t \leq 0$, note that

$$|R(g_i(t))| = \frac{1}{Q_i} |R(g(t_i + Q_i^{-1} t))| \leq 1$$

by the definition of Q_i. Also observe, for each i and times $t > 0$, that

$$\sup_M |R(g_i(t))| \leq \frac{1}{1 - C(n)t}$$

by the doubling-time estimate discussed in Sect. 7.2.2. Therefore $g_i(t)$ is defined with

$$\sup_i \sup_{M\times(a,b)} |R(g_i(t))| < \infty$$

for any $a < 0$ and some $b = b(n) > 0$. In which case, the sequence $\{(M, g_i(t), O_i), t \in (a,b)\}$ has uniform bounded geometry. So by Theorem 9.11, we can pass to a subsequence j such that $(M, g_j(t), O_j)$ converges to a complete pointed solution to the Ricci flow $(M_\infty, g_\infty(t), O_\infty)$, for all $t \in (a,b)$, *provided* we can establish a suitable lower bound on the injectivity radius.

Remark 9.16. It is precisely such a bound that is missing from our analysis. Historically this has been a major difficulty, except in special circumstances.[8] However, as we shall see in the next two chapters, this issue has been elegantly resolved by the work of Perel'man [Per02].

[8] One of the cases which can be handled is where the sectional curvatures are positive: Klingenberg [Kli59] proved that an even dimensional simply connected manifold with positive sectional curvatures has injectivity radius equal to its conjugate radius, which is at least $\pi/\sqrt{K_{\max}}$ by the Rauch comparison theorem. He also proved this for odd dimensions provided the sectional curvatures are globally 1/4-pinched [Kli61], and Abresch and Meyer [AM94] extended this result to allow global pinching with some explicit pinching ratio below 1/4.

Chapter 10
The $\mathcal{F}$-Functional and Gradient Flows

After Ricci flow was first introduced, it appeared for many years that there was no variational characterisation of the flow as the gradient flow of a geometric quantity. In particular, Bryant and Hamilton established that the Ricci flow is not the gradient flow of any functional on $\mathfrak{Met}$ – the space of smooth Riemannian metrics – with respect to the natural L^2 inner product (with the exception of the two-dimensional case, where there is indeed such an 'energy'). Considering the prominent role variational methods have played in geometric analysis, PDE's and mathematical physics, it seemed surprising that such a natural equation as Ricci flow should be an exception. One of the many important contributions Perel'man made was to elucidate a gradient flow structure for the Ricci flow, not on $\mathfrak{Met}$ but on a larger augmented space. Part of this structure was already implicit in the physics literature [Fri85]. In this chapter we discuss this structure, at the centre of which is Perel'man's $\mathcal{F}$-functional [Per02]. The analysis will provide the ground work for the proof of a lower bound on injectivity radius at the end of Chap. 11.

10.1 Introducing the Gradient Flow Formulation

We introduce 'the gradient flow' associated to an energy functional in general terms. The concept naturally arises in branches of physics, PDE's, numerical analysis and related areas.

Definition 10.1. If $\mathcal{H}$ is a Hilbert space with a smooth functional $E : \mathcal{H} \to \mathbb{R}$, the *gradient vector field* $\nabla E : \mathcal{H} \to \mathcal{H}$ is given at each $u \in \mathcal{H}$ by the unique vector $\nabla E(u) \in \mathcal{H}$, such that

$$\langle \nabla E(u), v \rangle = dE(u)(v) \tag{10.1}$$

for all $v \in \mathcal{H}$.

B. Andrews and C. Hopper, *The Ricci Flow in Riemannian Geometry*, Lecture Notes in Mathematics 2011, DOI 10.1007/978-3-642-16286-2_10,

A consequence of (10.1) is that $\|dE(u)\| = \|\nabla E(u)\|$ for any $u \in \mathcal{H}$. Moreover, ∇E defines a *gradient flow* Φ given by the ODE

$$\begin{aligned}\frac{d}{dt}\Phi_u(t) &= -\nabla E(\Phi_u(t))\\ \Phi_u(0) &= u\end{aligned}$$

for any fixed $u \in \mathcal{H}$. To interpret Φ, we observe that flow lines, with respect to the graph $\mathrm{gr}(E) = \{(u, E(u)) \in \mathcal{H} \times \mathbb{R}\}$, are paths of steepest decent. Indeed, for any time dependent $u = u(t) \in \mathcal{H}$, we have

$$\frac{d}{dt}E(u(t)) = dE(u)(\dot{u}) = \langle \nabla E(u), \dot{u}\rangle \geq -\|\nabla E(u)\|\|\dot{u}\|$$

with equality holding if and only if $\dot{u} = -\lambda \nabla E(u)$ for $\lambda > 0$.

The archetypical gradient flow in PDE is the one associated to the Dirichlet energy functional

$$E(u) = \frac{1}{2}\int_{\mathbb{R}^n} |\nabla u|^2 dx = \frac{1}{2}\|\nabla u\|^2_{L^2}$$

with the usual L^2-inner product. Indeed, by Proposition 2.69(c), if u is smooth in x and t then

$$\frac{d}{dt}E(u) = \int_{\mathbb{R}^n} \left\langle \frac{\partial}{\partial t}\nabla u, \nabla u\right\rangle dx = -\int_{\mathbb{R}^n} \frac{\partial u}{\partial t}\Delta u\, dx = \langle -\Delta u, \dot{u}\rangle.$$

Thus formally the gradient flow is the standard heat equation $\frac{\partial u}{\partial t} = \Delta u$. Note that this example (in common with most other examples arising in PDE and calculus of variations) does not quite fit into the gradient flow framework defined above, since the energy E is not even defined on the Hilbert space $L^2(\mathbb{R}^n)$. We can make sense of E on the smaller space $L^2 \cap C^\infty$, but on this space the inner product is not complete, and E is not differentiable in the Fréchet sense (see Appendix A – note that Proposition A.3 does not apply here, since the derivative $-\Delta u$ is not continuous with respect to the L^2 metric):

$$\begin{aligned}E(u+v) - E(u) - \delta_v(u) &= \frac{1}{2}\int |\nabla(u+v)|^2 - |\nabla u|^2 + 2v\Delta u\, dx\\ &= \frac{1}{2}\int |\nabla v|^2\, dx + \int (v\Delta u + \nabla v \cdot \nabla u)\, dx\\ &= \frac{1}{2}\int |\nabla v|^2\end{aligned}$$

which is not even bounded as $\|v\|_{L^2} \to 0$ in $L^2 \cap C^\infty$, let alone $o(\|v\|_{L^2})$. However, E is Gâteaux differentiable on this space, and the derivative $\delta_v u$ is a continuous linear map with respect to the L^2 metric for any smooth u, so the gradient vector is still well defined.

Remark 10.2. When examining the gradient flow formulation of the Ricci flow there is a natural L^2-inner product, induced by the fibrewise product on $\mathrm{Sym}^2 T^*M$, given by

$$\langle h, h' \rangle = \int_M g^{ik} g^{j\ell} h_{ij} h'_{k\ell} d\mu(g) \tag{10.2}$$

referred to as the *canonical Riemannian metric* on the space of Riemannian metrics $\mathfrak{Met}$.[1] Note that this metric, by construction, is invariant under the action of the diffeomorphism group; however, as in the previous example the inner product is not complete.

10.2 Einstein-Hilbert Functional

Let (M, g) be a closed[2] Riemannian manifold. One of the most natural functionals one can construct on $\mathfrak{Met}$ is the so-called Einstein-Hilbert functional $E : \mathfrak{Met} \to \mathbb{R}$, which is the integral of the scalar curvature:

$$E(g) = \int_M \mathrm{Scal}\, d\mu.$$

By computing the variation of E at g, in direction $h = \dot{g}$, we see that

$$\begin{aligned} \delta_h E(g) &= \int_M \delta_h \mathrm{Scal}(g)\, d\mu + \int_M \mathrm{Scal}\, \delta_h d\mu(g) \\ &= \int_M -\Delta \mathrm{tr}_g h + \delta^2 h - \langle h, \mathrm{Ric} \rangle + \frac{\mathrm{Scal}}{2} \mathrm{tr}_g h \, d\mu \end{aligned}$$

where we recall the variation equations Proposition 4.10 and 4.11. Now as M is closed, $\int_M \Delta \mathrm{tr}_g h d\mu = 0$, so by the divergence theorem (i.e. Theorem 2.68) we see that $\int_M \delta^2 h\, d\mu = \int_M g^{ij} g^{pq} \nabla^2_{q,j} h_{pi}\, d\mu = \int_M \mathrm{div}\, V d\mu = 0$, where $V = V^q \partial_q = (g^{pq} g^{ij} \nabla_j h_{ip}) \partial_q$. Thus the variation of E is

---

[1] Note that $\mathfrak{Met}$ is an infinite dimensional cone (in the vector space $\mathrm{Sym}^2 T^*M$) and so is highly non-compact. Arbitrary sequences of Riemannian metrics can degenerate in complicated ways; however there are two rather trivial but nonetheless important sources of non-compactness, namely those of diffeomorphism invariance and scaling (cf. Chap. 9).

[2] That is, compact without boundary.

$$\delta_h E(g) = \int_M \left\langle \frac{\mathrm{Scal}}{2} g - \mathrm{Ric}, h \right\rangle d\mu.$$

It is important to note that the $\frac{\mathrm{Scal}}{2} g$-term is due to variation of the volume element $d\mu$. From the formula above, (twice) the gradient flow of E is given by

$$\frac{\partial}{\partial t} g_{ij} = 2(\nabla E)_{ij} = \mathrm{Scal}\, g_{ij} - 2R_{ij}.$$

We note that this equation looks similar to the Ricci flow, but the extra term means that this equation is not parabolic: From (5.2) the symbol is given by

$$\begin{aligned}(\widehat{\sigma}[\mathrm{Scal}\, g - 2\mathrm{Ric}]\xi)(h) &= |\xi|^2 h + \mathrm{tr}\, h\, \xi \otimes \xi \\ &\quad - \xi \otimes h^\sharp(\xi) - h^\sharp(\xi) \otimes \xi + \big(h(\xi,\xi) - |\xi|^2 \mathrm{tr}\, h\big) g,\end{aligned}$$

where $h^\sharp$ is the linear operator corresponding to h (obtained by raising an index using the metric). Choosing $\xi = e_n$, this maps $h = \xi \otimes \xi - |\xi|^2 g$ to $(2-n)$ times itself, but maps $h = e_1 \otimes e_2 + e_2 \otimes e_1$ to itself. Of course there is also a kernel which can be removed using DeTurck's trick, but no such trick can make the symbol have definite real part. Therefore the gradient flow is not parabolic. Such equations do not generally have solutions even for a short time – that is, there do not exist paths of steepest descent for this functional.

10.3 $\mathcal{F}$-Functional

To overcome the problems associated with the Einstein-Hilbert functional, Perel'man considers a functional $\mathcal{F}$ on the enlarged space $\mathfrak{Met} \times C^\infty(M)$, defined by

$$\mathcal{F}(g, f) := \int_M (\mathrm{Scal} + |\nabla f|^2) e^{-f} d\mu. \tag{10.3}$$

We shall follow the physics literature and call f the dilaton. It is important to note that $\mathcal{F}$ can also be written as $\mathcal{F}(g,f) = \int_M (\mathrm{Scal} + \Delta f) e^{-f} d\mu$ on a closed manifold M.[3]

Remark 10.3. There are some elementary symmetry properties associated with $\mathcal{F}$. The first of these is that $\mathcal{F}$ is diffeomorphism invariant: If $\varphi \in \mathrm{Diff}(M)$ then $\mathcal{F}(\varphi^* g, f \circ \varphi) = \mathcal{F}(g, f)$. The second is the scaling behaviour: For any scalars b and $c > 0$ we have $\mathcal{F}(c^2 g, f + b) = e^{n-2} e^{-b} \mathcal{F}(g, f)$.

[3] Since $\Delta e^{-f} = (|\nabla f|^2 - \Delta f) e^{-f}$ implies that $\int_M |\nabla f|^2 e^{-f} d\mu = \int_M \Delta f e^{-f} d\mu$.

Proposition 10.4 (Variation of $\mathcal{F}$). *On a closed manifold M, the variation of $\mathcal{F}$ is equal to*

$$\begin{aligned}\delta_{(h,k)}\mathcal{F}(g,f) = &-\int_M \langle \mathrm{Ric} + \mathrm{Hess}(f), h\rangle\, e^{-f}\, d\mu \\ &+ \int_M \big(\frac{1}{2}\mathrm{tr}_g h - k\big)\big(2\Delta f - |\nabla f|^2 + \mathrm{Scal}\big)e^{-f}\, d\mu.\end{aligned}$$

Proof. The variation of $\mathcal{F}$ in direction (h,k) is defined by Appendix A to be $\delta_{(h,k)}\mathcal{F}(g,f) = \frac{d}{ds}\big|_{s=0}\mathcal{F}(g+sh, f+sk)$. This implies that

$$\begin{aligned}\delta_{(h,k)}\mathcal{F}(g,f) = &\int_M \big(\delta_{(h,k)}(\mathrm{Scal} + |\nabla f|^2)(g,f)\big)e^{-f}\, d\mu \\ &+ \int_M (\mathrm{Scal} + |\nabla f|^2)\big(\delta_{(h,k)}(e^{-f}\, d\mu)(g,f)\big).\end{aligned}$$

Now by Proposition 4.11 we find that $\delta_{(h,k)}(e^{-f}\, d\mu)(g,f) = (\frac{\mathrm{tr}_g h}{2} - k)\, e^{-f}\, d\mu(g)$, and by (4.9) that

$$\begin{aligned}\delta_{(h,k)}|\nabla f|^2(g,f) &= -g^{ik}g^{j\ell}h_{k\ell}\nabla_i f\nabla_j f + 2g^{ij}\nabla_i f\nabla_j k \\ &= -h^{ij}\nabla_i f\nabla_j f + 2\langle\nabla f, \nabla k\rangle\,.\end{aligned}$$

Putting this together gives the variation

$$\begin{aligned}\delta_{(h,k)}\mathcal{F}(g,f) = &\int_M \Big(-\Delta\mathrm{tr}_g h + \delta^2 h - \langle h, \mathrm{Ric}\rangle \\ &\qquad - h^{ij}\nabla_i f\nabla_j f + 2\langle\nabla f, \nabla k\rangle\Big)e^{-f}\, d\mu \\ &+ \int_M (\mathrm{Scal} + |\nabla f|^2)\Big(\frac{\mathrm{tr}_g h}{2} - k\Big)e^{-f}\, d\mu. \qquad (10.4)\end{aligned}$$

Now since $\Delta e^{-f} = (|\nabla f|^2 - \Delta f)e^{-f}$, Proposition 2.69(b) implies that $\int_M -(\Delta\mathrm{tr}_g h)e^{-f}\, d\mu = -\int_M \mathrm{tr}_g h\Delta e^{-f}\, d\mu = \int_M \mathrm{tr}_g h(\Delta f - |\nabla f|^2)e^{-f}\, d\mu$, and Proposition 2.69(c) implies that $\int_M \langle\nabla f, \nabla k\rangle\, e^{-f}\, d\mu = \int_M(\Delta e^{-f})k\, d\mu = \int_M(|\nabla f|^2 - \Delta f)ke^{-f}\, d\mu$. Putting these two identities together gives

$$\int_M (-\Delta\mathrm{tr}_g h + 2\langle\nabla f, \nabla k\rangle)\, e^{-f}\, d\mu = 2\int_M \Big(\frac{\mathrm{tr}_g h}{2} - k\Big)(\Delta f - |\nabla f|^2)e^{-f}\, d\mu.$$

Also, we find by the divergence theorem that

$$\int_M \delta^2 h\, e^{-f}\, d\mu = \int_M \mathrm{div} V\, e^{-f}\, d\mu = -\int_M V^q\nabla_q e^{-f}\, d\mu,$$

where $V = V^q \partial_q = (g^{pq} g^{ij} \nabla_j h_{ip}) \partial_q$. Now by using the identity

$$\nabla_j (g^{pq} g^{ij} h_{ip} \nabla_q e^{-f}) = V^q \nabla_q e^{-f} + g^{pq} g^{ij} h_{ip} \nabla_j \nabla_q e^{-f},$$

we see that $\int_M \nabla_j (g^{pq} g^{ij} h_{ip} \nabla_q e^{-f}) d\mu = 0$ by the divergence theorem again, so that

$$\begin{aligned}\int_M \delta^2 h \, e^{-f} d\mu &= \int_M g^{pq} g^{ij} h_{ip} \nabla_j \nabla_q e^{-f} d\mu \\ &= \int_M h^{qj} \partial_j f \partial_q f e^{-f} + (-\partial_j \partial_q f + \Gamma^r_{jq} \partial_r f) e^{-f} d\mu \\ &= \int_M \left(h^{ij} \nabla_i f \nabla_j f - \langle \mathrm{Hess}(f), h \rangle \right) e^{-f} d\mu.\end{aligned}$$

Therefore we find that

$$\begin{aligned}&\int_M \left(-\Delta \mathrm{tr}_g h + \delta^2 h - \langle h, \mathrm{Ric} \rangle - h^{ij} \nabla_i f \nabla_j f + 2 \langle \nabla f, \nabla k \rangle \right) e^{-f} d\mu \\ &= \int_M - \langle \mathrm{Ric} + \mathrm{Hess}(f), h \rangle \, e^{-f} d\mu + \int_M \left(\frac{\mathrm{tr}_g h}{2} - k \right) (2\Delta f - 2|\nabla f|^2) e^{-f} d\mu.\end{aligned}$$

Combining this with (10.4) yields the desired result. □

Corollary 10.5 (Measure-preserving Variation of $\mathcal{F}$). *For variations (h,k) satisfying $\delta_{(h,k)} e^{-f} d\mu(g,f) = 0$, the variation*

$$\delta_{(h,k)} \mathcal{F}(g,f) = - \int_M \langle \mathrm{Ric} + \mathrm{Hess}(f), h \rangle \, e^{-f} d\mu.$$

10.4 Gradient Flow of $\mathcal{F}^m$ and Associated Coupled Equations

With Proposition 10.4 in mind, Perel'man formulates an appropriate gradient flow, in the form of a coupled system of equations, that can be related to the Ricci flow. To see this, first fix a smooth positive background measure[4] $d\omega$ on M, and define a smooth graph $X : \mathfrak{Met} \to \mathfrak{Met} \times C^\infty(M)$ by letting

$$X : g \mapsto \left(g, \log \frac{d\mu(g)}{d\omega} \right).$$

[4] Here we take a measure to mean a positive n-form. Note if dm is any positive n-form, then for any open $U \subset M$ we can define $m(U) := \int_U dm$. Hence $m : \mathcal{B}(M) \to [0, \infty)$ is a positive measure (in the strict sense) defined on the Borel σ-algebra $\mathcal{B}(M)$.

The resulting composition $\mathcal{F}^m = \mathcal{F} \circ X : \mathfrak{Met} \to \mathbb{R}$ is a functional on $\mathfrak{Met}$ that modifies $\mathcal{F}$. It takes the form

$$\begin{aligned}\mathcal{F}^m(g) &= \int_M \left(\text{Scal} + \left| \nabla \log \frac{d\mu}{d\omega} \right|^2 \right) d\omega \\ &= \int_M (\text{Scal} + |\nabla f|^2) d\omega,\end{aligned}$$

where now $f := \log \frac{d\mu}{d\omega}$.[5] From Corollary 10.5, the variation of $\mathcal{F}^m$ is

$$\delta_h \mathcal{F}^m(g) = - \int_M h^{ij}(R_{ij} + \nabla_i \nabla_j f) d\omega,$$

since $d\omega$ is fixed and $d\omega = e^{-f} d\mu$ by the definition of f. So by (10.2) and Proposition A.3, the gradient vector field of $\mathcal{F}^m$ (if it exists) is given by

$$\nabla \mathcal{F}^m(g) = -(R_{ij} + \nabla_i \nabla_j f).$$

Hence (twice) the positive gradient flow of $\mathcal{F}^m$ on $\mathfrak{Met}$ is

$$\frac{\partial}{\partial t} g_{ij} = 2\nabla \mathcal{F}^m(g) = -2(R_{ij} + \nabla_i \nabla_j f),$$

and the associated evolution equation for $f = \log \frac{d\mu}{d\omega}$ equal to

$$\frac{\partial f}{\partial t} = -\Delta f - \text{Scal},$$

since $\frac{\partial f}{\partial t} = \frac{1}{2} g^{ij} \frac{\partial g_{ij}}{\partial t} = -(\text{Scal} + g^{ij} \nabla_i \nabla_j f)$.[6] In which case we consider the *coupled modified Ricci flow*:

$$\frac{\partial}{\partial t} g = -2(\text{Ric} + \text{Hess}(f)) \tag{10.5a}$$

$$\frac{\partial f}{\partial t} = -\text{Scal} - \Delta f \tag{10.5b}$$

---

[5] The quotient of two n-forms makes sense: For if dm_1 and dm_2 are two positive n-forms, we have that $dm_1 = \varphi_1 d\omega$ and $dm_2 = \varphi_2 d\omega$, where φ_1, φ_2 are the corresponding Randon-Nykodym derivatives with respect to the background measure $d\omega$. In which case we define $\frac{dm_1}{dm_2} := \frac{\varphi_1}{\varphi_2}$.

[6] Note that if $\omega_1(t)$ and $\omega_2(t)$ are time-dependent n-forms, then

$$\frac{\partial}{\partial t} \log \frac{\omega_1}{\omega_2} = \frac{\frac{\partial \omega_1}{\partial t}}{\omega_1} - \frac{\frac{\partial \omega_2}{\partial t}}{\omega_2}.$$

In general, with Sect. 10.1 in mind, any gradient flow of $\mathcal{F}^m$ is non-decreasing along the flow lines, with $\frac{\partial}{\partial_t}\mathcal{F}^m = \|2\nabla\mathcal{F}^m\|^2_{L^2}$. This now gives Perel'man's monotonicity formula for the gradient flow of $\mathcal{F}^m$.

Proposition 10.6. *If* $(g(t), f(t))$ *is a solution to the coupled modified Ricci flow* (10.5), *then*

$$\frac{d}{dt}\mathcal{F}^m(g(t)) = 2\int_M \left|R_{ij} + \nabla_i\nabla_j f\right|^2 d\omega.$$

10.4.1 Coupled Systems and the Ricci Flow

Remarkably the gradient flow (10.5) is (up to diffeomorphism) equivalent to Ricci flow. This is achieved simply by performing a time-dependent diffeomorphisms (similar to that seen in Sect. 10.5) that transforms the coupled modified system into the Ricci flow.

On an intuitive level, if the diffeomorphism is generated by flowing along the time-dependent vector field $V(t)$, then the new equations for g and f become $\dot{g}_{ij} = -2(R_{ij} + \nabla_i\nabla_j f) + \mathcal{L}_V g$ and $\dot{f} = -\Delta f - \mathrm{Scal} + \mathcal{L}_V f$. Using these together with the fact that $\mathcal{L}_{\nabla f} g = 2\nabla\nabla f$ and $\mathcal{L}_{\nabla f} f = |\nabla f|^2$,[7] we consider following *coupled Ricci flow*:

$$\frac{\partial}{\partial t} g = -2\mathrm{Ric} \tag{10.6a}$$

$$\frac{\partial f}{\partial t} = -\Delta f + |\nabla f|^2 - \mathrm{Scal}. \tag{10.6b}$$

We get a solution to this system first by solving $\dot{g} = -2\mathrm{Ric}$ forwards in time then by solving $\dot{f} = -\Delta f + |\nabla f|^2 - \mathrm{Scal}$ backwards in time. The rest of this section is devoted to confirming this intuition formally.

---

[7] To see this, we first observe:

Proposition 10.7. *For any 1-form* ω *and* X, Y *vector fields,*

$$(\mathcal{L}_{\omega^\sharp} g)(X, Y) = (\nabla\omega)(X, Y) + (\nabla\omega)(Y, X).$$

Proof. By compatibility of g, $X(g(\omega^\sharp, Y)) = g(\nabla_X\omega^\sharp, Y) + g(\omega^\sharp, \nabla_X Y)$. Hence $g(\nabla_X\omega^\sharp, Y) = X(\omega(Y)) - \omega(\nabla_X Y) = (\nabla\omega)(X, Y)$. However on the other hand, since $\nabla g = 0$, we have $\mathcal{L}_{\omega^\sharp} g(X, Y) = g(\nabla_X\omega^\sharp, Y) + g(X, \nabla_Y\omega^\sharp)$.

Thus when $\omega = (df)^\sharp$ we have that $(\mathcal{L}_{\nabla f} g)(X, Y) = (\nabla\nabla f)(X, Y) + (\nabla\nabla f)(Y, X) = 2\mathrm{Hess}(f)(X, Y)$, since the Hessian is symmetric.

10.4.1.1 Converting the Gradient Flow to a Solution of the Ricci Flow

Given a solution $(\bar{g}(t), \bar{f}(t))$ to the gradient flow (10.5), we show that there is a solution $(g(t), f(t))$ to the coupled Ricci flow (10.6) by flowing along the gradient of $\bar{f}$.

Lemma 10.8. *Let $(\bar{g}(t), \bar{f}(t))$, $t \in [0, T]$, be a solution to the system* (10.5). *Define the one-parameter family of diffeomorphism $\Phi(t) \in \mathrm{Diff}(M)$ by*

$$\frac{d}{dt}\Phi(t) = \nabla_{\bar{g}(t)}\bar{f}(t), \qquad \Phi(0) = \mathrm{id}_M. \tag{10.7}$$

Then the pullback metric $g(t) = \Phi(t)^*\bar{g}(t)$ and dilaton $f(t) = \bar{f} \circ \Phi(t)$ satisfy the system (10.6).

Proof. By the theory of time-dependent ODE's, (10.7) always has solution (cf. [Lee02, p. 451]). In which case[8]

$$\frac{\partial}{\partial t}g = \frac{\partial}{\partial t}(\Phi^* g) = \Phi^*\left(\frac{\partial \bar{g}}{\partial t}\right) + \Phi^*\left(\mathcal{L}_{\nabla_{\bar{g}}\bar{f}}\bar{g}\right) = -2\Phi^*\left(\mathrm{Ric}(\bar{g})\right) = -2\mathrm{Ric}(g).$$

So we find that

$$\begin{aligned}\frac{\partial}{\partial t}f = \frac{\partial}{\partial t}(\bar{f} \circ \Phi) &= \frac{\partial \bar{f}}{\partial t} \circ \Phi + \left\langle \bar{\nabla}\bar{f} \circ \Phi, \frac{\partial \Phi}{\partial t}\right\rangle_{\bar{g}} \\ &= \left(-\bar{\Delta} f - \bar{\mathrm{Scal}}\right) \circ \Phi + \left|\bar{\nabla}\bar{f} \circ \Phi\right|_{\bar{g}}^2 \\ &= -\Delta f - \mathrm{Scal} + \left|\nabla f\right|^2,\end{aligned}$$

where $\bar{\Delta}$ and $\bar{\nabla}$ are with reference to $\bar{g}(t)$. □

10.4.1.2 Converting the Ricci Flow to a Solution of the Gradient Flow

We now show the converse. Given a solution $(g(t), f(t))$ to the coupled Ricci flow (10.5), we show that there is a solution $(\bar{g}(t), \bar{f}(t))$ to the gradient flow (10.6) by flowing backwards along the gradient of f. To do this, we need to solve a backwards heat equation for f.

[8] Note, if ψ is a solution to the ODE: $\frac{\partial}{\partial t}\psi(x,t) = X(\psi(x,t),t)$. Then

$$\frac{\partial}{\partial t}\psi^* g = \frac{\partial g}{\partial t} + g(\nabla_i X, \partial_j) + g(\partial_i, \nabla_j X).$$

Lemma 10.9. *Let $g(t)$, $t \in [0, T)$, be a solution of the Ricci flow and let f_T be an arbitrary function on M.*

1. *Then there exists a unique solution to the backwards heat equation*

$$\frac{\partial f}{\partial t} = -\Delta f + |\nabla f|^2 - \mathrm{Scal}, \qquad t \in [0, T] \tag{10.8a}$$

$$f(T) = f_T \tag{10.8b}$$

2. *Furthermore, given a solution $f(t)$ to (10.8), define the one-parameter family of diffeomorphism $\Psi(t) \in \mathrm{Diff}(M)$ by*

$$\frac{d}{dt}\Psi(t) = -\nabla_{g(t)} f(t), \qquad \Psi(0) = \mathrm{id}_M. \tag{10.9}$$

Then the pullback metric $\bar{g}(t) = \Psi^*(t) g(t)$ and the pullback dilaton $\bar{f}(t) = f \circ \Psi(t)$ satisfy (10.5).

Proof. 1. Re-parametrise time by $\tau = T - t$ and set $u = e^{-f}$. Note that

$$\begin{aligned}\frac{\partial u}{\partial \tau} = -\frac{\partial u}{\partial t} &= \frac{\partial f}{\partial t} u = (\Delta f + |\nabla f|^2 - \mathrm{Scal}) u \\ &= \Delta u - \mathrm{Scal}\, u,\end{aligned}$$

and so u satisfies

$$\frac{\partial u}{\partial \tau} = \Delta u - \mathrm{Scal}\, u.$$

As this is a linear parabolic equation (forwards) in time with initial data at $\tau = 0$, there exists a unique solution on $[0, T]$.

2. One can verify that $\bar{g}$ and $\bar{f}$ satisfy (10.5) by the same procedure as in Lemma 10.8 except that here we flow along $-\nabla f$ rather than $\bar{\nabla} \bar{f}$. □

10.4.2 Monotonicity of $\mathcal{F}$ from the Monotonicity of $\mathcal{F}^m$

The diffeomorphism invariance of all quantities under consideration implies the monotonicity formula for the Ricci flow:

Proposition 10.10. *If $(g(t), f(t))$ is a solution to (10.6) on a closed manifold M, then*

$$\frac{d}{dt}\mathcal{F}(g(t), f(t)) = 2\int_M \left|R_{ij} + \nabla_i \nabla_j f\right|^2 e^{-f} d\mu.$$

Proof. As $(g(t), f(t))$ is a solution to (10.6), Lemma 10.9 implies that $\bar{g}(t) = \Psi^*(t)g(t)$ and $\bar{f}(t) = f(t) \circ \Psi(t)$ are a solution to (10.5).

By Remark 10.3, $\mathcal{F}$ is invariant under diffeomorphisms. Thus $\mathcal{F}(g, f) = \mathcal{F}(\bar{g}, \bar{f})$, and so $\frac{d}{dt}\mathcal{F}(g, f) = \frac{d}{dt}\mathcal{F}(\bar{g}, \bar{f})$. Moreover, by Proposition 10.6 we see that

$$\begin{aligned}\frac{d}{dt}\mathcal{F}(\bar{g}, \bar{f}) &= 2\int_M \left|\bar{R}_{ij} + \bar{\nabla}_i\bar{\nabla}_j\bar{f}\right|_{\bar{g}}^2 e^{-\bar{f}}\,d\mu(\bar{g}) \\ &= 2\int_M \left|R_{ij} + \nabla_i\nabla_j f\right|^2 e^{-f}\,d\mu(g). \qquad \square\end{aligned}$$

Chapter 11
The $\mathcal{W}$-Functional and Local Noncollapsing

The $\mathcal{F}$-functional provides a gradient flow formalism for the Ricci flow, discussed in Chap. 10. We hope to be able to use this to understand the singularities of Ricci flow, but the $\mathcal{F}$-functional is not yet enough to do this, because it does not behave well under the scaling transformations needed in the blow-up analysis. To overcome this, Perel'man introduced the $\mathcal{W}$-functional which includes a positive scale factor τ. The advantage of this functional is that it can be related to aspects of the local geometry, in particular volume ratios of balls with radius on the order of $\sqrt{\tau}$. As discussed at the end of Chap. 9, the missing ingredient in the singularity analysis is a suitable lower bound on the injectivity radius on the scale determined by the curvature, and Perel'man was able to prove such a bound using the $\mathcal{W}$-functional. The injectivity estimate is proved here in Sect. 11.4.

11.1 Entropy $\mathcal{W}$-Functional

On a closed n-dimensional manifold M, define Perel'man's entropy $\mathcal{W}$-functional $\mathcal{W} : \mathfrak{Met} \times C^\infty(M) \times \mathbb{R}^+ \to \mathbb{R}$ by

$$\mathcal{W}(g, f, \tau) = \int_M \big(\tau(\mathrm{Scal} + |\nabla f|^2) + f - n\big) u \, d\mu, \tag{11.1}$$

where $u := (4\pi\tau)^{-\frac{n}{2}} e^{-f}$ and $\tau > 0$ is the scale parameter. By inspection we note that this is related to $\mathcal{F}$-functional by

$$\mathcal{W}(g, f, \tau) = \frac{1}{(4\pi\tau)^{n/2}} \Big(\tau \, \mathcal{F}(g, f) + \int_M (f - n) e^{-f} d\mu\Big). \tag{11.2}$$

In a similar way to that of Chap. 10, we look to find a gradient flow for $\mathcal{W}$ that includes the Ricci flow which makes $\mathcal{W}$ monotone. With this, we prove the local noncollapsing result for the Ricci flow. Thereafter we prove the desired lower injectivity bounds.

B. Andrews and C. Hopper, *The Ricci Flow in Riemannian Geometry*, Lecture Notes in Mathematics 2011, DOI 10.1007/978-3-642-16286-2_11,

Remark 11.1. Like $\mathcal{F}$, the functional $\mathcal{W}$ is diffeomorphism-invariant: If $\Phi \in \mathrm{Diff}(M)$ then $\mathcal{W}(\Phi^* g, \Phi^* f, \tau) = \mathcal{W}(g, f, \tau)$, where $\Phi^* g$ is the pullback metric and $\Phi^* f = f \circ \Phi$. The scaling properties of $\mathcal{W}$ are rather nicer than those of $\mathcal{F}$: Under the scaling transformation $(g, f, \tau) \mapsto (cg, f, c\tau)$,we have $\mathcal{W}(cg, f, c\tau) = \mathcal{W}(g, f, \tau)$.

Proposition 11.2 (Variation of $\mathcal{W}$). *On a closed manifold M, the variation of $\mathcal{W}$ is equal to*

$$\delta_{(h,k,\zeta)}\mathcal{W}(g,f,\tau) = \int_M \Big\langle \mathrm{Ric} + \mathrm{Hess}(f) - \frac{1}{2\tau}g, -\tau h + \zeta g \Big\rangle u\, d\mu$$
$$+ \int_M \tau\Big(\frac{1}{2}\mathrm{tr}_g h - k - \frac{n}{2\tau}\zeta\Big)\Big(\mathrm{Scal} + 2\Delta f - |\nabla f|^2 + \frac{f-n-1}{\tau}\Big) u\, d\mu.$$

Proof. We calculate the variation $\delta_{(h,k,\zeta)}\mathcal{W}(g,f,\tau)$ of $\mathcal{W}$ at (g,f,τ) in the direction (h,k,ζ) via a two step process. This is done by separating out the variation in the scale parameter so that

$$\delta_{(h,k,\zeta)}\mathcal{W}(g,f,\tau) = \delta_{(h,k,0)}\mathcal{W}(g,f,\tau) + \delta_{(0,0,\zeta)}\mathcal{W}(g,f,\tau).$$

Now to compute $\delta_{(h,k,0)}\mathcal{W}(g,f,\tau)$ with τ fixed, we look to (11.2). By Proposition 10.4 we have that

$$\delta_{(h,k,0)}\Big(\frac{\tau}{(4\pi\tau)^{n/2}}\mathcal{F}(g,f)\Big)(g,f,\tau) = -\int_M \tau\langle \mathrm{Ric} + \mathrm{Hess}(f), h\rangle u\, d\mu$$
$$+ \int_M \tau\left(\frac{1}{2}\mathrm{tr}_g h - k\right)\left(2\Delta f - |\nabla f|^2 + \mathrm{Scal}\right) u\, d\mu,$$

and by direct computation

$$\delta_{(h,k,0)}\Big(\frac{1}{(4\pi\tau)^{n/2}}\int_M (f-n)e^{-f}\, d\mu\Big)(g,f,\tau)$$
$$= \int_M \left(k + \left(\frac{1}{2}\mathrm{tr}_g h - k\right)(f-n)\right) u\, d\mu.$$

To compute $\delta_{(0,0,\zeta)}\mathcal{W}(g,f,\tau)$ with g and f fixed, we see directly that

$$\delta_{(0,0,\zeta)}\mathcal{W}(g,f,\tau) = \int_M \left(\zeta\left(1-\frac{n}{2}\right)\left(\mathrm{Scal} + |\nabla f|^2\right) - \frac{n\zeta}{2\tau}(f-n)\right) u\, d\mu.$$

By combining all of the terms, we find that the variation of $\mathcal{W}$ equals

$$\delta_{(h,k,\zeta)}\mathcal{W}(g,f,\tau) = \int_M \Big[\langle \mathrm{Ric} + \mathrm{Hess}(f), -\tau h + \zeta g\rangle + \tau\left(\frac{1}{2}\mathrm{tr}_g h - k\right)\left(2\Delta f - |\nabla f|^2 + \mathrm{Scal} + \frac{f-n}{\tau}\right) + k + \zeta(|\nabla f|^2 - \Delta f) - \frac{n\zeta}{2\tau}(f-n) - \frac{n\zeta}{2}(\mathrm{Scal} + |\nabla f|^2)\Big] u\, d\mu.$$

Now absorb $-\frac{n}{2\tau}\zeta$ into the first bracket of the terms on the second line to get

$$\delta_{(h,k,\zeta)}\mathcal{W}(g,f,\tau) = \int_M \Big[\langle \mathrm{Ric} + \mathrm{Hess}(f), -\tau h + \zeta g\rangle + \tau\left(\frac{1}{2}\mathrm{tr}_g h - k - \frac{n}{2\tau}\zeta\right)\left(\mathrm{Scal} + 2\Delta f - |\nabla f|^2 + \frac{f-n}{\tau}\right) + k + (n-1)\zeta(\Delta f - |\nabla f|^2)\Big] u\, d\mu.$$

Also absorb $-\frac{1}{2\tau}g$ into the angled bracket terms, together with the fact that $\langle \frac{-1}{2\tau}g, -\tau h + \zeta g\rangle = \frac{1}{2}\mathrm{tr}_g h - \frac{n}{2\tau}\zeta$, so that finally

$$\delta_{(h,k,\zeta)}\mathcal{W}(g,f,\tau) = \int_M \Big[\langle \mathrm{Ric} + \mathrm{Hess}(f) - \frac{1}{2\tau}g, -\tau h + \zeta g\rangle + \tau\Big(\frac{1}{2}\mathrm{tr}_g h - k - \frac{n}{2\tau}\zeta\Big)\Big(\mathrm{Scal} + 2\Delta f - |\nabla f|^2 + \frac{f-n-1}{\tau}\Big) + (n-1)\zeta(\Delta f - |\nabla f|^2)\Big] u\, d\mu.$$

The desired result now follows since the last term of the latter equation vanishes because $\int_M (\Delta f - |\nabla f|^2)e^{-f} d\mu = \int_M \Delta e^{-f} d\mu = 0$. □

Corollary 11.3 (Measure-preserving Variation of $\mathcal{W}$). *For variations (h,k,ζ) satisfying $\delta_{(h,k,\zeta)} u\, d\mu(g,f,\tau) = 0$, the variation*

$$\delta_{(h,k,\zeta)}\mathcal{W}(g,f,\tau) = \int_M \langle \mathrm{Ric} + \mathrm{Hess}(f) - \frac{1}{2\tau}g, -\tau h + \zeta g\rangle u\, d\mu.$$

11.2 Gradient Flow of $\mathcal{W}$ and Monotonicity

In this section we want to formulate an appropriate gradient flow for $\mathcal{W}$ that makes the functional monotone. Whereas in Sect. 10.4 it was possible to do this formally for the $\mathcal{F}$-functional, here we derive the gradient flow for $\mathcal{W}$ heuristically.

We do this first by fixing the measure $dm = (4\pi\tau)^{-n/2}e^{-f}d\mu$ so that the variation $\delta_{(h,k,\zeta)}dm(g,f,\tau)$ vanishes (i.e. $-\frac{n}{2\tau}\zeta - k + \frac{1}{2}\mathrm{tr}_g h = 0$). By solving this for f we find that

$$f = \log\frac{d\mu}{dm} - \frac{n}{2}\log(4\pi\tau).$$

Now by taking the gradient flow for the metric g_{ij} as done in (10.5), we obtain the following coupled gradient flow:

$$\frac{\partial}{\partial t}g = -2(\mathrm{Ric} + \mathrm{Hess}(f)) \tag{11.3a}$$

$$\frac{\partial f}{\partial t} = -\Delta f - \mathrm{Scal} + \frac{n}{2\tau} \tag{11.3b}$$

$$\frac{d\tau}{dt} = -1 \tag{11.3c}$$

where the last condition $\dot{\tau} = -1$ is imposed in order to ensure monotonicity, since

$$\begin{aligned}\frac{d}{dt}\mathcal{W} &= \int_M \left(R_{ij} + \nabla_i\nabla_j f - \frac{1}{2\tau}g_{ij}\right)(-\tau\dot{g}_{ij} + \dot{\tau}g_{ij})\,dm \\ &= 2\tau\int_M \left|R_{ij} + \nabla_i\nabla_j f - \frac{1}{2\tau}g_{ij}\right|^2 dm\end{aligned}$$

whenever $dm = u\,d\mu$ is fixed, $\dot{g}_{ij} = -2(R_{ij} + \nabla_i\nabla_j f)$ and $\dot{\tau} = -1$.

By performing the same diffeomorphism change that was outlined in Sect. 10.4.1, we obtain the following coupled system of equations:

$$\frac{\partial}{\partial t}g = -2\mathrm{Ric} \tag{11.4a}$$

$$\frac{\partial f}{\partial t} = -\Delta f + |\nabla f|^2 - \mathrm{Scal} + \frac{n}{2\tau} \tag{11.4b}$$

$$\frac{d\tau}{dt} = -1 \tag{11.4c}$$

which includes the Ricci flow. It now follows, by a similar argument to that of Sect. 10.4.2, that we have the following monotonicity result for the $\mathcal{W}$-functional:

Proposition 11.4. *If $(g(t), f(t), \tau(t))$ be a solution to the coupled system (11.4) on a closed manifold M, then*

$$\frac{d}{dt}\mathcal{W}(g(t), f(t), \tau(t)) = \int_M 2\tau\left|\mathrm{Ric} + \mathrm{Hess}(f) - \frac{1}{2\tau}g\right|^2 u\,d\mu. \tag{11.5}$$

11.2.1 Monotonicity of $\mathcal{W}$ from a Pointwise Estimate

We note that there is an alternative approach to proving the entropy monotonicity of $\mathcal{W}$ based on the conjugate heat operator.

By defining the heat operator $\Box = \frac{\partial}{\partial t} - \Delta$ acting on $C^\infty(M \times [0,T))$ we see, by evaluating $\frac{d}{dt}\int vw$, that the conjugate heat operator $\Box^* = -\frac{\partial}{\partial t} - \Delta + \mathrm{Scal}$ is conjugate to $\Box$ in the following sense.

Lemma 11.5. *If $g(t)$, $t \in [0,T)$ is a solution to the Ricci flow and $v, w \in C^\infty(M \times [0,T))$, then*

$$\int_0^T \left(\int_M (\Box v)\, w \, d\mu\right) dt = \left[\int_M vw \, d\mu\right]_0^T + \int_0^T \left(\int_M v\,(\Box^* w)\, d\mu\right) dt.$$

Now by defining

$$w := \left(\tau(R + 2\Delta f - |\nabla f|^2) + f - n\right) u$$

so that

$$\mathcal{W} = \int_M w \, d\mu,$$

as $\int_M (\Delta f - |\nabla f|^2) u \, d\mu = \int_M \Delta u \, d\mu = 0$, we see that the monotonicity of $\mathcal{W}$ follows immediately from the following proposition since

$$\frac{d}{dt}\mathcal{W} = \frac{d}{dt}\int_M w \, d\mu = -\int_M \Box^* w \, d\mu$$

by Lemma 11.5 with $v = 1$.

Proposition 11.6 **([Per02, Proposition 9.1], [Top06, p. 77]).** *Suppose (g, f, τ) evolve according to (11.4). Then the function w satisfies*

$$\Box^* w = -2\tau \left|\mathrm{Ric} + \mathrm{Hess}(f) - \frac{1}{2\tau} g\right|^2 u.$$

11.3 μ-Functional

We now look at the functional $\mu : \mathfrak{Met} \times \mathbb{R}^+ \to \mathbb{R}$ defined by

$$\mu(g, \tau) = \inf\{\mathcal{W}(g, f, \tau) : f \in C^\infty(M) \text{ compatible with } g \text{ and } \tau\},$$

where we say that the tuple (g, f, τ) is *compatible* if

$$(4\pi\tau)^{-\frac{n}{2}} \int_M e^{-f} \, d\mu = \int_M u \, d\mu = 1.$$

As we shall see, the functional μ plays an important role in proving the local noncollapsing result of the next section. Although before we can do so, we need to check that μ is monotone and bounded from below, and that the infimum is attained.

Remark 11.7. μ is a homogeneous function of degree 0, i.e. $\mu(cg, c\tau) = \mu(g, \tau)$ for any scalar c, and has the diffeomorphism invariance property: $\mu(\Phi^* g, \tau) = \mu(g, \tau)$ for any $\Phi \in \mathrm{Diff}(M)$, inherited from $\mathcal{W}$.

Proposition 11.8 (μ is Bounded Below). *For any given g and $\tau > 0$ on a closed manifold M, there exists $c \in \mathbb{R}$ such that $\mathcal{W}(g, f, \tau) \geq c$ for all compatible $f \in C^\infty(M)$. Consequently $\mu(g, \tau) \geq c$.*

Proof. By our scaling property $\mu(g, 1) = \mu(\tau g, \tau)$, so without loss of generality let $\tau = 1$.

Let $w = \sqrt{u} = (4\pi)^{-n/4} e^{-f/2} > 0$ with $\int w^2 d\mu = 1$. From this we find that $f = -2\log w - \frac{n}{2}\log 4\pi$ and $\nabla f = -2\nabla w / w$. Hence we can write $\mathcal{W}(g, f, 1)$ in terms of w:

$$\begin{aligned}\mathcal{W}(g, f, 1) &= \int_M \left(4|\nabla w|^2 + \left(\mathrm{Scal} - 2\log w - \frac{n}{2}\log 4\pi - n\right) w^2\right) d\mu \\ &=: \mathcal{H}(g, w). \qquad (11.6)\end{aligned}$$

Since M is closed, $\mathrm{Scal} - n - \frac{n}{2}\log 4\pi \geq \inf_{x \in M} \mathrm{Scal} - n - \frac{n}{2}\log 4\pi > C > -\infty$. So the only problem term is the $w^2 \log w$ one. Fortunately, the Log Sobolev inequality allows this to be bounded by a Dirichet type term. It is typically stated (for instance see [CLN06, p. 184]) as follows.

Lemma 11.9 (Log Sobolev Inequality on a Manifold). *Let (M, g) be closed Riemannian manifold. For any $a > 0$ there exists a constant $C(a, g)$ such that if $\varphi > 0$ satisfies $\int \varphi^2 d\mu = 1$, then*

$$\int_M \varphi^2 \log \varphi \, d\mu \leq a \int_M |\nabla \varphi|^2 d\mu + C(a, g).$$

From the Lemma we finally get

$$\mathcal{H}(g, w) \geq 2 \int_M |\nabla w|^2 d\mu + C - C(1, g) \geq C - C(1, g). \qquad (11.7)$$

□

Proposition 11.10 (Existence of a Smooth Minimiser). *For any smooth metric g on a closed M with $\tau > 0$, the infimum of $\mathcal{W}$ over all compatible f is attained by a smooth compatible minimiser f_∞.*

Proof. Once again, without loss of generality, let $\tau = 1$ and define $\mathcal{H}(g, w)$ as in (11.6) above. We will use direct methods in the calculus of variations to show that $\mathcal{H}$ has a minimiser. From the estimate (11.7), any minimizing sequence $\{w_k\}$ of compatible functions for $\mathcal{H}(g, \cdot)$ (that is, functions which are positive and satisfy $\int_M w^2\,d\mu = 1$) has bounded Dirichlet energy, and there exists a subsequence with a weak limit w in $W^{1,2}$. Since the Dirichlet energy is weakly lower semicontinuous (see [Dac04, p. 82]), we have $\int |\nabla w|^2 \leq \liminf_{k\to\infty} \int |\nabla w_k|^2$. By the Rellich compactness theorem,[1] the sequence also converges in L^p for $p < \frac{2n}{n-2}$, so $\int w^2 = 1$, and $\int \mathrm{Scal}\, w_k^2$ converges to $\int \mathrm{Scal}\, w^2$.

We want to show that the term involving the integral of $w^2 \log w$ in the definition of $\mathcal{H}$ also converges. To see this, set $\rho = w^2 \log w$, and note that $\nabla\rho = (2w\log w + w)\nabla w$. Since $|w \log w| \leq c_1 + c_2 w^{1+\varepsilon/2}$ for any $\varepsilon > 0$, we find that

$$\begin{aligned} \int |\nabla \rho| &\leq \int \big|(w + 2w\log w)\nabla w\big| \\ &\leq \Big(\int |2c_1 + w|^2\Big)^{\frac{1}{2}} \Big(\int |\nabla w|^2\Big)^{\frac{1}{2}} + 2c_2\Big(\int |w|^{2+\varepsilon}\Big)^{\frac{1}{2}} \Big(\int |\nabla w|^2\Big)^{\frac{1}{2}}. \end{aligned}$$

Trivially $(\int |\nabla w|^2)^{1/2}$ is bounded by $\|w\|_{W^{1,2}}$ and using Sobolev inequalities[2] one can show the other terms are also bounded by $\|w\|_{W^{1,2}}$ as well.

Using the Rellich compactness theorem again, we have that the sequence ρ_k is precompact in L^1, so passing to a subsequence we have $\int \rho_k \to \int \rho$. It follows that $\mathcal{H}(w) \leq \lim_{n\to\infty} \mathcal{H}(w_k) = \inf \mathcal{H}$, so the limit w is a minimiser of $\mathcal{H}$ and is compatible. Hence by definition we have $\mathcal{H}(g, w) = \mu(g, 1)$.

The limit w is clearly non-negative. The first variation formula shows that it is a weak solution of the Euler–Lagrange equation for $\mathcal{H}$, which is the elliptic equation $\Delta w + (2\log w + n + \frac{n}{2}\log 4\pi - \mathrm{Scal} + \mu)w/4 = 0$. We wish to prove that w is smooth and positive, using techniques from PDE theory. This is slightly subtle due to the presence of the logarithmic nonlinearity.

We first show that w has continuous derivatives up to second order: Since $|w\log w| \leq C(\varepsilon) + \varepsilon w^{1+\varepsilon}$, and $w \in L^{\frac{2n}{n-2}}$ (for $n > 2$), we have that $w \log w \in L^p$ for any $p < \frac{2n}{n-2}$. But then L^p estimates (such as [GT83, Theorem 9.11]) imply that $w \in W^{2,p}$, and hence $w \in L^q$ for any $q < \frac{2n}{n-6}$ (or $w \in L^\infty$ if $n < 6$). But then L^p estimates imply $w \in W^{2,q}$, so that $w \in L^q$ for $q < \frac{2n}{n-10}$ (or $w \in L^\infty$ if $n < 10$). Continuing in this way, we find that $w \in L^\infty$ for $n < 2 + 4k$ after k iterations. In fact we have more, since if k is large

[1] If $\Omega \subset \mathbb{R}^n$ is a regular bounded domain with $1 \leq p < n$ and $1 \leq q < p^*$, the bounded sets on $W^{1,p}(\Omega)$ are precompact in $L^q(\Omega)$. In particular, if (u_k) is a sequence of functions in $W^{1,p}(\Omega)$ such that $\|u_k\|_{W^{1,p}} \leq C$, where C is independent of k, then there is a subsequence of (u_k) which converges in $L^q(\Omega)$ (cf. [Eva98, p. 272] or [Jos08, p. 549]).

[2] That is, $\|w\|_{L^{p^*}} \leq C\|w\|_{W^{1,p}}$ for any $p^* = np/(n-p)$ and $1 \leq p < n$ (cf. [Eva98, p. 265]).

enough so that $n < 2+4k$, then we have $w \in W^{2,q}$ for $q > n/2$, which implies that $w \in C^{0,\alpha}$ by the Sobolev embedding theorem [GT83, Corollary 7.11].

We now show that $w \log w$ is in $C^{0,\beta}$ for any $\beta < \alpha$: Since $0 \leq w \leq K$ for some K, we have for any $\varepsilon > 0$ a constant $C(\varepsilon)$ such that $|\log w| \leq C(\varepsilon) + w^{-\varepsilon}$. But then we have (writing $w_s = (1-s)w(x) = sw(y)$ and assuming $w(y) > w(x)$) that

$$\begin{aligned} |(w \log w)(x) - (w \log w)(y)| &= \left| \int_0^1 (1 + \log w_s)\, ds(w(y) - w(x)) \right| \\ &\leq \int_0^1 (C(\varepsilon) + w_s^{-\varepsilon})\, ds(w(y) - w(x)) \\ &= C(\varepsilon)|w(y) - w(x)| + \frac{1}{1-\varepsilon}|w(y)^{1-\varepsilon} - w(x)^{1-\varepsilon}|. \end{aligned}$$

But now we observe that

$$w(y)^{1-\varepsilon} - w(x)^{1-\varepsilon} = (1-\varepsilon) \int_0^1 w_s^{-\varepsilon}\, ds(w(y) - w(x)),$$

so we have (since w is Hölder continuous with exponent α) that

$$\begin{aligned} |(w \log w)(x) - (w \log w)(y)| &\leq C(\varepsilon)|w(y) - w(x)| + \frac{1}{1-\varepsilon}|w(y) - w(x)|^{1-\varepsilon} \\ &\leq C d(y,x)^{\alpha(1-\varepsilon)}. \end{aligned}$$

Thus $w \log w$ is Hölder continuous with exponent $\alpha(1-\varepsilon)$, for any $\varepsilon > 0$, as claimed. Schauder estimates [GT83, Theorem 6.2] then imply that $w \in C^{2,\beta}$. In particular w is a classical solution of the equation.

A strong maximum principle (precisely, one such as is proved in [Váz84, Theorem 1]) implies that w has a positive bound below, so that $w \log w$ is a smooth function of w, and so is $C^{2,\beta}$. Higher regularity now follows by Schauder estimates, so $w \in C^\infty(M)$. □

Proposition 11.11 (Monotonicity of μ). *If $(g(t), \tau(t))$, for $t \in [0,T)$, is a solution to*

$$\begin{aligned} \frac{\partial g}{\partial t} &= -2\mathrm{Ric} \\ \frac{\partial \tau}{\partial t} &= -1 \end{aligned}$$

on a closed manifold M with $\tau(t) > 0$. Then

$$\mu(g(t_2), \tau(t_2)) \leq \mu(g(t_1), \tau(t_1))$$

for all times $0 \leq t_1 \leq t_2 \leq T$.

Proof. For any $t_0 \in (0, T]$, let f solve (11.4b) backwards in time on $[0, t_0]$ with

$$f\big|_{t=t_0} = \arg\min\{\mathcal{W}(g(t_0), \hat{f}, \tau(t_0)) : \hat{f} \in C^\infty(M) \text{ compatible with } g \text{ and } \tau\}.$$

By Proposition 11.4, the monotonicity of $\mathcal{W}$ implies that

$$\frac{d}{dt}\mathcal{W}(g(t), f(t), \tau(t)) \geq 0,$$

for all $t \in [0, t_0]$. We note that compatibility is preserved by the flow (11.4) since

$$\begin{aligned} \frac{d}{dt}\int_M (4\pi\tau)^{-n/2} e^{-f} d\mu &= \int \Big(\frac{n}{2\tau} - \frac{\partial f}{\partial t} + \frac{1}{2}\mathrm{tr}_g \frac{\partial g}{\partial t}\Big) u \, d\mu \\ &= \int \Big(\frac{\partial u}{\partial t} - \mathrm{Scal}\, u\Big) d\mu \\ &= \int (-\Box^* u - \Delta u) d\mu = 0. \end{aligned}$$

Hence for all times t we have that

$$\begin{aligned} \mu(g(t), \tau(t)) &\leq \mathcal{W}(g(t), f(t), \tau(t)) \\ &\leq \mathcal{W}(g(t_0), f(t_0), \tau(t_0)) \\ &= \mu(g(t_0), \tau(t_0)) \end{aligned}$$

where the last equality is by construction. □

In particular, by letting $\tau(s) = -s + r^2 + t_2$, where $t_1 = 0$ and $t_2 = t$ we get the following useful inequality.

Corollary 11.12. *If $g(t)$, $t \in [0, T)$, is solution to the Ricci flow on a closed manifold M. Then for all $t \in [0, T)$ and $r > 0$ we have*

$$\mu(g(0), r^2 + t) \leq \mu(g(t), r^2). \tag{11.8}$$

11.4 Local Noncollapsing Theorem

In order to prove the desired injectivity bounds, it is enough to work with volume ratios. This is both analytically convenient and necessary since it is rather difficult to work with the injectivity radius directly. We show that there exists a lower bounds on the volume collapsing ratio under the Ricci flow by establishing an upper bound on μ in terms of various local geometric quantities. From this we prove a stronger version of Perel'man's 'local noncollapsing' result [Per02, Sect. 4] where only a pointwise bounds on the scalar curvature Scal is required rather than the full curvature tensor R.

Proposition 11.13 ([Top05]). *Let (M, g) be a closed Riemannian manifold. Then for any point p and $r > 0$ we have*

$$\mu(g, r^2) \leq \log \frac{\operatorname{Vol} B(p,r)}{r^n} + \left(36 + r^2 \fint_{B(p,r)} |\mathrm{Scal}| d\mu\right) \frac{\operatorname{Vol} B(p,r)}{\operatorname{Vol} B(p,r/2)}. \quad (11.9)$$

Remark 11.14. Note that the first term on the right-hand side of the inequality is the desired volume ratio. We shall look to bound the other terms under reasonable conditions on the initial data.

Proof. Choose $\tau = r^2$ and let $w = \sqrt{u}$ so that $\int w^2 d\mu = 1$. Note that this implies that $f = -2\log w - \frac{n}{2}\log(4\pi r^2)$ and $\nabla f = -2\nabla w/w$. By taking the infimum over all $f \in C^\infty(M)$ compatible with g and τ, we get

$$\mu(g, r^2) \leq \int_M \left(r^2(\mathrm{Scal}\, w^2 + 4|\nabla w|^2) + f - n\right) w^2 d\mu. \quad (11.10)$$

We now make a judicious choice of compatible f so that the right-hand side of (11.10) reflects the local geometry at a point p with respect to the metric g. In particular let

$$f(x) = c - \log \phi \left(\frac{d_g(x,p)}{r}\right)^2$$

or alternatively

$$w(x)^2 = (4\pi r^2)^{-n/2} \phi \left(\frac{d_g(x,p)}{r}\right)^2 e^{-c}$$

where $c = c(n, g, x, r)$ is chosen so that $\int w^2 d\mu = 1$, $d_g(x,p)$ is the distance between x and p with respect to the metric g, and $\phi : [0,\infty) \to [0,1]$ is the smooth cut-off function distributed so that $\phi(y) = 1$ for $y \in [0, 1/2]$; $\phi(y) = 0$ for $y \in [1, \infty)$ with a slope chosen so that $|\phi'| \leq 3$ for $1/2 \leq y \leq 1$.

Claim 11.15. The constant c satisfies the following inequality:

$$\frac{1}{\operatorname{Vol} B(p,r)} \leq (4\pi r^2)^{-n/2} e^{-c} \leq \frac{1}{\operatorname{Vol} B(p, r/2)}. \quad (11.11)$$

Proof of Claim. As $|\phi| \leq 1$ and $\operatorname{supp} w \subset B(p,r)$, then

$$1 = \int w^2 d\mu \leq (4\pi r^2)^{-n/2} e^{-c} \operatorname{Vol} B(p,r).$$

Furthermore, as $\phi(y) = 1$ for $0 \leq y \leq 1/2$ we get

$$\int_M w^2 d\mu \geq \int_{B(p,r/2)} w^2 d\mu = (4\pi r^2)^{-n/2} e^{-c} \operatorname{Vol} B(p,r/2). \qquad \square$$

We now estimate each of the terms in (11.10) separately.

Term 1. By letting $\psi(x) = \phi(d_g(x,p)/r)$ we note that $|\nabla \psi| \leq \frac{1}{r} \sup |\phi'| \leq 3/r$ and as the gradient of ψ is supported on $B(p,r)/B(p,r/2)$ we have by (11.11) that

$$\begin{aligned} 4r^2 \int_M |\nabla w|^2 d\mu &= 4r^2 \int_M (4\pi r^2)^{-n/2} e^{-c} |\nabla \psi|^2 d\mu \\ &\leq \frac{4r^2}{\operatorname{Vol} B(p,r/2)} \int_M |\nabla \psi|^2 d\mu \\ &\leq 36 \frac{\operatorname{Vol} B(p,r)}{\operatorname{Vol} B(p,r/2)}. \end{aligned}$$

Term 2. As $\phi \equiv 0$ outside $B(p,r)$ then (11.11) implies that

$$\begin{aligned} r^2 \int_M \operatorname{Scal} w^2 d\mu &= r^2 \int_M \operatorname{Scal} (4\pi r^2)^{-n/2} e^{-c} \psi^2 d\mu \\ &\leq \frac{r^2}{\operatorname{Vol} B(p,r)} \int_{B(p,r)} |\operatorname{Scal}| d\mu. \end{aligned}$$

Term 3. As the support $\operatorname{supp} w \subset B(p,r)$ and $\log(4\pi)^{-n/2} < 0$ we have that

$$\begin{aligned} \int_M f w^2 d\mu &= \int_M \left(-\frac{n}{2} \log(4\pi r^2) - \log w^2\right) w^2 d\mu \\ &= \log(4\pi r^2)^{-n/2} \int_M w^2 d\mu - \int_M w^2 \log w^2 d\mu \\ &= \log(4\pi)^{-n/2} + \log r^{-n} - \int_{B(p,r)} w^2 \log^2 d\mu \\ &\leq \log r^{-n} + \log \operatorname{Vol} B(p,r) \end{aligned}$$

where the last line follows from Jensen's inequality.[3] □

Motivated by the right-hand side of (11.9), we define

$$\nu_r(g) = \inf_{\tau \in (0,r^2]} \mu(g,\tau)$$

$$M_R(p,r) = \sup_{0<s\leq r} s^2 \fint_{B(p,s)} |\mathrm{Scal}| d\mu.$$

So by (11.9) we have that

$$\frac{\mathrm{Vol}\, B(p,s)}{s^n} \geq e^{\mu(g,s^2)} \exp\left(-(36 + M_R(p,s)) \frac{\mathrm{Vol}\, B(p,s)}{\mathrm{Vol} B(p,s/2)} \right). \tag{11.12}$$

Remark 11.16. Since $s^2 \fint_{B(p,s)} |\mathrm{Scal}| d\mu \to 0$ as $s \to 0$, the quantity M_R is a well defined finite number for all $r > 0$. Trivially, if $r_1 \leq r_2$ we get $M_R(p, r_1) \leq M_R(p, r_2)$.

We now look to bound the volume ratios by ν_r and M_R.

Corollary 11.17. *If (M, g) is a closed manifold and $0 < s \leq r$ then*

$$\frac{\mathrm{Vol}\, B(p,s)}{s^n} \geq e^{\nu_r(g)} \exp\big(-3^n(36 + M_R(p,r))\big).$$

Proof. Firstly if $\frac{\mathrm{Vol}\, B(p,s)}{\mathrm{Vol} B(p,s/2)} \leq 3^n$, that is if at the point p the volume doubling property holds at scale s. Then by (11.12) and $\nu_r(g) \leq \mu(g, s^2)$, the desired estimate follows.

Now suppose $\frac{\mathrm{Vol}\, B(p,s)}{\mathrm{Vol} B(p,s/2)} \geq 3^n$, then it is clear that

$$\frac{\mathrm{Vol}\, B(p, s/2^k)}{\mathrm{Vol} B(p, s/2^{k+1})} \to 2^n$$

as $k \to \infty$.[4] So there is a $k > 0$ such that $\frac{\mathrm{Vol}\, B(p,s/2^k)}{\mathrm{Vol}\, B(p,s/2^{k+1})} \leq 3^n$. However $\frac{\mathrm{Vol}\, B(p,s/2^k)}{\mathrm{Vol}\, B(p,s/2^{k+1})} > 3^n$ for all $0 \leq i < k$ (i.e. it is the first such k where ratio

---

[3] Which states that if on a manifold N, φ is s convex function on $\mathbb{R}$ and $v \in L^1(N)$ then

$$\fint_N \varphi \circ v \, d\mu \geq \varphi\Big(\fint_N v \, d\mu \Big).$$

In particular, if $\varphi(x) = x \log x$ and $v \geq 0$ with $\int_N v \, d\mu = 1$ then

$$\int_N v \log v \, d\mu \geq -\log \mathrm{Vol}\, N.$$

[4] This is intuitively clear since small balls will, up to first order, have volumes close to that of balls of $\mathbb{R}^n$ (cf. Theorem 2.65).

below 3^n). Now using Proposition 11.13 with radius of $\frac{s}{2^k}$ we find that

$$\mu(g,(s/2^k)^2) \leq \log \frac{\operatorname{Vol} B(p,s/2^k)}{\left(\frac{s}{2^k}\right)^n} + 3^n(36 + M_R(p,s/2^k)).$$

Hence we find that

$$\begin{aligned}\frac{\operatorname{Vol} B(p,s)}{s^n} &\geq \left(\frac{3}{2}\right)^n \frac{\operatorname{Vol} B(p,s/2)}{(\frac{s}{2})^n} \\ &\geq \left(\frac{3}{2}\right)^{nk} \frac{\operatorname{Vol} B(p,s/2^k)}{\left(\frac{s}{2^k}\right)^n} \\ &\geq \left(\frac{3}{2}\right)^{nk} e^{\nu_r(g)} e^{-3^n(36+M_R(p,r))}\end{aligned}$$

from which the result now follows. □

Corollary 11.18. *Let (M,g) be a closed manifold with $0 < s \leq r$. If* $\operatorname{Scal} \leq K(n)r^{-2}$ *in $B(p,r)$, then $M_R(p,r) \leq K(n)$ so that*

$$\frac{\operatorname{Vol} B(p,s)}{s^n} \geq e^{\nu_r(g)} \exp\big(-3^n(36 + K(n))\big). \tag{11.13}$$

We are now in a position to prove the following local noncollapsing theorem:

Theorem 11.19 (Local Noncollapsing). *Let $(M, g(t))$, for $t \in [0,T)$, be a solution to the Ricci flow on a closed manifold with $T < \infty$ and let $\rho \in (0,\infty)$. There exists a constant $\kappa = \kappa(n, g(0), T, \rho) > 0$ such that for $p \in M$, $t \in [0,T)$ and $r \in (0,\rho]$ with*

$$\operatorname{Scal} \leq \frac{1}{r^2}$$

in $B_{g(t)}(p,r)$, the volume ratio

$$\frac{\operatorname{Vol}_{g(t)} B_{g(t)}(p,s)}{s^n} \geq \kappa$$

for all $0 < s < r$.

Proof. By (11.8) we have

$$\nu_{\sqrt{\rho^2+T}}(g(0)) \leq \nu_r(g(t))$$

for $r \in (0,\rho]$ and $t \in [0,T)$. So by (11.13) the result follows. □

11.4.1 Local Noncollapsing Implies Injectivity Radius Bounds

From Theorem 11.19 we can now deduce a positive lower bound on the injectivity radius. This will complete our discussion on the blowing up at singularities that started in Sect. 9.5 of Chap. 9. The main theorem is the following:

Theorem 11.20. *There exists $\rho > 0$ and $K = K(n) > 0$ such that if (M, g) is a closed Riemannan manifold satisfying $|R| \leq 1$ then there exists $p \in M$ such that*

$$\frac{\operatorname{Vol} B(p,r)}{r^n} \leq \frac{K}{r} \operatorname{inj}(M)$$

for all $r \in (0, \rho]$.[5]

In the work of Perelman [Per02] this point was considered too obvious to warrant mention. Indeed the idea is rather standard in comparison geometry: For example, predecessors to Hamilton's compactness theorem, such as the results of Greene–Wu [GW88] and Peters [Pet87], had no explicit assumption on injectivity radius, instead assuming an upper bound on diameter and a lower bound on volume. They then inferred a lower bound on injectivity radius from a result of Cheeger [Che70, Corollary 2.2], which is similar to the above result but somewhat trickier since it uses a global rather than local lower bound on volume. Here we present an argument similar to that in [Top05, Lemma 8.4.1], which is itself an adaptation of the original argument of Cheeger (see also [HK78], where a similar result is proved by a quite different argument). We give a sketch of the main argument only.

We need only consider the case where the injectivity radius is small (on the scale of the curvature), since otherwise there is nothing to prove. In such a setting one can apply the following lemma of Klingenberg:

Lemma 11.21. *If M is a compact Riemannian manifold with sectional curvature satisfying $\operatorname{sect}(M) \leq 1$ and $\operatorname{inj}(M) \leq \pi$ then there exists a closed unit speed geodesic $\gamma : \mathbb{R}/(\lambda\mathbb{Z}) \to M$ with $\lambda = 2\operatorname{inj}(M)$.*

Now we show that for any point p in γ, the volume of $B(p, r)$ is small if the injectivity radius is small. We observe that $\operatorname{Vol} B(p, r) \leq \operatorname{Vol} B(\gamma, r)$, where $B(\gamma, r)=\{q \in M : d(q, \gamma(s)) \leq r \text{ for some } s\}$. Now any point $q \in B(\gamma, r)$ can be reached by following a geodesic from a point of γ in an orthogonal direction for a distance at most r (see Fig. 11.1). Let $\{E_1(s), \ldots, E_{n-1}(s), E_n(s)=\gamma'(s)\}$ be

[5] Recall that the injectivity radius $\operatorname{inj}(p)$ of a point p is defined to be the supremum of all $r > 0$ such that $\exp_p$ is an embedding when restricted to $B_r(0)$ (i.e. $\operatorname{inj}(p) = \sup\{r > 0 : \exp_p$ is defined on $d_r(0) \subset T_pM$ and is injective$\}$), and that the injectivity radius of a Riemannian manifold M is $\operatorname{inj}(M) = \inf_{p \in M} \operatorname{inj}(p)$.

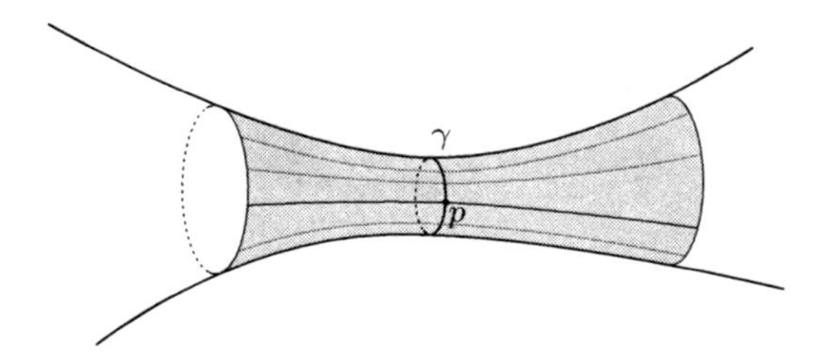

Fig. 11.1 Points within distance r of γ are obtained by following orthogonal geodesics from γ for distance at most r

an orthonormal frame for $T_{\gamma(s)}M$ obtained by parallel transport, for $0 \le s < \lambda$. Then define $f:\ [0,\lambda)\times B_r^{n-1}(0)\to M$ by

$$f(s,V^1,\ldots,V^{n-1}) = \exp_{\gamma(s)}\left(\sum_{i=1}^{n-1} V^i E_i(s)\right).$$

That is, $f(s,V)$ is obtained by following the geodesic from $\gamma(s)$ in direction $V^i E_i(s)$ orthogonal to $\gamma'(s)$ for time 1, and so travelling distance less than r. The map f is a smooth, so we can write by the change of variables formula

$$\operatorname{Vol} B(\gamma,r) \le \int_{[0,\lambda)\times B_r^{n-1}(0)} |\det Df|.$$

The Jacobian Df can be expressed in terms of Jacobi fields, the behaviour of which is controlled once curvature is controlled.[6] In particular, for r small compared to 1 we have $|\det DF|$ comparable to 1, so the right-hand side is comparable to $\lambda \operatorname{Vol}(B_r^{n-1}(0)) = c_n \lambda r^{n-1}$. Dividing through by r^n implies that $\operatorname{Vol} B(p,r)/r^n \le C_n\lambda/r$, for some $C_n > 0$, which gives the result.

11.5 The Blow-Up of Singularities and Local Noncollapsing

We are now in a position to complete the blow-up of singularities result by using the local noncollapsing theorem to obtain the desired injectivity bound discussed in Sect. 9.5. We follow a similar discussion presented in [Top05, Sect. 8.5].

As before, suppose there is a solution $(M, g(t))$ of the Ricci flow on a maximal time interval $t \in [0,T)$ with $T < \infty$. By Theorem 8.4 there exists

[6] The estimate required is provided by the Rauch comparison theorem, see for example [dC92, Theorem 2.3].

points $O_i \in M$ and times $t_i \nearrow T$ such that

$$|R|(O_i, t_i) = \sup_{M \times [0,t_i]} |R|(x,t).$$

With this sequence, define the blow-up metrics

$$g_i(t) = Q_i g(t_i + Q_i^{-1} t),$$

where $Q_i = |R|(O_i, t_i)$. As discussed in Sect. 9.5, for all $a < 0$ and some $b = b(n) > 0$, the curvature of g_i is bounded. In particular,

$$\sup_{M \times (a,b)} |\mathrm{Scal}(g_i(t))| < C$$

for sufficiently large $i = i(a)$ and some constant $C = C(n) < \infty$. Now if $0 < r < 1/\sqrt{C}$, then $|\mathrm{Scal}(g_i(0))| \leq r^{-2}$. So if $0 < r \leq 1/\sqrt{CQ_i}$, then $|\mathrm{Scal}(g(t_i))| \leq r^{-2}$.

By Theorem 11.19 for all $p \in M$, $0 < r \leq 1/\sqrt{CQ_i}$ and sufficiently large i, there is a lower bound

$$\frac{\mathrm{Vol}\, B_{g(t_i)}(p,r)}{r^n} > \kappa$$

where $\kappa = \kappa(n, g(0), T) > 0$. Returning to the blow-up flows g_i, we find for all $p \in M$ that

$$\frac{\mathrm{Vol}\, B_{g_i(0)}(p,r)}{r^n} > \kappa$$

for all $0 < r \leq 1/\sqrt{C}$. So by Theorem 11.20, fix $r = \min\{\frac{1}{\sqrt{C}}, \rho\} > 0$ so that

$$\mathrm{inj}(M, g_i(0)) \geq \frac{r}{K} \frac{\mathrm{Vol}\, B_{g_i(0)}(p,r)}{r^n} > \frac{\kappa}{K} r > 0.$$

In particular $\kappa r/K$ is a positive bound depending only on n, $g(0)$ and T. Therefore the compactness theorem (i.e. Theorem 9.11) gives the following:

Theorem 11.22 (Blow-up of Singularities). *Suppose $(M, g(t))$ is a solution to the Ricci flow on a maximal time interval $[0,T)$ with finite final time $T < \infty$. Then there exist sequences $O_i \in M$ and $t_i \nearrow T$ with*

$$|R|(O_i, t_i) = \sup_{M \times [0,t_i]} |R|(x,t) \to \infty$$

such that by defining $g_i(t) := Q_i g(t_i + Q_i^{-1} t)$, where $Q_i = |R|(O_i, t_i)$, there exists $b = b(n) > 0$ and a complete Ricci flow $(M_\infty, g_\infty(t))$, for $t \in (-\infty, b)$, and a point O_∞ such that

$$(M, g_i(t), O_i) \longrightarrow (M_\infty, g_\infty(t), O_\infty)$$

as $i \to \infty$. Moreover $|R(g_\infty(0))|(O_\infty) = 1$ and $|R(g_\infty(t))| \leq 1$ for $t \leq 0$.

11.6 Remarks Concerning Perel'man's Motivation From Physics

It is remarked by Perel'man [Per02, p. 3] that:

> 'The Ricci flow has also been discussed in quantum field theory, as an approximation to the renormalisation group (RG) flow for the two-dimensional nonlinear σ-model... this connection between the Ricci flow and the RG flow suggests that Ricci flow must be gradient-like; the present work confirms this expectation.'

In this section we expand on this comment. By doing so we see, at least in part, where the motivation for the $\mathcal{F}$ and $\mathcal{W}$ functionals originate from.

As [Gro99] explains, the method of renormalisation group is a method designed to describe how dynamics of a physical system change as the scale (i.e. distance or energies) at which we probe changes.

Physics is heavily scale dependent, for instance in fluid dynamics each scale of distance has a different theory. For instance at length scales of 1 cm, classical continuous mechanics (Navier–Stokes equations) is appropriate. However at length scales 10^{-13}–10^{-18} cm quantum chromodynamics (quarks) is a more suitable theory.

Moreover, physics at larger scales decouples from the physics at a smaller scale. The theory at a larger scale remembers only finitely many parameters from the theories at smaller scales and throws the rest of the details away. Passing from smaller scales to larger scales involves averaging over irrelevant degrees of freedom. For instance, it is possible to reconstruct the thermodynamics of a gas from molecular theory (by averaging over all configurations) but it is impossible to reconstruct the behaviour of molecules from the macroscopic behaviour of the gas its self. This decoupling is the reason why we are able to do physics. The aim of renormalisation group method is to explain how this decoupling takes place and why exactly information is transmitted from scale to scale through finitely many parameters.

In the context of particle physics, renormalisation is a process of making sense of ultraviolet and infrared divergences arising in the Feynman diagram integrals. It is done in a two stage process. The first step is *regularisation*. It consists in introducing a cut-off which makes the integrals converge but depend on the cut-off. There are many types of cut-off schemes, for instance momentum cut-off, dimensional regularisation and so on. The integrals will usually tend to infinity as the cut-off goes to infinity. Because of this we need a second step called *renormalisation*. This step consists in making finitely many parameters of the Lagrangian depend on the regularisation, so that they go to infinity as the regularisation goes to infinity but all the correlation functions remain finite.

The behaviour of a quantum field theory under renormalisation is governed by the so-called β-function. Formally, it is a vector field on the space of dimensionless parameters for a particular quantum theory. The differential equation associated to this vector field is called the *renormalisation group equation*. Fixed points of this equation correspond the ultraviolet and infrared limits of the theory. In the case of a single coupling constant g, the vector field generating the renormalisation flow is written as

$$\mu \frac{\partial}{\partial \mu} + \beta(g) \frac{\partial}{\partial g}$$

where μ is the scale parameter.[7] When there are n such couplings $g_1, \ldots, g_n$, the renormalisation group equation for g_i takes the form

$$\mu \frac{dg_i}{\partial \mu} = \beta_i(g_1, \ldots, g_n). \tag{11.14}$$

In general we know nothing about the flow except that the point $g_i = 0$ (the free theory) is a fixed point of this flow. It is conjectured that topologically, the flow behaves as a gradient flow.

Conjecture 11.23. There exists a function $F(g_1, \ldots, g_n)$ such that its total derivative with respect to μ is positive at any nonsingular point of the flow (11.14).[8]

The conjecture prohibits various complicated dynamical patterns which could occur in a multi-dimensional dynamical systems. Thus it implies that any trajectory is either driven to a fixed point or to infinity.

According to [Gro99, p. 578], the intuition behind the conjecture is the Wilsonian point of view that the (backwards) renormalisation group flow comes from erasing degrees of freedom. The meaning of F is 'the measure of the number of degrees of freedom'. Unfortunately it is unknown how to make mathematical sense of this idea in general but is proved in special case 2-dimensional field theories.

For instance, in the case of 2-dimensional Bosonic non-linear σ-models, we are interested in mappings $\phi = \phi(\sigma, \tau) : \Sigma \to M$ where Σ is a 2-dimensional Riemann surface with metric h_{ab} and M is a D-dimensional closed Riemannian manifold with metric $g_{\mu\nu}$. The action S is of the form

---

[7] A coupling constant is a coefficient in a Lagrangian that measures the strength of a particular interaction among elementary fields.

[8] By taking a logarithmic scale $\lambda = \log \mu$, (11.14) can be re-written as

$$\frac{dg_i}{\partial \lambda} = \beta_i(g_1, \ldots, g_n).$$

The choice of the scale parameter μ is just a convention.

$$S = \frac{1}{4\pi\alpha'} \int_\Sigma \Big(g_{\mu\nu}(\phi) \frac{\partial\phi^\mu}{\partial x^a} \frac{\partial\phi^\nu}{\partial x^b} + \epsilon^{ab} B_{\mu\nu}(\phi) \frac{\partial\phi^\mu}{\partial x^a} \frac{\partial\phi^\nu}{\partial x^b} + \alpha' \Phi(\phi) R^{(2)} \Big) h^{ab} \sqrt{h}\, d\sigma d\tau$$

where $R^{(2)}$ is the world-sheet Ricci scalar with respect to h_{ab}, Φ is the so-called dilaton field, $B_{\mu\nu}$ is a 2-form antisymmetic tensor field and ϵ^{ab} is the antisymmetric tensor (taking values $\epsilon^{01} = -\epsilon^{10} = 1$ and $\epsilon^{00} = \epsilon^{11} = 0$). The constant α' is needed to make action dimensionless. In this case the couplings for S are $g_{\mu\nu}$, $B_{\mu\nu}$ and Φ. To first order the respective β-functions (see [GSW88, Sect. 3.4] and [Pol98, Sect. 3.7]) are

$$\begin{aligned}
\beta^g_{\mu\nu} &= -\alpha'(R_{\mu\nu} + 2\nabla_\mu\nabla_\nu\Phi) + \frac{\alpha'}{4} H_\mu{}^{\lambda\rho} H_{\nu\lambda\rho} + O(\alpha'^2) \\
\beta^B_{\mu\nu} &= \alpha' \left(\frac{1}{2} \nabla_\lambda H^\lambda{}_{\mu\nu} - H^\lambda{}_{\mu\nu} \nabla_\lambda \Phi \right) + O(\alpha'^2) \\
\beta^\Phi_{\mu\nu} &= \frac{26 - D}{6} + \alpha' \left(\frac{1}{2} \nabla^2\Phi - \nabla_\mu\Phi\nabla^\mu\Phi + \frac{1}{24} H_{\mu\nu\rho} H^{\mu\nu\rho} \right) + O(\alpha'^2)
\end{aligned}$$

where $H_{\mu\nu\rho} = \partial_\mu B_{\nu\rho} + \partial_\rho B_{\mu\nu} + \partial_\mu B_{\rho\mu}$ or simply $H = dB$.

From this system of couplings, the formulas (10.5) and (11.3) seem to be strongly motivated by this example. Furthermore, Perel'man's comments indicate his emphasis on the gradient flow approach (even though it is not used strongly in many of the key arguments in his work on the geometrisation conjecture, as the more robust reduced volume functional is employed instead). For further discussion on the connection between Perel'man's work and renormalisation group (RG) flows see [OSW06, Car10].

Chapter 12
An Algebraic Identity for Curvature Operators

In this chapter and the next we look at one of the most important recent developments in the theory of Ricci flow: The work of Böhm and Wilking [BW08] which gives a method for producing whole families of preserved convex sets for the Ricci flow from a given one. This remarkable new method has broken through what was an enormous barrier to further applications of Ricci flow: In particular the proof of the differentiable sphere theorem relies heavily on this work.

The problem to be dealt with is the following: There are numerous examples of convex cones in the space of curvature tensors which are known to be preserved by the Ricci flow. Examples which have been known for a long time include the cone of positive curvature operators, and the cone of 2-positive curvature operators. We will see other examples in Chap. 14, such as the cone defined by the positive isotropic curvature condition. One would like to be able to find a whole family of preserved convex cones interpolating between the given cone and the 'degenerate' cone consisting of constant positive sectional curvature operators. If this can be done, then one can argue using the maximum principle (Theorem 7.15) that solutions of the Ricci flow which have their curvature in the given cone at the initial time will evolve to have constant curvature as they approach their maximal time of existence. In the next chapter we will give the details of the construction of the family of cones and the argument required for their application (see Sect. 7.5.3 for a discussion of how this works in the simple case of Ricci flow for 3-manifolds).

In the present chapter we discuss a fundamental identity, Theorem 12.33, due to Böhm and Wilking. It is the basis for their cone construction and is the result of a detailed and delicate analysis of how the various parts of the curvature tensor (first discussed in Sect. 4.5) combine to produce the reaction terms in the evolution equation for curvature seen in Theorem 6.21.

B. Andrews and C. Hopper, *The Ricci Flow in Riemannian Geometry*, Lecture Notes in Mathematics 2011, DOI 10.1007/978-3-642-16286-2_12,

12.1 A Closer Look at Tensor Bundles

We begin with an examination of the framework needed to apply the vector bundle maximum principle to the evolution equation for the curvature tensor.

In Sect. 7.5.3 we discussed the construction of subsets of the bundle of symmetric 2-tensors which are convex in the fibre and invariant under parallel transport, from any $O(n)$-invariant convex subset of the vector space of symmetric $n \times n$ matrices. In that situation further simplification is possible since symmetric matrices are diagonalisable, and the question reduces to convex subsets of $\mathbb{R}^n$ invariant under interchange of coordinates.

In higher dimensions the curvature tensor is no longer simply a symmetric 2-tensor, but lives in a vector bundle with considerably more complicated structure. We will investigate this structure in detail in the next section, however before doing so we discuss the construction of suitable subsets of vector bundles in a more general context.

12.1.1 Invariant Tensor Bundles

In three dimensions we applied the vector bundle maximum principle on the bundle of symmetric 2-tensors of the spatial tangent bundle $\mathfrak{S}$, which we know is a parallel subbundle of the bundle of 2-tensors, and so (by the arguments of Sect. 2.5.3) inherits a canonical metric and compatible connection. In order to carry out this step in higher dimensions we consider an important method of constructing parallel subbundles of tensor bundles.

Suppose we have a vector bundle E over a manifold M, with a metric g and compatible connection ∇ supplied. Then the construction given in Sect. 2.5.3 provides metrics and compatible connections on each of the tensor bundles $T^p_q(E)$. Let $O(E)$ be the orthonormal frame bundle of E, so that the fibre of $O(E)$ at a point $p \in M$ consists of the isometries from $\mathbb{R}^k$ (with the standard inner product) to E_p (with inner product g_p). Recall that $O(k)$ acts on $O(E)$ by

$$O(k) \times O(E) \to O(E);\ (\mathbb{O}, Y) \mapsto Y^{\mathbb{O}},$$

where $Y^{\mathbb{O}}(u) = Y(\mathbb{O}u)$ for each $u \in \mathbb{R}^k$. For each frame $Y \in O(E)_x$ there exists a dual frame Y_* for E^*_x, defined by $(Y_*(\omega))(Y(u)) = \omega(u)$ for all $\omega \in (\mathbb{R}^k)^*$ and $u \in \mathbb{R}^k$. There is a corresponding action of $O(k)$ on the dual frames, defined by $(Y_*)^{\mathbb{O}} = (Y^{\mathbb{O}})_*$.

As noted in Sect. 6.2.1, a (p,q)-tensor T at x gives rise to a function $\tilde{T}$ from $O(E)_x$ to $T^p_q(\mathbb{R}^k)$, defined by

$$(\tilde{T}(Y))(v_1, \dots, v_p, \omega_1, \dots, \omega_q) = T(Y(v_1), \dots, Y(v_p), Y_*(\omega_1), \dots, Y_*(\omega_q))$$

for any $v_1, \ldots, v_p \in \mathbb{R}^k$ and $\omega_1, \ldots, \omega_q \in (\mathbb{R}^k)^*$. The function $\tilde{T}$ is equivariant with respect to the action of $O(k)$ on $T^p_q(\mathbb{R}^k)$ defined by

$$O(k) \times T^p_q(\mathbb{R}^k) \to T^p_q(\mathbb{R}^k); \ (\mathbb{O}, T) \mapsto T^{\mathbb{O}},$$

where $T^{\mathbb{O}}(v_1, \ldots, v_p, \omega_1, \ldots, \omega_q) = T(\mathbb{O}v_1, \ldots, \mathbb{O}v_p, \mathbb{O}^*\omega_1, \ldots, \mathbb{O}^*\omega_q)$ for all $v_1, \ldots, v_p \in \mathbb{R}^k$ and $\omega_1, \ldots, \omega_q \in (\mathbb{R}^k)^*$, and $\mathbb{O}^*$ is the dual transformation on $(\mathbb{R}^k)^*$ corresponding to $\mathbb{O}$, defined by $(\mathbb{O}^*\omega)(\mathbb{O}v) = \omega(v)$, for all ω and v. Here equivariance means that

$$\tilde{T}(Y^{\mathbb{O}}) = (\tilde{T}(Y))^{\mathbb{O}}$$

where the action of $\mathbb{O}$ on the left-hand side is on the frame $Y \in O(E)$, and that on the right-hand side is on $\tilde{T}(Y) \in T^p_q(\mathbb{R}^k)$.

An *invariant subset* K of $T^p_q(\mathbb{R}^k)$ is a subset which is invariant under the action of $O(k)$ – in the sense that $T \in K$ implies that $T^{\mathbb{O}} \in K$ for every $\mathbb{O} \in O(k)$. In particular, a vector subspace V of $T^p_q(\mathbb{R}^k)$ which is invariant is called an *invariant subspace.*[1] From any invariant subset we can construct a subset $K(E)$ of $T^p_q(E)$ as follows:

$$K(E) \cap T^p_q(E)_x = \{T \in T^p_q(E)_x : \ \forall Y \in O(E)_x, \ \tilde{T}(Y) \in K\}.$$

In particular from an invariant subspace V we construct a subbundle $V(E)$ of $T^p_q(E)$, by

$$V(E)_x = \{T \in T^p_q(E)_x : \ \forall Y \in O(E)_x, \ \tilde{T}(Y) \in V\}.$$

We call $V(E)$ the *invariant subbundle* corresponding to the invariant subspace V. The main result we need is the following:

Proposition 12.1. *If K is an invariant subset of $T^p_q(\mathbb{R}^k)$, then the subset $K(E)$ of $T^p_q(E)$ is invariant under parallel transport.*

Proof. The argument is essentially the same as that used in Sect. 7.5.3.1 for the case of the 2-tensor bundle. Let $\sigma : \ I \to M$ be a smooth curve, and let T be a parallel section of $\sigma^*(T^p_q(E))$ such that $T_0 \in K(E)_{\sigma(0)}$. That is, for any frame $Y_0 \in O(E)_{\sigma(0)}$, we have $\tilde{T}_0(Y_0) \in K$. Fix any such frame, and define a frame $Y_s \in F(E)_{\sigma(s)}$ for each $s \in I$ by parallel transport, i.e. for each $u \in \mathbb{R}^k$,

$$^{\sigma}\nabla_{\partial_s}(T_s(u)) = 0.$$

[1] Note that the action of $O(k)$ on $T^p_q(\mathbb{R}^k)$ defined above is an orthogonal representation, i.e. a Lie group homomorphism from $O(k)$ to $O(T^p_q(\mathbb{R}^k))$. Representation theory tells us that $T^p_q(\mathbb{R}^k)$ decomposes as an orthogonal direct sum of irreducible invariant subspaces. An arbitrary invariant subspace can be written as a direct sum of these irreducible ones.

Then $Y_s \in O(E)_{\sigma(s)}$ since the compatibility of ∇ gives

$$\frac{\partial}{\partial s} g(Y_s(u), Y_s(v)) = g({}^\sigma\nabla_{\partial_s}(Y_s(u)), Y_s(v)) + g(Y_s(u), {}^\sigma\nabla_{\partial_s}(Y_s(v))) = 0.$$

Note that the dual frame Y_s^* is also parallel by definition of the dual connection. But we also have, for all $v_1, \ldots, v_p \in \mathbb{R}^k$ and $\omega_1, \ldots, \omega_q \in (\mathbb{R}^k)^*$,

$$\begin{aligned} &\frac{\partial}{\partial s}\left(\tilde{T}_s(Y_s)\right)(v_1, \ldots, v_p, \omega_1, \ldots, \omega_q) \\ &\quad = \frac{\partial}{\partial s}\left(T_s\left(Y_s(v_1), \ldots, Y_s(v_k), Y_s^*(\omega_1), \ldots, Y_s^*(\omega_q)\right)\right) \\ &\quad = 0, \end{aligned}$$

since T_s, $Y_s(v_i)$ and $Y_s^*(\omega_j)$ are all parallel. Therefore $\tilde{T}_s(Y_s)$ is constant, and so is in K for all $s \in I$, and so $T_s \in K(E)$ for all s. □

The particular case where K is a subspace immediately gives the following:

Corollary 12.2. *Invariant subspaces of $T_q^p(E)$ are parallel.*

It follows from the results of Sect. 2.8.7 that invariant subbundles inherit metrics and compatible connections from the tensor bundle.

Example 12.3. There are no nontrivial invariant subbundles of E itself: Given any non-zero vector v in $\mathbb{R}^k$, there is an $O(k)$ element which takes v to any other given vector of the same length, so the linear span of the orbit of v under the action is all of $\mathbb{R}^k$.

Example 12.4. Since the metric tensor is invariant, the operations of raising and lowering indices are also, and the invariant subspaces of $T_q^p(\mathbb{R}^k)$ are in one-to-one correspondence with those of $T_0^{p+q}(\mathbb{R}^k)$. In particular, E^* has no nontrivial invariant subspaces.

Example 12.5. The invariant subbundles of the bundle of 2-tensors on E are given by finding the invariant subspaces of the action of $O(n)$ on $T_0^2(\mathbb{R}^k)$ (i.e. the space of $k \times k$ matrices) with the action $(\mathbb{O}, M) \mapsto \mathbb{O}^T M \mathbb{O}$. We have already found one invariant subspace, namely the subspace of symmetric matrices: In this case $\left(\mathbb{O}^T M \mathbb{O}\right)^T = \mathbb{O}^T M^T \left(\mathbb{O}^T\right)^T = \mathbb{O}^T M \mathbb{O}$ if $M = M^T$. Similarly the subspace of antisymmetric matrices is preserved.[2] Another invariant subspace is the trace-free matrices, $\{M : \operatorname{tr}(M) = 0\}$, since $\operatorname{tr}\left(\mathbb{O}^T M \mathbb{O}\right) = \operatorname{tr}\left(M \mathbb{O}\mathbb{O}^T\right) = \operatorname{tr}(M)$ for $\mathbb{O} \in O(k)$. And finally, there is the subspace consisting of multiples of the identity matrix, which is invariant

[2] We get this for free: The orthogonal complement of an invariant subspace is always invariant.

since $\mathbb{O}^T I_k \mathbb{O} = \mathbb{O}^T \mathbb{O} = I_k$. This gives a decomposition of $T_0^2(\mathbb{R}^k)$ as the following direct sum of invariant subspaces:

$$T_0^2(\mathbb{R}^k) = \bigwedge^2(\mathbb{R}^k) \oplus \mathrm{Sym}_0^2(\mathbb{R}^k) \oplus \mathbb{R} I_k$$

where $\bigwedge^2(\mathbb{R}^k)$ is the antisymmetric matrices, and $\mathrm{Sym}_0^2(\mathbb{R}^k)$ is given by the intersection of the symmetric matrices with the traceless ones.

12.1.2 Constructing Subsets in Invariant Subbundles

We now have a method constructing subsets of invariant subspaces which are invariant under parallel transport, since if K is an invariant subset of $T_q^p(\mathbb{R}^k)$ which is contained in an invariant subspace V, then by Proposition 12.1 $K(E)$ is a subset of $V(E)$ which is invariant under parallel transport. Furthermore, if K is also convex then the subset $K(E)$ is convex in the fibres of $K(V)$, since any frame $Y \in O(E)_x$ gives a linear isomorphism of $V(E)_x$ to V which takes $K(E)_x$ to K, and linear isomorphisms preserve convexity.

Unlike the situation discussed in Sect. 7.5.3, the action of $O(k)$ on higher tensor spaces does not allow reduction to a simple form such as the diagonal matrices, so the construction of invariant convex subsets of $T_q^p(\mathbb{R}^k)$ is not as straightforward. However, there is a recipe which will prove quite convenient and which fits in nicely with our formulation of the maximum principle: Let $c \in \mathbb{R}$ and let ℓ be an arbitrary nontrivial linear function on $T_q^p(\mathbb{R}^k)$ (that is, an element of the dual space $\left(T_q^p\right)^* \simeq T_p^q(\mathbb{R}^k)$). Then we can construct an invariant convex set $K_{\ell,c}$ in $T_q^p(\mathbb{R}^k)$ as follows:

$$K_{\ell,c} = \bigcap_{\mathbb{O} \in O(k)} \left\{T \in T_q^p(\mathbb{R}^k) : \ \ell(T^{\mathbb{O}}) \le c\right\}.$$

This is an intersection of half-spaces, hence convex, and is invariant by construction. An arbitrary invariant convex set may be constructed by taking intersections of sets of this form.

Example 12.6. The cone of positive definite matrices in $\mathrm{Sym}^2(\mathbb{R}^k)$ can be expressed in exactly the form above: Define $\ell(T) := -T(e_1, e_1)$, and choose $c = 0$. Then we have

$$\begin{aligned} K_{\ell,c} &= \bigcap_{\mathbb{O} \in O(k)} \left\{T \in \mathrm{Sym}^2(\mathbb{R}^k) : \ T(\mathbb{O}e_1, \mathbb{O}e_1) \ge 0\right\} \\ &= \left\{T \in \mathrm{Sym}^2(\mathbb{R}^k) : \ \forall v \in \mathbb{R}^k, \ T(v,v) \ge 0\right\}. \end{aligned}$$

The other cones constructed in Sect. 7.5.3 may be similarly described: The cone of positive Ricci curvature (for $k=3$) corresponds to $\ell(T)=-T(e_1,e_1)$ $-T(e_2,e_2)$ and $c=0$; and the cones of pinched Ricci curvature correspond to $\ell(T)=\varepsilon T(e_3,e_3)-T(e_1,e_1)-T(e_2,e_2)$ and $c=0$.

12.1.3 Checking that the Vector Field Points into the Set

To apply the maximum principle there is one further ingredient required: We must check that the vector field in the reaction-diffusion equation points into the set. We will discuss the specific situation corresponding to Ricci flow in the next section, but here we discuss more generally a situation where checking this condition can be reduced to a rather concrete question.

Let E be a vector bundle over $M\times\mathbb{R}$ with a given metric and compatible connection. Let u be a section of an invariant tensor bundle $V(E)$ which satisfies the reaction-diffusion equation of the form

$$\nabla_{\partial_t}u=\Delta u+F(u).$$

Recall that for any orthonormal frame $Y\in O(E)$, we have a map which take $T\in V(E)$ to $\tilde{T}(Y)$ in the invariant subspace V. Assume that there exists a vector field $\psi:\ V\to V$, equivariant in the sense that $(\psi(T))^{\mathbb{O}}=\psi(T^{\mathbb{O}})$ for all $T\in V$ and $\mathbb{O}\in O(k)$. We require that the vector field F maps to ψ, in the sense that for any $u\in V(E)_{(x,t)}$ and $Y\in O(E)_{(x,t)}$,

$$\widetilde{F(u)}(Y)=\psi(\tilde{u}(Y)).$$

In this case it is straightforward to check the following:

Proposition 12.7. *Let K be a convex invariant subset of V, and $K(E)$ the corresponding subset of $V(E)$. Then the vector field F on $V(E)$ points (strictly) into $K(E)$ if and only if the vector field ψ on V points (strictly) into K.*

Proof. The definition of F pointing into K is given in Definition 7.12. Let s_K be the support function of K, so that

$$s_K(\ell)=\sup\{\ell(w):\ w\in K\}$$

for each $\ell\in V^*$. Also let s be the support function of $K(E)$, so that

$$s(x,t,\ell)=\sup\{\ell(w):\ w\in K(E)_{(x,t)}\}$$

for each $(x,t,\ell) \in V(E)^*$. These are related as follows: Fix $(x,t) \in M \times \mathbb{R}$, and let $Y \in F(E)_{(x,t)}$. Then for any $\ell \in V(E)^*_{(x,t)}$ we have $\tilde{\ell}(Y) \in V^*$, and

$$\begin{aligned} s(x,t,\ell) &= \sup\{\ell(w) : \ w \in K(E)_{(x,t)}\} \\ &= \sup\{(\tilde{\ell}(Y))(\tilde{w}(Y)) : \ \tilde{w}(Y) \in K\} \\ &= \sup\{(\tilde{\ell}(Y))(w) : \ w \in K\} \\ &= s_K(\tilde{\ell}(Y)). \end{aligned}$$

From the definitions of tangent and normal cones in Appendix B it follows that

$$\mathcal{N}_{\tilde{v}(Y)}K = \{\tilde{\ell}(Y) : \ \ell \in \mathcal{N}_v K_{(x,t)}\}$$

and

$$\mathcal{T}_{\tilde{v}(Y)}K = \{\tilde{z}(Y) : \ z \in \mathcal{T}_v K_{(x,t)}\}.$$

It follows that $\psi(\tilde{v}(Y)) = (\widetilde{F(v)})(Y)$ is in (the interior of) $\mathcal{T}_{\tilde{v}(Y)}K$ if and only if $F(v)$ is in (the interior of) $\mathcal{T}_v K_{(x,t)}$, and the Proposition is proved. □

Remark 12.8. As we will see below, the evolution equation for the curvature tensor has exactly this property. This is a drastic simplification: To check that the vector field points into the set we now have only to consider an explicit vector field on a fixed finite-dimensional vector space, and ask whether it points into a given convex invariant set. In effect the maximum principle has removed the geometry from the analysis, and all that remains is a very concrete and rather algebraic question. This fact has been well understood since at least the early 1990s, so it is perhaps surprising that useful applications of these ideas in dimensions $n \geq 4$ have only been possible recently (following the appearance of [BW08]).

12.2 Algebraic Curvature Operators

The (spatial) curvature tensor can be considered a $(4,0)$-tensor on the spatial tangent bundle – that is, a section of the tensor space $T^4_0(\mathfrak{S})$. However, the symmetries of the curvature tensor mean that the curvature tensor in fact lies inside an invariant subbundle, which we call the bundle of algebraic curvature operators $\mathrm{Curv}(\mathfrak{S})$, corresponding to an invariant subspace Curv of the vector space $T^4_0(\mathbb{R}^n)$ defined to be the set of all $R \in T^4_0(\mathbb{R}^n)$ such that

$$\begin{aligned} R(u,v,w,z) + R(v,u,w,z) &= 0 \\ R(u,v,w,z) + R(u,v,z,w) &= 0 \\ R(u,v,w,z) - R(w,z,u,v) &= 0 \\ R(u,v,w,z) + R(v,w,u,z) + R(w,u,v,z) &= 0 \end{aligned}$$

for all $u, v, w, z \in \mathbb{R}^n$. One can check directly that this subspace is $O(n)$-invariant (in fact $GL(n)$-invariant).

Since elements of Curv are antisymmetric in the first and last pairs of arguments, and symmetric under interchange of the first pair with the last pair, it is natural to interpret them as symmetric bilinear forms acting on the $\frac{n(n+1)}{2}$-dimensional space $\bigwedge^2(\mathbb{R}^n)$, with each antisymmetric pair of arguments in $\mathbb{R}^n$ corresponding to a single $\bigwedge^2(\mathbb{R}^n)$-valued argument: That is, $\mathrm{Curv} \subset \mathrm{Sym}^2(\bigwedge^2 \mathbb{R}^n)$. Accordingly, the curvature tensor can be viewed as a self-adjoint linear endomorphism of $\bigwedge^2(\mathfrak{S})$, defined by the formula

$$g(R(x,y)u, v) = \langle R(x \wedge y), u \wedge v \rangle = R(x \wedge y, u \wedge v),$$

for all $x, y, u, v \in \mathfrak{S}_{(p,t)} \simeq \mathbb{R}^n$, where $\langle \cdot, \cdot \rangle$ is defined by (C.2).[3]

By letting $(e_i)_{i=1}^n$ denote an orthonormal basis for T_pM, the curvature operator R maps $e_i \wedge e_j$ to $\frac{1}{2}\sum_{p,q} A_{pq} e_p \wedge e_q$ under our adopted summation convention discussed in Appendix C. By definition $A_{k\ell} = \langle R(e_i \wedge e_j), e_k \wedge e_\ell \rangle = R_{ijk\ell}$, in which case the curvature operator can be written in the form

$$R(e_i \wedge e_j) = \sum_{k<\ell} R_{ijk\ell} e_k \wedge e_\ell \tag{12.1}$$

with respect to the orthonormal basis $(e_i \wedge e_j)_{i<j}$. Likewise one can express R when interpreted as a symmetric bilinear form, i.e. an element of $\mathrm{Sym}^2(\bigwedge^2 T_p^*M)$, by $R = \sum_{i<j,k<\ell} R_{ijk\ell}(e^i \wedge e^j) \otimes (e^k \wedge e^\ell)$.[4]

Now consider a different orthonormal basis $(\sigma_\alpha)_{\alpha=1}^N$ for $\bigwedge^2 T_pM$ with corresponding dual (φ^α). With respect to this basis, the curvature R (as a bilinear form) can be expressed as $R = R_{\alpha\beta}\varphi^\alpha \otimes \varphi^\beta$, where $R_{\alpha\beta} = R(\sigma_\alpha, \sigma_\beta)$. Since $\varphi^\alpha = \sum_{i<j} \varphi^\alpha_{ij} e^i \wedge e^j$ and $\sigma_\alpha = \sum_{i<j} \sigma^{ij}_\alpha e_i \wedge e_j$ we find the components of R, with respect to the two orthonormal bases, are related by

$$R_{ijk\ell} = R_{\alpha\beta}\varphi^\alpha_{ij}\varphi^\beta_{k\ell} \qquad \text{and} \qquad R_{\alpha\beta} = R_{ijk\ell}\sigma^{ij}_\alpha \sigma^{k\ell}_\beta.$$

12.2.1 Interpreting the Reaction Terms

We first observe that the evolution equation for the curvature tensor is exactly of the form described in Sect. 12.1.3, in the sense that the reaction terms are

[3] Note our definition here differs in sign convention from, say [Pet06, p. 36].

[4] Every symmetric bilinear form $\beta : V \times V \to \mathbb{R}$ defines a mapping $V \to V^*; v \mapsto (w \mapsto \beta(v,w))$ and visa versa. Here we have implicitly identified $V \simeq V^*$ to conform to our adopted summation convention.

described by an invariant vector field. Recall the evolution equation derived in Theorem 6.21:

$$\nabla_{\partial_t} R_{ijkl} = \Delta R_{ijkl} + 2(B_{ijkl} - B_{ijlk} + B_{ikjl} - B_{iljk}).$$

We seek to understand the quadratic reaction terms

$$Q_{ijk\ell} = B_{ijk\ell} - B_{ij\ell k} + B_{ikj\ell} - B_{i\ell jk}, \tag{12.2}$$

which we introduced in Chap. 4.[5] The first observation is that these terms are indeed invariant, since they are given simply by contracting with the metric. Thus according to Proposition 12.7 we need only consider the corresponding vector field Q on the invariant subspace Curv. By appealing to the inherent Lie algebra structure associated with the curvature operator we will rewrite these as the sum of two natural quadratic vector fields on Curv.

To achieve this formally, first identify $U \otimes U \simeq \mathfrak{gl}(U)$ where the isomorphism given by (C.1), with $U = \bigwedge^2 \mathbb{R}^n$. Then identify $\bigwedge^2 \mathbb{R}^n \simeq \mathfrak{so}(n)$ with the isomorphism given by (C.3). The quadratic curvature terms, referred to as the squared and sharp products respectively, are defined by:

$$\begin{aligned} R^2_{\alpha\beta} &= R_{\alpha\lambda} R_{\lambda\beta} \\ R^\#_{\alpha\beta} &= \frac{1}{2} c_\alpha^{\gamma\eta} c_\beta^{\delta\theta} R_{\gamma\delta} R_{\eta\theta} \end{aligned}$$

where $c_\alpha^{\beta\gamma}$ are the structure constants for $\mathfrak{so}(n)$ with respect to the basis (φ^α). One can easily show that if $R \geq 0$ then $R^2 \geq 0$ and $R^\# \geq 0$.

Remark 12.9. The sharp product is a contraction with the structure constants and the squared product is 'natural' in the sense that:

$$\begin{aligned} R^2 &= (R_{\alpha\beta} \varphi^\alpha \otimes \varphi^\beta)(R_{\gamma\delta} \varphi^\gamma \otimes \varphi^\delta) \\ &= R_{\alpha\beta} R_{\gamma\delta} \delta^{\beta\gamma} \varphi^\alpha \otimes \varphi^\delta \\ &= R_{\alpha\lambda} R_{\lambda\beta} \varphi^\alpha \otimes \varphi^\beta \end{aligned}$$

where the second equality is due to (C.1).

With these identifications, we find that the vector field Q on Curv can be written in the form $Q(R) = R^2 + R^\#$. This is an immediate consequence of the following two lemmas.

[5] Recall by (4.2) that $B_{ijk\ell} = R_{ipjq} R_{pkq\ell}$ where, in this algebraic context, we suppress the index raising.

Lemma 12.10. *For any* $R \in \mathrm{Curv}$,

$$R^2_{ijk\ell} = B_{ijk\ell} - B_{ij\ell k}.$$

Proof. Since $B_{ijk\ell} = B_{ji\ell k} = B_{k\ell ij}$ we find that

$$\begin{aligned} R_{ijpq}R_{k\ell pq} &= (R_{ipqj} + R_{iqjp})(R_{kpq\ell} + R_{kq\ell p}) \\ &= (R_{iqjp} - R_{ipjq})(R_{kq\ell p} - R_{kp\ell q}) \\ &= B_{\ell kji} - B_{jik\ell} - B_{ij\ell k} + B_{ijk\ell} \\ &= 2(B_{ijk\ell} - B_{ij\ell k}). \end{aligned}$$

In which case

$$\begin{aligned} \sum_{p,q} R_{ijpq}R_{pqk\ell} &= \sum_{p,q} R_{\alpha\beta}\varphi^\alpha_{ij}\varphi^\beta_{pq}R_{\gamma\delta}\varphi^\gamma_{pq}\varphi^\delta_{k\ell} \\ &= R_{\alpha\beta}R_{\gamma\delta}\varphi^\alpha_{ij}\varphi^\delta_{k\ell}\sum_{p,q}\varphi^\beta_{pq}\varphi^\gamma_{pq} \\ &= 2R_{\alpha\beta}R_{\gamma\delta}\varphi^\alpha_{ij}\varphi^\delta_{k\ell}\delta^{\beta\gamma} \\ &= 2R^2_{ijk\ell} \end{aligned}$$

since $\delta^{\beta\gamma} = \langle \varphi^\beta, \varphi^\gamma \rangle = \frac{1}{4}\varphi^\beta_{ij}\varphi^\gamma_{k\ell}\langle e^i \wedge e^j, e^k \wedge e^\ell \rangle = \frac{1}{2}\sum_{p,q}\varphi^\beta_{pq}\varphi^\gamma_{pq}$. □

Lemma 12.11. *For* $R \in \mathrm{Curv}$,

$$R^\#_{ijk\ell} = B_{ikj\ell} - B_{i\ell jk}.$$

Proof. From (C.5) we find that

$$\begin{aligned} B_{ikj\ell} - B_{i\ell jk} &= R_{ipkq}R_{jp\ell q} - R_{ip\ell q}R_{jpkq} \\ &= R_{\alpha\beta}\varphi^\alpha_{ip}\varphi^\beta_{kq}R_{\gamma\delta}\varphi^\gamma_{jp}\varphi^\delta_{\ell q} - R_{\alpha\beta}\varphi^\alpha_{ip}\varphi^\beta_{\ell q}R_{\gamma\delta}\varphi^\gamma_{jp}\varphi^\delta_{kq} \\ &= \varphi^\alpha_{ip}\varphi^\gamma_{jp}(\varphi^\beta_{\ell q}\varphi^\delta_{qk} - \varphi^\delta_{\ell q}\varphi^\beta_{qk})R_{\alpha\beta}R_{\gamma\delta} \\ &= \varphi^\alpha_{ip}\varphi^\gamma_{jp}[\varphi^\beta, \varphi^\delta]_{\ell k}R_{\alpha\beta}R_{\gamma\delta} \\ &= \varphi^\alpha_{ip}\varphi^\gamma_{jp}c^{\beta\delta}_\eta\varphi^\eta_{\ell k}R_{\alpha\beta}R_{\gamma\delta} \\ &= \frac{1}{2}(\varphi^\alpha_{ip}\varphi^\gamma_{jp} - \varphi^\gamma_{ip}\varphi^\alpha_{jp})\varphi^\eta_{\ell k}c^{\beta\delta}_\eta R_{\alpha\beta}R_{\gamma\delta} \\ &= \frac{1}{2}c^{\gamma\alpha}_\theta c^{\beta\delta}_\eta R_{\alpha\beta}R_{\gamma\delta}\varphi^\theta_{ij}\varphi^\eta_{k\ell} = R^\#_{ijk\ell} \end{aligned}$$

□

Remark 12.12. We note that neither R^2 nor $R^\#$ satisfy the first Bianchi identity. However, their sum does. In particular:

Proposition 12.13. *For any* $R \in \mathrm{Curv}$, $Q(R) \in \mathrm{Curv}$.

Proof. The symmetry and antisymmetry identities are trivial. To check the Bianchi symmetry, we see that

$$\begin{aligned}
2(Q_{ijk\ell} + Q_{ik\ell j} + Q_{i\ell jk}) &= R_{ijpq}R_{k\ell pq} + 2R_{ipjq}R_{\ell pkq} - 2R_{ipjq}R_{kp\ell q} \\
&\quad + R_{ikpq}R_{\ell jpq} + 2R_{ipkq}R_{jp\ell q} - 2R_{ipkq}R_{\ell pjq} \\
&\quad + R_{i\ell pq}R_{jkpq} + 2R_{ip\ell q}R_{kpjq} - 2R_{ip\ell q}R_{jpkq} \\
&= R_{ijpq}R_{k\ell pq} - 2R_{ipjq}R_{k\ell pq} \\
&\quad + R_{ikpq}R_{\ell jpq} - 2R_{ipkq}R_{\ell jpq} \\
&\quad + R_{i\ell pq}R_{jkpq} - 2R_{ip\ell q}R_{jkpq}
\end{aligned}$$

since R satisfies the first Bianchi identity. Thus

$$\begin{aligned}
2(Q_{ijk\ell} + Q_{ik\ell j} + Q_{i\ell jk}) &= (R_{ijpq} - R_{ipjq} + R_{iqjp})R_{k\ell pq} \\
&\quad + (R_{ikpq} - R_{ipkq} + R_{iqkp})R_{\ell jpq} \\
&\quad + (R_{i\ell pq} - R_{ip\ell q} + R_{iq\ell p})R_{jkpq} \\
&= 0.
\end{aligned}$$

□

Example 12.14. To get a feel for the quadratic structure, in particular that of the problematic sharp term, we look at the special case when $n = 3$. In such a situation $N = n(n-1)/2 = 3$, and the Lie algebra $\mathfrak{so}(3) \simeq \mathbb{R}^3$ – via the isomorphism ι defined in Sect. 7.5.3 – so that $[X, Y]_i = (X \times Y)_i = \varepsilon_{ijk}X_jY_k$ (cf. Example C.1). In which case (as described in Sect. 7.5.3) the curvature operator becomes the tensor $\Lambda \in \mathrm{Sym}^2(\mathbb{R}^3)$, and the structure constants

$$c_\gamma^{\alpha\beta} = (\varphi^\alpha \times \varphi^\beta) \cdot \varphi^\gamma = \varepsilon_{\alpha\beta\gamma}$$

equal elementary alternating 3-tensors (i.e. the volume form). Thus we have

$$\Lambda^\#_{\alpha\beta} = \frac{1}{2}c_\alpha^{\gamma\eta}c_\beta^{\delta\theta}\Lambda_{\gamma\delta} = \frac{1}{2}\varepsilon_{\alpha\gamma\eta}\varepsilon_{\beta\delta\theta}\Lambda_{\gamma\delta}\Lambda_{\eta\theta}$$

and we claim that:

Claim 12.15. The sharp term equals the cofactor matrix. That is,

$$\Lambda^\#_{\alpha\beta} = (\mathrm{adj}\,\Lambda)_{\beta\alpha}.$$

To see this, note that $\Lambda^{-1}_{\alpha\beta} = (2\det\Lambda)^{-1}\varepsilon_{\alpha pq}\varepsilon_{\beta ab}\Lambda_{ap}\Lambda_{bq}$.[6] So by Lemma 4.13 we find that $\Lambda^\# = \det\Lambda \cdot {}^t\Lambda^{-1} = {}^t(\mathrm{adj}\,\Lambda)$. Therefore if

[6] Since $(\det\Lambda)\varepsilon_{ijk} = \varepsilon_{pqr}\Lambda_{ip}\Lambda_{jq}\Lambda_{kr}$.

$$\Lambda = \begin{pmatrix} a & b & c \\ d & e & f \\ g & h & k \end{pmatrix}$$

we see that Λ^2 corresponds to the usual matrix product and

$$\Lambda^{\#} = \det \Lambda \cdot {}^t\Lambda^{-1} = \begin{pmatrix} ek - fh & fg - dk & dh - eg \\ ch - bk & ak - cg & bg - ah \\ bf - cd & cd - af & ae - bd \end{pmatrix}.$$

In particular if $\Lambda = \operatorname{diag}(a, e, k)$, then $\Lambda^{\#} = \operatorname{diag}(ek, ak, ae)$.

12.2.2 Algebraic Relationships and Generalisations

To present the results of Böhm and Wilking we need to extend the above squared and sharp products a little. By setting $V = \mathbb{R}^n$, we define the *circle* and *sharp* operators $\circ, \# : S^2(\bigwedge^2 V^*) \times S^2(\bigwedge^2 V^*) \to S^2(\bigwedge^2 V^*)$ to be

$$R \circ S = \frac{1}{2}(RS + SR)$$
$$(R\#S)_{\alpha\beta} = (R\#S)(\sigma_\alpha, \sigma_\beta) = \frac{1}{2} c_\alpha^{\gamma\eta} c_\beta^{\delta\theta} R_{\gamma\delta} S_{\eta\theta}$$

respectively.

Remark 12.16. From the antisymmetry of $c_\gamma^{\alpha\beta}$, $R\#S = S\#R$. Moreover, in what is to come we shall let I denote the identity element in $S^2(\bigwedge^2 V^*)$. However, one should note $R\#I$ is not equal to R in general.

Now with these definitions in place, Lemma 12.10 and 12.11 can easily be extended to the following result.

Lemma 12.17. *For $R, S \in S^2(\bigwedge^2 V^*)$,*

$$(R \circ S)_{ijk\ell} = \frac{1}{2} R_{ijpq} S_{pqk\ell}$$
$$(R\#S)_{ijk\ell} = R_{ipkq} S_{jp\ell q} - R_{ip\ell q} S_{jpkq}.$$

Furthermore, the quadratic term defines a bilinear operator $Q : S^2(\bigwedge^2 V^*) \times S^2(\bigwedge^2 V^*) \to S^2(\bigwedge^2 V^*)$ by polarisation:

$$Q(R, S) = R \circ S + R\#S. \tag{12.3}$$

Related to this is the operator $Q : S^2(\bigwedge^2 V^*) \to S^2(\bigwedge^2 V^*)$ which is taken to be $Q(R) = Q(R,R) = R^2 + R^\#$.

12.2.2.1 Huisken's Trilinear Form

There is a trilinear form on $S^2(\bigwedge^2 V^*)$ defined by

$$\operatorname{tri}(R,S,T) = 2\,\langle Q(R,S),T\rangle = 2\operatorname{tr}\big((R\circ S + R\#S)T\big).$$

The trilinear form is symmetric in all three components, since it is straightforward to check that

$$\operatorname{tr}\big((R\#S)T\big) = \frac{1}{2}\sum_{\alpha,\beta,\gamma=1}^{N} \langle [R(\sigma_\alpha),S(\sigma_\beta)],T(\sigma_\gamma)\rangle\,\langle [\sigma_\alpha,\sigma_\beta],\sigma_\gamma\rangle\,,$$

which is clearly symmetric in all three components.

The vector bundle maximum principle (together with Corollary 7.17) allows us to produce curvature conditions preserved by the Ricci flow PDE from sets preserved by flow of the vector field Q on Curv. So it is of great interest to construct functions which are monotone under this flow.

Proposition 12.18. *The* ODE $\frac{d}{dt}R = Q(R)$ *is the gradient flow of*

$$P(R) = \frac{1}{3}\operatorname{tr}(R^3 + RR^\#) = \frac{1}{6}\operatorname{tri}(R,R,R).$$

Proof. Let $\frac{\partial}{\partial t}R = S$ and observe – by the total symmetry of $\operatorname{tr}(R\#S)T$ – that

$$\begin{aligned}\frac{\partial}{\partial t}P(R) &= \frac{1}{3}\operatorname{tr}(SR^2 + RSR + R^2S + (R\#S + S\#R)\circ R + R^\#\circ S)\\ &= \operatorname{tr}\big((R^2+R^\#)S\big) = \big\langle R^2 + R^\#, S\big\rangle\,.\end{aligned}$$ □

What is more, this can be improved to give a *scaling-invariant* monotone function:

Proposition 12.19. *Under the* ODE $\frac{d}{dt}R = Q(R)$, *the radial projection* $\widehat{R} = R/|R|$ *onto the unit sphere evolves in the direction of the gradient of the scaling-invariant function* $P(\widehat{R})$. *In particular* $P(\widehat{R}) = P(R)/|R|^3$ *is strictly increasing except at points where* $Q(R) = \lambda R$ *for some* $\lambda \in \mathbb{R}$.

Proof. We directly compute:

$$\frac{d}{dt}P(\widehat{R}) = \frac{1}{|R|^3}\Big\langle Q(R), \frac{d}{dt}R\Big\rangle - \frac{6P(R)}{|R|^5}\Big\langle R, \frac{d}{dt}R\Big\rangle$$

$$= \frac{1}{|R|^3}\Big(|Q(R)|^2 - \frac{\mathrm{tri}(R,R,R)}{|R|^2}\langle R, Q(R)\rangle\Big)$$
$$= \frac{1}{|R|^3}\Big|Q(R) - \frac{\langle Q(R), R\rangle}{|R|^2}R\Big|^2,$$

since $\langle Q(R), R\rangle = \mathrm{tri}(R,R,R)$. □

In particular, this says that the super-level sets of $P(\widehat{R})$ are cones in the space of curvature operators, and the vector field $Q(R)$ points into them. The bad news is, however, that these cones are in general *not* convex (see Fig. 12.1 and Fig. 12.2) so that the maximum principle unfortunately does not apply.

12.2.2.2 The Wedge Product

The $\circ$ and $\#$ products enable one to construct a new element in $S^2(\bigwedge^2 V^*)$ from two given elements. It is important to note there is another method of constructing elements in $S^2(\bigwedge^2 V^*)$ from two elements of $S^2(V)$. Following

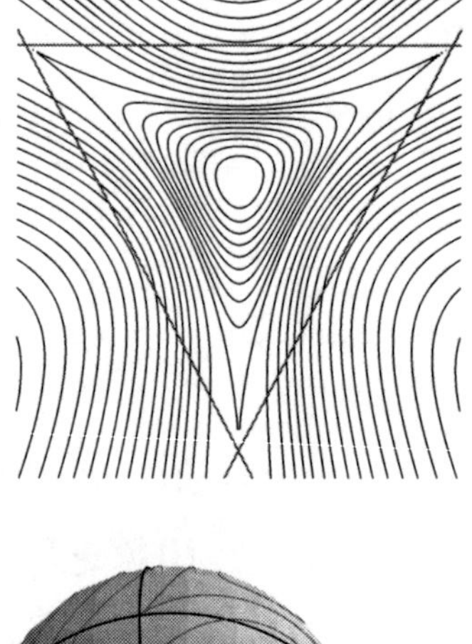

Fig. 12.1 Level sets for $P(\widehat{R})$ restricted to the plane $\{R: \mathrm{tr}R = 1\}$ for $n = 3$. The centre point corresponds to curvature of S^3, and the triangle corresponds to the cone of positive sectional curvature, with the vertices corresponding to the curvature of $S^2 \times \mathbb{R}$

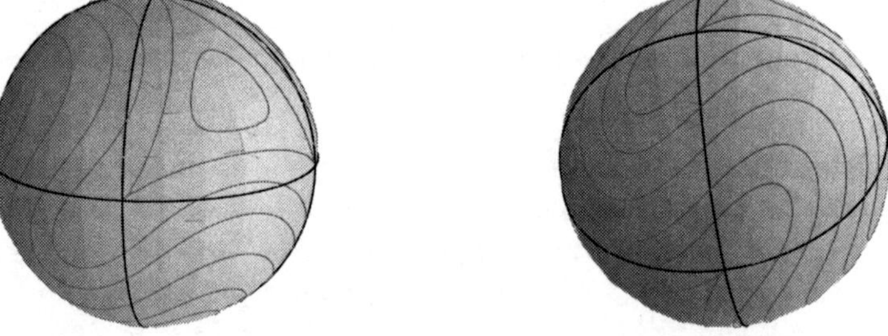

Fig. 12.2 Top and side views of the level sets of $P(\widehat{R})$ restricted to the unit sphere, showing the lines where one of the sectional curvatures vanishes. The intersection points of these lines are the curvature operators of $S^2 \times \mathbb{R}$ and $\mathbb{H}^2 \times \mathbb{R}$, and the curvature of S^3 and of $\mathbb{H}^3$ are at the 'top' amd 'bottom' of the sphere

[BW08, p. 1082], we define the *wedge product* $A \wedge B : \bigwedge^2 V \to \bigwedge^2 V$ of two elements $A, B \in \mathfrak{gl}(n, \mathbb{R}) \simeq V \otimes V$ by

$$(A \wedge B)(x \wedge y) := \frac{1}{2}\big(A(x) \wedge B(y) + B(x) \wedge A(y)\big). \tag{12.4}$$

It is easy to see that $A \wedge B = B \wedge A$ if $A, B \in S^2(V)$. By using the wedge, there is a natural inclusion map $\mathrm{id}_\wedge : S^2(V) \to S^2(\bigwedge^2 V^*)$ given by

$$A \mapsto A \wedge \mathrm{id}.$$

Note that $I = \mathrm{id} \wedge \mathrm{id}$, so $I_{ijk\ell} = \delta_{ik}\delta_{j\ell} - \delta_{i\ell}\delta_{jk}$ is the curvature tensor on the standard sphere. When restricted to the space of symmetric 2-tensors, the wedge product is (up to a constant) equal to the Kulkarni–Nomizu product, namely:

$$\begin{aligned} \langle (A \wedge B)(e_i \wedge e_j), e_k \wedge e_\ell \rangle &= \frac{1}{2} \langle A_{ip}B_{jq}e_p \wedge e_q + B_{ip}A_{jq}e_p \wedge e_q, e_k \wedge e_\ell \rangle \\ &= \frac{1}{2}(A_{ik}B_{j\ell} + A_{j\ell}B_{ik} - A_{i\ell}B_{jk} - A_{jk}B_{i\ell}) \\ &= \frac{1}{2}(A \owedge B)_{ijk\ell}. \end{aligned} \tag{12.5}$$

It is advantageous to define the *Ricci operator* $\mathrm{Rc} : S^2(\bigwedge^2 V^*) \to S^2(V)$ by

$$\langle \mathrm{Rc}(R)(e_i), e_j \rangle = \sum_{k=1}^{n} \langle R(e_i \wedge e_k), e_j \wedge e_k \rangle = R_{ikjk},$$

for all $R \in S^2(\bigwedge^2 V^*)$. Note that $\mathrm{Rc}(R) = \mathrm{Ric}$ as expected. Similarly, the *scalar operator* scal is defined by

$$\mathrm{scal}(R) := \mathrm{tr}\,\mathrm{Rc}(R) = R_{ikik},$$

for all $R \in S^2(\bigwedge^2 V^*)$, where $\mathrm{scal}(R) = \mathrm{Scal}$ as expected.

It is interesting to note:

Proposition 12.20. *The operator* $2\,\mathrm{id}_\wedge$ *is the adjoint of* Rc.

Proof. Using the trace norms on $S^2(V)$ and $S^2(\bigwedge^2 V^*)$ we compute:

$$\begin{aligned} \langle 2\,\mathrm{id}_\wedge(A), R \rangle &= (2A \wedge \mathrm{id})_{\alpha\lambda} R_{\lambda\alpha} \\ &= \frac{1}{4}(A \owedge \mathrm{id})_{ijk\ell} R_{k\ell ij} \\ &= \frac{1}{4}(A_{ik}\delta_{j\ell} + A_{j\ell}\delta_{ik} - A_{i\ell}\delta_{jk} - A_{jk}\delta_{i\ell}) R_{k\ell ij} \\ &= A_{ik}\mathrm{Ric}_{ki} = \langle A, \mathrm{Rc}(R) \rangle \end{aligned}$$

□

For future computations we also observe: If $A, B \in S^2(V)$ then

$$\begin{aligned}\langle \mathrm{Rc}(A \wedge B)(e_i), e_j\rangle &= \sum_k \langle (A \wedge B)(e_i \wedge e_k), e_j \wedge e_k\rangle \\ &= \frac{1}{2}(A_{ij}B_{kk} + B_{ij}A_{kk} - A_{ik}B_{kj} - B_{ik}A_{kj}) \\ &= \frac{1}{2}(A_{ij}\mathrm{tr}(A) - A_{ik}B_{kj} + B_{ij}\mathrm{tr}(A) - B_{ik}A_{kj}),\end{aligned}$$

which implies

$$\mathrm{Rc}(A \wedge B) = \frac{1}{2}A\big(\mathrm{tr}(B)\,\mathrm{id} - B\big) + \frac{1}{2}B\big(\mathrm{tr}(A)\,\mathrm{id} - A\big).$$

In particular we find that

$$\mathrm{Rc}(A \wedge A) = \mathrm{tr}(A)\,A - A^2 \quad \text{and} \quad \mathrm{Rc}(A \wedge \mathrm{id}) = \frac{n-2}{2}A + \frac{\mathrm{tr}(A)}{2}\mathrm{id}. \qquad (12.6)$$

So if $\mathrm{tr}(A) = 0$ then

$$\mathrm{Rc}(A \wedge A) = -A^2 \quad \text{and} \quad \mathrm{Rc}(A \wedge \mathrm{id}) = \frac{n-2}{2}A. \qquad (12.7)$$

12.3 Decomposition of Algebraic Curvature Operators

The space of curvature operators can be decomposed in a directly analogous way to Sect. 4.5. By explicitly identifying $\bigwedge^2 V^* \simeq \mathfrak{so}(n)$, there is a natural decomposition of

$$S^2(\mathfrak{so}(n)) = \langle I\rangle \oplus \langle \mathring{\mathrm{Ric}}\rangle \oplus \langle W\rangle \oplus \bigwedge^4(\mathbb{R}^n)$$

into $O(n)$-invariant, irreducible, pairwise inequivalent subspaces, whenever $n \geq 4$. The non-trivial aspect of the statement is the irreducibility of the decomposition, however this plays only a minor role in our discussion. A detailed algebraic proof of this fact can be found in [GW09, Sect. 10.3] (one could also consult [Bes08, p. 45]).

An endomorphism $R \in S^2(\mathfrak{so}(n))$ is satisfies the first Bianchi identity if and only if it has no component in the last factor, so we have

$$\mathrm{Curv} = \langle I\rangle \oplus \langle \mathring{\mathrm{Ric}}\rangle \oplus \langle W\rangle\,.$$

Now given a curvature operator $R \in \operatorname{Curv}$ let R_I, $R_{\overset{\circ}{\operatorname{Ric}}}$, R_W denote the projections onto $\langle I \rangle$, $\langle \overset{\circ}{\operatorname{Ric}} \rangle$ and $\langle W \rangle$ respectively, so that

$$R = R_I + R_{\overset{\circ}{\operatorname{Ric}}} + R_W.$$

Here, naturally, $\langle I \rangle$ denotes the subspace generated by the identity, $\langle \overset{\circ}{\operatorname{Ric}} \rangle$ denote the traceless Ricci subspace and $\langle W \rangle$ denotes the Weyl part. For such an R, also define

$$\bar{\lambda} = \frac{\operatorname{scal}(R)}{n} \qquad \text{and} \qquad \sigma = \frac{\|\operatorname{Rc}_0(R)\|^2}{n},$$

where $\operatorname{Rc}_0(R) = \operatorname{Rc}(R) - \frac{\operatorname{scal}(R)}{n}\operatorname{id}$.

In a direct comparison with (4.14) and (4.15), an algebraic curvature operator can be decomposed as

$$R = \frac{\bar{\lambda}}{n-1}\operatorname{id} \wedge \operatorname{id} + \frac{2}{n-2}\operatorname{Rc}_0(R) \wedge \operatorname{id} + R_W \tag{12.8}$$

or

$$R = \frac{-\operatorname{scal}(R)}{(n-1)(n-2)}\operatorname{id} \wedge \operatorname{id} + \frac{2}{n-2}\operatorname{Rc}(R) \wedge \operatorname{id} + R_W.$$

Proposition 12.21. *For any $A \in S^2(V)$ with* $\operatorname{tr}(A) = 0$,

$$A \wedge A = -\frac{\operatorname{tr}(A^2)}{n(n-1)}\operatorname{id} \wedge \operatorname{id} - \frac{2}{n-2}(A^2)_0 \wedge \operatorname{id} + (A \wedge A)_{\mathrm{W}}$$

where $(A^2)_0$ is the trace-free part of A^2.

Proof. Using (12.8) with (12.7) one finds

$$A \wedge A = \frac{\operatorname{tr}(-A^2)}{n(n-1)}\operatorname{id} \wedge \operatorname{id} + \frac{2}{n-2}\Big(- A^2 - \frac{\operatorname{scal}(-A^2)}{n}\operatorname{id}\Big) \wedge \operatorname{id} + (A \wedge A)_W,$$

from which the result follows. □

12.3.1 Schur's Lemma

With equation (12.8) in mind, we relate manifolds with constant sectional curvature to the decomposition of the curvature endomorphism. We will use Schur's Lemma in the final argument discussed in Chap. 15, since we seek to deform our manifold into a space form.

Lemma 12.22. *A Riemannian manifold M has constant sectional curvature, equal to κ_0, if and only if $R = \kappa_0 \operatorname{id} \wedge \operatorname{id}$, where R is the curvature of M.*

Proof. Suppose, for each $p \in M$, that $K(\Pi) = \kappa_0$ for all 2-planes $\Pi \leq T_pM$. From Sect. 2.7.6 we have that

$$\begin{aligned} R(x, y, x, y) &= \kappa_0(|x|^2|y|^2 - \langle x, y\rangle^2) \\ &= \langle \kappa_0(\operatorname{id} \wedge \operatorname{id})(x \wedge y), x \wedge y\rangle , \end{aligned}$$

for any $x, y \in T_pM$ linearly independent. By (12.5), $\operatorname{id} \wedge \operatorname{id}$ has the same symmetry properties as the curvature R, so uniqueness gives the result. The converse is immediate. □

We shall say that a manifold M is *isotropic* if, for each $p \in M$, the sectional curvature $K_p(\Pi)$ is independent of the 2-plane $\Pi \leq T_pM$. The following lemma of Schur [Sch86] show the distinction between isotropic manifolds and manifolds of constant curvature is artificial.

Lemma 12.23 (Schur). *Let M^n be a connected Riemannian manifold with $n \geq 3$. If $K_p(\Pi) = f(p)$, for all 2-planes $\Pi \leq T_pM$ and $p \in M$, then f must be constant. In other words, all isotropic manifolds M have constant sectional curvature.*

Proof. For each $p \in M$, our hypothesis implies that

$$R\big|_p(x, y, x, y) = f(p)(|x|^2|y|^2 - g(x, y)^2)$$

for all $x, y \in T_pM$ linearly independent. So by Lemma 12.22 we must have $R = \frac{1}{2} f\, g \owedge g$. In which case $2\nabla R = (\nabla f)\, g \owedge g + f\, (\cancel{\nabla g}) \owedge g + f\, g \owedge (\cancel{\nabla g}) = (\nabla f)\, g \owedge g$, since $\nabla g \equiv 0$. Using the second Bianchi identity for R gives

$$\begin{aligned} 0 &= (Uf)\big(g(W, X)g(Z, Y) - g(Z, X)g(W, Y)\big) \\ &\quad + (Xf)\big(g(W, Y)g(Z, U) - g(Z, Y)g(W, U)\big) \\ &\quad + (Yf)\big(g(W, U)g(Z, X) - g(Z, U)g(W, X)\big) \end{aligned}$$

for all $X, Y, W, Z, U \in \mathscr{X}(M)$.

If $X \in T_pM$ is an arbitrary tangent vector to the manifold, then it is possible to choose Y and Z such that X, Y, Z are orthogonal (since $\dim M \geq 3$). By setting $U = Z$ at p, the above equation implies that $g(X(f)Y - Y(f)X, W) = 0$ for all W. Since X and Y are linearly independent, we have $X(f) = Y(f) = 0$. Since X is arbitrary, the derivative of f vanishes, and f is locally constant. Since M is connected, f is constant. □

12.3.2 The Q-Operator and the Weyl Subspace

An important step of Böhm and Wilking is to establish the relationship between Q, defined by (12.3), and the various curvature subspace discussed in the above.

Proposition 12.24. *For any $R \in \mathrm{Curv}$,*

$$\begin{aligned} \mathrm{Rc}(R^2 + R^\#)_{ij} &= \mathrm{Rc}(R)_{k\ell} R_{ikj\ell} \\ \mathrm{scal}(R^2 + R^\#) &= \mathrm{Rc}(R)_{k\ell} \mathrm{Rc}(R)_{k\ell}. \end{aligned}$$

In particular, if $R \in \langle \mathrm{W} \rangle$ then $Q(R) = R^2 + R^\# \in \langle \mathrm{W} \rangle$.

Proof. From Lemma 12.10 and 12.11,

$$\begin{aligned} (R^2 + R^\#)_{ijkj} = Q_{ijkj} &= B_{ijkj} - B_{ijjk} + B_{ikjj} - B_{ijjk} \\ &= R_{piqj} R_{pkqj} - 2 R_{piqj} R_{pjqk} + R_{ipkq} R_{pjqj}. \end{aligned}$$

By the first Bianchi identity,

$$\begin{aligned} \sum_{p,j} R_{iqpj} R_{krpj} &= \sum_{p,j} (R_{ipqj} R_{krpj} + R_{ijpq} R_{krpj}) \\ &= \sum_{p,j} (R_{ipqj} R_{krpj} + R_{ipjq} R_{krjp}) = 2 \sum_{p,j} R_{ipqj} R_{krpj}. \end{aligned}$$

In which case,

$$R_{piqj} R_{pjqk} = R_{ipqj} R_{kqpj} = \frac{1}{2} R_{iqpj} R_{kqpj} = \frac{1}{2} R_{qipj} R_{qkpj}.$$

Thus $(R^2 + R^\#)_{ijkj} = R_{ipkq} R_{pjqj} = R_{ipkq} \mathrm{Rc}(R)_{pq}$.

Now if $R \in \langle W \rangle$ then $\mathrm{Rc}(R) = 0$, so $\mathrm{Rc}(R^2 + R^\#)_{ik} = R_{ipkq} \mathrm{Rc}(R)_{pq} = 0$ and $\mathrm{scal}(R^2 + R^\#) = 0$. Hence $R^2 + R^\# \in \langle W \rangle$. □

Proposition 12.25. *If $R \in \langle \mathring{\mathrm{Ric}} \rangle$ and $S \in \langle W \rangle$ then $Q(R, S) \in \langle \mathring{\mathrm{Ric}} \rangle$.*

Proof. For $S, W \in \langle W \rangle$ and $R \in \langle \mathring{\mathrm{Ric}} \rangle$, it suffices to show:

$$\begin{aligned} \mathrm{tri}(S, R, W) &= 0 \\ \mathrm{tri}(S, R, I) &= 0 \end{aligned}$$

To prove the first, recall that tri is totally symmetric, so that $\mathrm{tri}(S, R, W) = \mathrm{tri}(W, S, R)$. By Proposition 12.24, $(S + W)^2 + (S + W)^\# \in \langle W \rangle$ and so $WS + SW + 2W\#S \in \langle W \rangle$. Therefore as $R \in \langle \mathring{\mathrm{Ric}} \rangle$,

$$0 = \langle WS + SW + 2W\#S, R\rangle = \mathrm{tr}\big((WS + SW + 2W\#S)R\big) = \mathrm{tri}(W, S, R).$$

To prove the second, we find that

$$\mathrm{tri}(S, R, I) = \mathrm{tri}(S, I, R) = 2\mathrm{tr}\big((S + S\#I)R\big) = 0,$$

where the last equality is due to Lemma 12.26 below.

Since the identities can be rewritten as

$$\langle SR + RS + 2R\#S, W\rangle = 0$$
$$\langle SR + RS + 2R\#S, I\rangle = 0$$

one must have $SR + RS + 2S\#R \in \langle \overset{\circ}{\mathrm{Ric}}\rangle$. The result now follows. □

12.3.3 Algebraic Lemmas of Böhm and Wilking

For the computation purposes of the next section, it is necessary to examine the following lemmas of [BW08, Sect. 2].

Lemma 12.26. *For* $R \in \mathrm{Curv}$,

$$R + R\#I = (n-1)R_I + \frac{n-2}{2}R_{\overset{\circ}{\mathrm{Ric}}} = \mathrm{Ric} \wedge \mathrm{id}.$$

Remark 12.27. The lemma implies $Q(R, I) = R + R\#I$ has no Weyl part, i.e. $(R + R\#I)_W = 0$.

Proof. By Lemma 12.17,

$$\begin{aligned}(R\#I)_{ijk\ell} &= R_{ipkq}I_{jp\ell q} - R_{ip\ell q}I_{jpkq} \\ &= R_{ipkp}\delta_{j\ell} - R_{i\ell kj} - R_{ip\ell p}\delta_{jk} + R_{ik\ell j} \\ &= \mathrm{Ric}_{ik}\delta_{j\ell} - \mathrm{Ric}_{i\ell}\delta_{jk} - R_{ijk\ell}\end{aligned}$$

since $I_{ijk\ell} = \langle I(e_i \wedge e_j), e_k \wedge e_\ell\rangle = \langle e_i \wedge e_j, e_k \wedge e_\ell\rangle = \delta_{ik}\delta_{j\ell} - \delta_{i\ell}\delta_{jk}$. Using this we find that

$$\begin{aligned}(\mathrm{Ric} \wedge \mathrm{id})_{ijk\ell} &= \frac{1}{2}(\mathrm{Ric}_{ik}\delta_{j\ell} + \mathrm{Ric}_{j\ell}\delta_{ik} - \mathrm{Ric}_{i\ell}\delta_{jk} - \mathrm{Ric}_{jk}\delta_{i\ell}) \\ &= \frac{1}{2}\big((R + R\#I)_{ijk\ell} - (R + R\#I)_{jik\ell}\big) \\ &= R_{ijk\ell} + (R\#I)_{ijk\ell},\end{aligned}$$

where the last equality is due to

$$\begin{aligned}(R\#I)_{ijk\ell} = (R\#I)_{\alpha\beta}\varphi_{ij}^{\alpha}\varphi_{k\ell}^{\beta} &= \frac{1}{2}c_{\alpha}^{\gamma\eta}c_{\beta}^{\delta\theta}R_{\gamma\delta}I_{\eta\theta}\varphi_{ij}^{\alpha}\varphi_{k\ell}^{\beta} \\ &= -(R\#I)_{\alpha\beta}\varphi_{ji}^{\alpha}\varphi_{k\ell}^{\beta} = -(R\#I)_{jik\ell}. \qquad \square\end{aligned}$$

We say a curvature operator R is of *Ricci type* if $R = R_I + R_{\overset{\circ}{\mathrm{Ric}}}$, i.e. $R_W = 0$.

Lemma 12.28. *If $R \in \mathrm{Curv}$ be a curvature operator of Ricci type, then*

$$\begin{aligned}R^2 + R^{\#} = &\frac{1}{n-2}\overset{\circ}{\mathrm{Ric}} \wedge \overset{\circ}{\mathrm{Ric}} + \frac{2\bar{\lambda}}{n-1}\overset{\circ}{\mathrm{Ric}} \wedge \mathrm{id} - \frac{2}{(n-2)^2}(\overset{\circ}{\mathrm{Ric}}{}^2)_0 \wedge \mathrm{id} \\ &+ \left(\frac{\bar{\lambda}^2}{n-1} + \frac{\sigma}{n-2}\right) I.\end{aligned}$$

Proof. Recall that $R_I = \frac{\bar{\lambda}}{n-1}I$ and $R_0 := R_{\overset{\circ}{\mathrm{Ric}}} = \frac{2}{n-2}\overset{\circ}{\mathrm{Ric}} \wedge \mathrm{id}$, so that

$$\begin{aligned}R^2 + R^{\#} &= (R_I + R_0)^2 + (R_I + R_0)^{\#} \\ &= R_0^2 + R_0^{\#} + R_I^2 + R_I^{\#} + 2(R_I R_0 + R_I \# R_0) \\ &= R_0^2 + R_0^{\#} + \frac{\bar{\lambda}^2}{(n-1)^2}(I + I\#I) + \frac{2\bar{\lambda}}{n-1}(R_0 + R_0\#I).\end{aligned}$$

As the last two summands are known by Lemma 12.26, it suffices to show:

Claim 12.29.

$$R_0^2 + R_0^{\#} = \frac{1}{n-2}\overset{\circ}{\mathrm{Ric}} \wedge \overset{\circ}{\mathrm{Ric}} - \frac{2}{(n-2)^2}(\overset{\circ}{\mathrm{Ric}}{}^2)_0 \wedge \mathrm{id} + \frac{\sigma}{n-2}I$$

To prove this claim, choose an orthonormal basis $(e_i)_{i=1}^n$ of eigenvectors of $\overset{\circ}{\mathrm{Ric}}$ with corresponding eigenvalues $(\lambda_i)_{i=1}^n$. The curvature operator R_0 is diagonal with respect to $(e_i \wedge e_j)_{i<j}$, with eigenvalues given by

$$R_0(e_i \wedge e_j) = \frac{1}{n-2}\left(\overset{\circ}{\mathrm{Ric}}(e_i) \wedge e_j + e_i \wedge \overset{\circ}{\mathrm{Ric}}(e_j)\right) = \frac{\lambda_i + \lambda_j}{n-2}e_i \wedge e_j.$$

On the left-hand side of the claim this implies that

$$R_0^2(e_i \wedge e_j) = \left(\frac{\lambda_i + \lambda_j}{n-2}\right)^2 e_i \wedge e_j.$$

Furthermore, with

$$(R_0)_{ijk\ell} = \langle R_0(e_i \wedge e_j), e_k \wedge e_\ell \rangle = \frac{\lambda_i + \lambda_j}{n-2}(\delta_{ik}\delta_{j\ell} - \delta_{jk}\delta_{i\ell})$$

and Lemma 12.17, one has

$$\begin{aligned}
R_0^\#(e_i \wedge e_j) &= \frac{1}{2}(R_0^\#)_{ijk\ell} e_k \wedge e_\ell \\
&= \frac{1}{2}(R_0)_{ipkq}(R_0)_{jp\ell q} e_k \wedge e_\ell - (k \leftrightarrow \ell) \\
&= \frac{1}{2}\sum_{p,q} \frac{\lambda_i + \lambda_p}{n-2}\frac{\lambda_j + \lambda_p}{n-2}(\delta_{ik}\delta_{pq} - \delta_{iq}\delta_{pk})(\delta_{j\ell}\delta_{pq} - \delta_{jq}\delta_{p\ell}) e_k \wedge e_\ell \\
&\quad - (k \leftrightarrow \ell) \\
&= \frac{1}{2}\Big(\sum_{p} \frac{\lambda_i + \lambda_p}{n-2}\frac{\lambda_j + \lambda_p}{n-2} e_i \wedge e_j \\
&\quad - \frac{\lambda_i + \lambda_j}{n-2}\frac{\lambda_j + \lambda_j}{n-2} e_i \wedge e_j - \frac{\lambda_i + \lambda_i}{n-2}\frac{\lambda_j + \lambda_i}{n-2} e_i \wedge e_j\Big) - (k \leftrightarrow \ell) \\
&= \sum_{p \neq i,j} \frac{\lambda_i + \lambda_p}{n-2}\frac{\lambda_j + \lambda_p}{n-2} e_i \wedge e_j.
\end{aligned}$$

Meanwhile, on the right-hand side one finds that

$$\begin{aligned}
\big((\mathring{\mathrm{Ric}}^2)_0 \wedge \mathrm{id}\big)(e_i \wedge e_j) &= \frac{1}{2}\big((\mathring{\mathrm{Ric}}^2)_0(e_i) \wedge e_j + e_i \wedge (\mathring{\mathrm{Ric}}^2)_0(e_j)\big) \\
&= \frac{\lambda_i^2 + \lambda_j^2 - 2\sigma}{2} e_i \wedge e_j,
\end{aligned}$$

since $(\mathring{\mathrm{Ric}}^2)_0(e_i) = (\mathring{\mathrm{Ric}}^2(e_i) - \sigma)e_i = (\lambda_i^2 - \sigma)e_i$. From this, we see that

$$\begin{aligned}
&\left(\frac{1}{n-2}\mathring{\mathrm{Ric}} \wedge \mathring{\mathrm{Ric}} - \frac{2}{(n-2)^2}(\mathring{\mathrm{Ric}}^2)_0 \wedge \mathrm{id} + \frac{\sigma}{n-2} I\right)(e_i \wedge e_j) \\
&\quad = \left(\frac{\lambda_i \lambda_j}{n-2} + \frac{n\sigma - \lambda_i^2 - \lambda_j^2}{(n-2)^2}\right) e_i \wedge e_j.
\end{aligned}$$

Since λ_k is the k-th eigenvalue of the traceless Ricci tensor $\mathring{\mathrm{Ric}}$, the sum $\sum_k \lambda_k$ is the trace (which is zero). Hence

$$\sum_{k \neq i,j} \lambda_k = -\lambda_i - \lambda_j$$

$$\sum_{k \neq i,j} \lambda_k^2 = n\sigma - \lambda_i^2 - \lambda_j^2.$$

Thus we find that

$$\left(\frac{\lambda_i + \lambda_j}{n-2}\right)^2 + \sum_{p \neq i,j} \frac{\lambda_i + \lambda_p}{n-2} \frac{\lambda_j + \lambda_p}{n-2} = \frac{1}{(n-2)^2}\left((n-2)\lambda_i\lambda_j + n\sigma - \lambda_i^2 - \lambda_j^2\right)$$

from which the claim is proved. □

12.4 A Family of Transformations for the Ricci flow

So far we have established the relationship between curvature conditions preserved by the Ricci flow and convex subsets of the vector space Curv which the vector field Q points into. We call such a set a *preserved set*, and in particular a cone which is preserved is called a *preserved cone*. We seek to find a whole family of preserved convex cones interpolating between a given initial cone C and the 'degenerate' cone consisting of constant positive sectional curvature operators. This is done by considering an appropriate linear transformation of Curv which maps the cone C to a new one. We show that the new cone is preserved if and only if the vector field Q pulls back by the given transformation to a vector field which points into C. So if the original cone C is preserved by the Ricci flow, in the sense that Q points into it, then it suffices to show that the difference between Q and the pulled back vector field points into C. It is this difference which we explicitly compute in Theorem 12.33, and hence control in the next chapter.

The question remains as to what transformation will work. Since we want to map convex $O(n)$-invariant cones to convex $O(n)$-invariant cones, it makes sense to consider linear maps $l : \mathrm{Curv} \to \mathrm{Curv}$ which are $O(n)$-equivariant, so that $l(G(R)) = G(l(R))$ for all $G \in O(n)$ and $R \in \mathrm{Curv}$. The equivariance implies that every eigenspace of ℓ is $O(n)$-invariant,[7] and hence is a direct sum of irreducible components of Curv. The image and kernel of ℓ on each component is again invariant, and the components are pairwise inequivalent, so the only possibility is that ℓ maps each irreducible component either to zero or to itself. Since the eigenspaces of ℓ are invariant subspaces, ℓ is necessarily a real multiple of the identity on each component.

With this insight, a naïve idea would be to consider a linear transformation $l_c(R) = R + cR_I$ which magnifies only the scalar curvature part. It is easy

[7] Since eigenspaces can be complex, we must use the fact that the complexifications of our real irreducible representations are irreducible representations of $O(n, \mathbb{C})$, and so have no nontrivial invariant *complex* subspaces. See [GW09, Sect. 10.3.2] for details.

to see that $l_c(C)$ converges to $\mathbb{R}^+ I$ whenever $c \to \infty$ (since $\frac{1}{c} l_c(R) \to R_I$ as $c \to \infty$). However, it is not true in general that the vector field $Q(R)$ points into these cones. It turns out to be important to allow a more general family of linear operators, to give more freedom to select a family of transformations which have the required behaviour (in fact, since scalings of a cone do not change it, the family defined below are in effect *all* of the $O(n)$-invariant linear transformations of Curv, though it is convenient to choose the overall scaling to leave the Weyl part unchanged):

Definition 12.30. For any $a, b \in \mathbb{R}$, define the $O(n)$-invariant linear transformation $l_{a,b} : \mathrm{Curv} \to \mathrm{Curv}$ by

$$l_{a,b}(R) = R + 2(n-1)aR_I + (n-2)bR_{\mathring{\mathrm{Ric}}}. \tag{12.9}$$

Remark 12.31. Note that each transformation $l_{a,b}$ preserves the Weyl part and is a multiple of the identity on each of the other two irreducible components. By the decomposition (12.8), the transformation can be rewritten as

$$l_{a,b}(R) = (1 + 2(n-1)a)R_I + (1 + (n-2)b)R_{\mathring{\mathrm{Ric}}} + R_W.$$

It is easy to see that $l_{0,0}(R) = R$ and $l_{a,b}$ is invertible provided both $a \neq -\frac{1}{2(n-1)}$ and $b \neq -\frac{1}{(n-2)}$ (we will always assume this is the case). Note if $a \to \infty$ while $b/a \to 0$, $l_{a,b}(R)/|l_{a,b}(R)| \to I/|I|$ with $\mathrm{scal}(R) > 0$, so $l_{a,b}(C)$ converges to the line of positive constant sectional curvature operators.

Böhm and Wilking work with these transformations by bringing the dynamics back to the original cone C. Thus we need to determine the vector field obtained by pulling back the vector field Q from the transformed cone $\ell_{a,b}(C)$ to C. This is given by a vector field $X_{a,b}$ defined as follows:

$$\begin{aligned} X_{a,b}(R) &= l_{a,b}^{-1}\left(l_{a,b}(R)^2 + l_{a,b}(R)^{\#}\right) \\ &= (l_{a,b}^{-1} \circ Q \circ l_{a,b})(R). \end{aligned} \tag{12.10}$$

The difference, denoted by $D_{a,b}$, between $X_{a,b}$ and the original vector field for the Ricci flow is defined by

$$\begin{aligned} D_{a,b}(R) &= (l_{a,b}^{-1} \circ Q \circ l_{a,b})(R) - Q(R) \\ &= X_{a,b}(R) - Q(R). \end{aligned}$$

From this we find that:

Lemma 12.32. *The vector field Q strictly points into $l_{a,b}(C)$ (in the sense of Definition 7.12) if and only if the vector field $X_{a,b}$ strictly points into C.*

Proof. Note that $\partial(l_{a,b}C) = l_{a,b}(\partial C)$, so that points in $\partial(l_{a,b}C)$ have the form $l_{a,b}R$ for some $R \in \partial C$. By definition, Q strictly points into $l_{a,b}(C)$ if for any $l_{a,b}R \in \partial l_{a,b}(C)$ we have $Q(l_{a,b}R) \in \operatorname{Int}\left(\mathcal{T}_{l_{a,b}R} l_{a,b}(C)\right)$. Now

$$\begin{aligned}
\mathcal{T}_{l_{a,b}(R)}\left(l_{a,b}C\right) &= \bigcup_{h>0} h^{-1}\left(l_{a,b}C - l_{a,b}R\right) \\
&= l_{a,b}\Big(\bigcup_{h>0} h^{-1}\left(C - R\right)\Big) \\
&= l_{a,b}\left(\mathcal{T}_R C\right),
\end{aligned}$$

and the same holds for the interiors. Thus

$$Q(l_{a,b}R) \in \operatorname{Int}(\mathcal{T}_{l_{a,b}R} l_{a,b}(C)) \quad \text{if and only if} \quad l_{a,b}^{-1}\left(Q(l_{a,b}R)\right) \in \operatorname{Int}\left(\mathcal{T}_R C\right)$$

which says precisely that $X_{a,b}(R)$ points strictly into C. □

The idea now is as follows: Suppose we know that C is preserved by Ricci flow, in the sense that the vector field Q points into C. We want $X_{a,b} = Q + D_{a,b}$ to point strictly into C, so it suffices to prove that $D_{a,b}$ points strictly into C.

The miraculous discovery of Böhm and Wilking is that the operator $D_{a,b}(R)$ is independent of the Weyl component of R, and so can be simply computed in terms of the eigenvalues of the Ricci curvature of R. This greatly simplifies the task of understanding how $R^2 + R^{\#}$ changes under the linear map $l_{a,b}$.

Theorem 12.33 (Main Formula for $D_{a,b}$). *For any $a, b \in \mathbb{R}$,*

$$\begin{aligned}
D_{a,b}(R) = \big((n-2)b^2 - 2(a-b)\big)\overset{\circ}{\mathrm{Ric}} \wedge \overset{\circ}{\mathrm{Ric}} + 2a\,\mathrm{Ric} \wedge \mathrm{Ric} + 2b^2\,\overset{\circ}{\mathrm{Ric}}{}^2 \wedge \mathrm{id} \\
+ \frac{\operatorname{tr}(\overset{\circ}{\mathrm{Ric}}{}^2)}{n + 2n(n-1)a}\big(nb^2(1-2b) - 2(a-b)(1-2b+nb^2)\big)I.
\end{aligned}$$

Proof. We first establish that $D_{a,b}(R)$ is independent of the Weyl part of R. The precise form of $D_{a,b}(R)$ can then be explicitly computed from the Weyl free part of R.

Claim 12.34. The algebraic curvature operator $D_{a,b}(R)$ is independent of the Weyl component of R. That is,

$$D_{a,b}(R + S) = D_{a,b}(R)$$

for all $S \in \langle W \rangle$ and $R \in \mathrm{Curv}$.

Proof of Claim. First we note that if $S \in \langle W \rangle$ then $l_{a,b}(S) = S$, and hence

$$D_{a,b}(S) = l_{a,b}^{-1}Q(S) - Q(S) = 0 \tag{12.11}$$

since $Q(S) \in \langle W \rangle$ by Proposition 12.24. We view $D = D_{a,b}$ as a quadratic form in R, and let

$$B(R,S) := \frac{1}{4}\big(D(R+S) - D(R-S)\big) = l_{a,b}^{-1} Q(l_{a,b}(R), S) - Q(R,S)$$

be the corresponding bilinear form.

It suffices to show $B(R,S) = 0$ for all $S \in \langle W \rangle$ and $R \in \text{Curv}$: In this case $D(R+S) - D(R) = (D(R) + 2B(R,S) + D(S)) - D(R) = 0$, since $D(S) = 0$ for any $S \in \langle W \rangle$ by (12.11).

We fix $S = W \in \langle W \rangle$ and prove $B(S, \cdot\,) = 0$ by considering $B(R,S)$ for R in each of the $O(n)$-irreducible components of Curv:

1. Suppose $R \in \langle W \rangle$. Then by (12.11) we have $D(R+S) = D(R-S) = 0$, so $B(R,S) = 0$.
2. When $R = I$ is the identity, $l_{a,b}(I) = (1 + 2(n-1)a)I =: (1+\alpha)I$ by Definition 12.30, so

   $$B(W,I) = l_{a,b}^{-1}\left((1+\alpha)Q(I,W)\right) - Q(I,W).$$

   However we have by the polarisation identity

   $$\begin{aligned} Q(I,W) &= \frac{1}{4}\left(Q(R+I) - Q(R-I)\right) \\ &= \frac{1}{4}\left((W+I)^2 + (W+I)^{\#} - (W-I)^2 - (W-I)^{\#}\right) \\ &= (W + W\#I)\,, \end{aligned}$$

   which vanishes by Lemma 12.26.
3. It remains to consider the case when $R \in \langle \mathring{\text{Ric}} \rangle$. From the definition of $B(W,R)$ it suffices to prove that $Q(W, l(R))$=$l(Q(W,R)$ To do this, note that since $R \in \langle \mathring{\text{Ric}} \rangle$, we have $l(R)$=$(1 + (n-2)b)R =: (1+\beta)R$ by Definition 12.30, and so

   $$Q(W, l(R)) = (1+\beta)Q(W,R) = l(Q(W,R)),$$

   since $Q(W,R) \in \langle \mathring{\text{Ric}} \rangle$ by Proposition 12.25.

This completes the proof that $Q(W, \cdot\,)$=0, and hence that $D(R)$ is independent of the Weyl part of R. □

Now we can compute the proof of Theorem 12.33 by assuming that $R_W = 0$, so that $R = R_I + R_{\mathring{\text{Ric}}}$. Let $\alpha = 2(n-1)a$, $\beta = (n-2)b$ and $R_0 = R_{\mathring{\text{Ric}}} = \frac{2}{n-2}\mathring{\text{Ric}} \wedge \text{id}$ so that

$$l_{a,b}(R) = \frac{(1+\alpha)\bar{\lambda}}{n-1} I + (1+\beta)R_0.$$

Using this with Lemma 12.26 yields

$$\begin{aligned}
D_{a,b}(R) &= l_{a,b}^{-1}\Big(\frac{(1+\alpha)^2\bar\lambda^2}{(n-1)^2}(I^2+I\#I)+\frac{2(1+\alpha)(1+\beta)\bar\lambda}{n-1}(R_0+I\#R_0)\\
&\qquad +(1+\beta)^2(R_0^2+R_0^\#)\Big)-Q(R)\\
&= l_{a,b}^{-1}\Big(\frac{(1+\alpha)^2\bar\lambda^2}{n-1}I+\frac{2(1+\alpha)(1+\beta)\bar\lambda}{n-1}\mathring{\mathrm{Ric}}\wedge\mathrm{id}\\
&\qquad +(1+\beta)^2Q(R_0)\Big)-Q(R).
\end{aligned}$$

By Proposition 12.21, with $A=\mathring{\mathrm{Ric}}$, we have

$$\mathring{\mathrm{Ric}}\wedge\mathring{\mathrm{Ric}}=\frac{-\sigma}{n-1}I-\frac{2}{n-2}(\mathring{\mathrm{Ric}}^2)_0\wedge\mathrm{id}+(\mathring{\mathrm{Ric}}\wedge\mathring{\mathrm{Ric}})_W. \tag{12.12}$$

So by Claim 12.29 we get

$$\begin{aligned}
Q(R_0) &= \frac{1}{n-2}\mathring{\mathrm{Ric}}\wedge\mathring{\mathrm{Ric}}-\frac{2}{(n-2)^2}(\mathring{\mathrm{Ric}}^2)_0\wedge\mathrm{id}+\frac{\sigma}{n-2}I\\
&= \frac{\sigma}{n-1}I-\frac{4}{(n-2)^2}(\mathring{\mathrm{Ric}}^2)_0\wedge\mathrm{id}+\frac{1}{n-2}(\mathring{\mathrm{Ric}}\wedge\mathring{\mathrm{Ric}})_W.
\end{aligned}$$

In this form we know how $\ell_{a,b}^{-1}$ acts on each term. Also, by Lemma 12.28 we have

$$\begin{aligned}
Q(R) &= \Big(\frac{\bar\lambda^2}{n-1}+\frac{\sigma}{n-2}\Big)I+\frac{2\bar\lambda}{n-1}\mathring{\mathrm{Ric}}\wedge\mathrm{id}\\
&\quad +\frac{1}{n-2}\mathring{\mathrm{Ric}}\wedge\mathring{\mathrm{Ric}}-\frac{2}{(n-2)^2}(\mathring{\mathrm{Ric}}^2)_0\wedge\mathrm{id}.
\end{aligned}$$

Using these formulas for $Q(R_0)$ and $Q(R)$ with the fact that $l_{a,b}^{-1}(R)=\frac{1}{1+\alpha}\frac{\bar\lambda}{n-1}I+\frac{1}{1+\beta}R_0$ results in

$$\begin{aligned}
D_{a,b}(R) &= \frac{(1+\alpha)^2\bar\lambda^2}{n-1}l_{a,b}^{-1}(I)+\frac{2(1+\alpha)(1+\beta)\bar\lambda}{n-1}l_{a,b}^{-1}(\mathring{\mathrm{Ric}}\wedge\mathrm{id})\\
&\quad +(1+\beta)^2l_{a,b}^{-1}(Q(R_0))-Q(R)\\
&= \frac{(1+\alpha)\bar\lambda^2}{n-1}I+\frac{2(1+\alpha)\bar\lambda}{n-1}\mathring{\mathrm{Ric}}\wedge\mathrm{id}\\
&\quad +\frac{(1+\beta)^2}{1+\alpha}\frac{\sigma}{n-1}I-\frac{4(1+\beta)}{(n-2)^2}(\mathring{\mathrm{Ric}}^2)_0\wedge\mathrm{id}+\frac{(1+\beta)^2}{n-2}(\mathring{\mathrm{Ric}}\wedge\mathring{\mathrm{Ric}})_W\\
&\quad -\Big(\frac{\bar\lambda^2}{n-1}+\frac{\sigma}{n-2}\Big)I-\frac{2\bar\lambda}{n-1}\mathring{\mathrm{Ric}}\wedge\mathrm{id}\\
&\quad -\frac{1}{n-2}\mathring{\mathrm{Ric}}\wedge\mathring{\mathrm{Ric}}+\frac{2}{(n-2)^2}(\mathring{\mathrm{Ric}}^2)_0\wedge\mathrm{id}.
\end{aligned}$$

Substituting $(\overset{\circ}{\mathrm{Ric}} \wedge \overset{\circ}{\mathrm{Ric}})_W$ from (12.12) and simplifying gives

$$\begin{aligned}
D_{a,b}(R) &= \frac{\alpha\bar{\lambda}^2}{n-1} I + \frac{2\alpha\bar{\lambda}}{n-1} \overset{\circ}{\mathrm{Ric}} \wedge \mathrm{id} + \frac{2\beta+\beta^2}{n-2} \overset{\circ}{\mathrm{Ric}} \wedge \overset{\circ}{\mathrm{Ric}} \\
&\quad + \frac{2\beta^2}{(n-2)^2} (\overset{\circ}{\mathrm{Ric}}{}^2)_0 \wedge \mathrm{id} + \Big(\frac{(1+\beta)^2}{1+\alpha} - \frac{n-1}{n-2} + \frac{(1+\beta)^2}{n-2} \Big) \frac{\sigma}{n-1} I \\
&= \frac{\alpha}{n-1} (\mathrm{Ric} \wedge \mathrm{Ric} - \overset{\circ}{\mathrm{Ric}} \wedge \overset{\circ}{\mathrm{Ric}}) + \frac{2\beta+\beta^2}{n-2} \overset{\circ}{\mathrm{Ric}} \wedge \overset{\circ}{\mathrm{Ric}} \\
&\quad + \frac{2\beta^2}{(n-2)^2} (\overset{\circ}{\mathrm{Ric}}{}^2)_0 \wedge \mathrm{id} + \Big(\frac{(1+\beta)^2}{1+\alpha} - 1 + \frac{2\beta+\beta^2}{n-2} \Big) \frac{\sigma}{n-1} I
\end{aligned}$$

since $\mathrm{Ric} \wedge \mathrm{Ric} - \overset{\circ}{\mathrm{Ric}} \wedge \overset{\circ}{\mathrm{Ric}} = \bar{\lambda}^2 I + 2\bar{\lambda} \overset{\circ}{\mathrm{Ric}} \wedge \mathrm{id}$. Substituting a and b back in for α and β together with $(\overset{\circ}{\mathrm{Ric}}{}^2)_0 = \overset{\circ}{\mathrm{Ric}}{}^2 - \sigma \mathrm{id}$ and $\mathrm{tr}(\overset{\circ}{\mathrm{Ric}}{}^2) = n\sigma$ gives the desired equation. □

Corollary 12.35 (Eigenvalues of $D_{a,b}$ and $\mathrm{Rc}(D_{a,b})$). *Suppose $(e_i)_{i=1}^n$ is an orthonormal basis of eigenvectors corresponding to the eigenvalues $(\lambda_i)_{i=1}^n$ of $\overset{\circ}{\mathrm{Ric}}$. Then $e_i \wedge e_j$, for $i < j$, is an eigenvector of $D_{a,b}(R)$ corresponding to the eigenvalue*

$$\begin{aligned}
d_{ij} &= \big((n-2)b^2 - 2(a-b)\big)\lambda_i\lambda_j + 2a(\bar{\lambda}+\lambda_i)(\bar{\lambda}+\lambda_j) + b^2(\lambda_i^2+\lambda_j^2) \\
&\quad + \frac{\sigma}{1+2(n-1)a}\big(nb^2(1-2b) - 2(a-b)(1-2b+nb^2)\big). \qquad (12.13)
\end{aligned}$$

Furthermore, e_i is an eigenvector of $\mathrm{Rc}(D_{a,b}(R))$ with eigenvalue

$$\begin{aligned}
r_i &= -2b\lambda_i^2 + 2a\bar{\lambda}(n-2)\lambda_i + 2a(n-1)\bar{\lambda}^2 \\
&\quad + \frac{\sigma}{1+2(n-1)a}\big(n^2b^2 - 2(n-1)(a-b)(1-2b)\big). \qquad (12.14)
\end{aligned}$$

Remark 12.36. Notice that $\lambda_i + \bar{\lambda}$ are the eigenvalues of the Ricci tensor Ric.

Proof. Choose an orthonormal basis $(e_i)_{i=1}^n$ of eigenvectors of $\overset{\circ}{\mathrm{Ric}}$ with corresponding eigenvalues $(\lambda_i)_{i=1}^n$. By (12.4) we find that $(\overset{\circ}{\mathrm{Ric}} \wedge \overset{\circ}{\mathrm{Ric}})(e_i \wedge e_j) = \lambda_i\lambda_j e_i \wedge e_j$, $(\overset{\circ}{\mathrm{Ric}}{}^2 \wedge \mathrm{id})(e_i \wedge e_j) = \frac{1}{2}(\lambda_i^2 + \lambda_j^2) e_i \wedge e_j$, and $(\mathrm{Ric} \wedge \mathrm{Ric})(e_i \wedge e_j) = (\lambda_i + \bar{\lambda})(\lambda_j + \bar{\lambda}) e_i \wedge e_j$. From this it easily follows that $e_i \wedge e_j$ is an eigenvector of $D_{a,b}(R)$ with eigenvalue d_{ij} given by (12.13).

For the second expression, using (12.6), (12.7) and (12.5) we find that $\mathrm{Rc}(\mathrm{Ric}\wedge\mathrm{Ric}) = (n-1)\bar{\lambda}^2\mathrm{id} + (n-2)\bar{\lambda}\overset{\circ}{\mathrm{Ric}} - \overset{\circ}{\mathrm{Ric}}{}^2$, $\mathrm{Rc}(\mathrm{id}\wedge\mathrm{id}) = (n-1)\mathrm{id}$, $\mathrm{Rc}(\overset{\circ}{\mathrm{Ric}} \wedge \overset{\circ}{\mathrm{Ric}}) = -\overset{\circ}{\mathrm{Ric}}{}^2$, and $\mathrm{Rc}(\overset{\circ}{\mathrm{Ric}}{}^2 \wedge \mathrm{id}) = \left(\frac{n}{2} - 1\right) \overset{\circ}{\mathrm{Ric}}{}^2 + \frac{n\sigma}{2}\mathrm{id}$. From which we have

$$
\begin{aligned}
\operatorname{Rc}(D_{a,b}(R)) &= -2b \overset{\circ}{\operatorname{Ric}}{}^2 + 2(n-2)a\bar{\lambda} \overset{\circ}{\operatorname{Ric}} + 2(n-1)a\bar{\lambda}^2 \operatorname{id} \\
&\quad + \frac{2(n-1)b + (n-2)^2 b^2 - 2(n-1)a(1-2b)}{1+2(n-1)a} \sigma \operatorname{id}
\end{aligned}
$$

So by the same diagonalisation argument for $\overset{\circ}{\operatorname{Ric}}$, (12.14) follows. □

Chapter 13
The Cone Construction of Böhm and Wilking

13.1 New Invariant Sets

In this section the remarkable formulas derived in the previous section, particularly the identities (12.13) and (12.14), will be applied to construct a family of cones preserved by the Ricci flow. We follow the argument presented by Böhm and Wilking who applied it to produce a family of preserved cones interpolating between the cone of positive curvature operators and the line of constant positive curvature operators. The construction applies much more generally, so that given any preserved cone satisfying a few conditions, there is a family of cones linking that one to the ray of constant positive curvature operators. As we will see, this is a crucial step in proving that solutions the Ricci flow converge to spherical space forms.

Definition 13.1 (Pinching Family of Convex Cones). We call a continuous family $C(s) \subset \mathrm{Curv}$ of top-dimensional closed convex cones, parametrised by $s \in [0, \infty)$, a *pinching family* (with respect to the vector field $Q(R) = R^2 + R^\#$) if

1. $C(s)$ is an $O(n)$-invariant cone for each $s \geq 0$.
2. Each $R \in C(s)\backslash\{0\}$ has positive scalar curvature.
3. $Q(R)$ strictly points into[1] $C(s)$ at every point $R \in \partial C(s)\backslash\{0\}$, for all $s > 0$.
4. $C(s)$ converges (in compact sets in the Hausdorff topology) to the one-dimensional cone $\mathbb{R}^+ I$ as $s \to \infty$.

The motivation for making this definition comes from previous work, particularly that by Hamilton [Ham82b] in which he proved that the cones

$$C(s)_{s\in[0,1)} = \left\{ R \in S^2(\mathfrak{so}(3)) : \mathrm{Ric} \geq s\frac{\mathrm{tr}\,\mathrm{Ric}}{3}\mathrm{id} \right\}$$

[1] Recall that this means $Q(R)$ is in the interior of the tangent cone $\mathcal{T}_R C(s)$.

B. Andrews and C. Hopper, *The Ricci Flow in Riemannian Geometry*, Lecture Notes in Mathematics 2011, DOI 10.1007/978-3-642-16286-2_13,

are preserved by Ricci flow, so that compact manifolds with positive Ricci maintain a bound on the ratio of the eigenvalues of the Ricci curvature as long as the solution exists. This provides a continuous family of closed convex cones with $C(0)$ equal to the cone of 3-dimensional curvature operators with nonnegative Ricci curvature, and $C(1)$ equal to the cone of constant positive curvature.

13.1.1 Initial Cone Assumptions

We will assume that we have an initial preserved closed convex cone $C(0)$ which is $O(n)$-invariant, contained in the cone of positive sectional curvature, and contains the cone of positive curvature operators. We will show that:

Theorem 13.2. *There exists a pinching family $C(s)$, for $0 \leq s < \infty$, of closed convex cones starting at the cone $C(0)$.*

The construction uses the identities proved in the previous chapter to produce a pinching family of cones, given by applying the operators $l_{a,b}$ (for carefully chosen a and b) to the intersection of $C(0)$ with a cone of operators with pinched Ricci tensor (defined by a pinching ratio p). The definition of this family is in two stages (see Fig. 13.1): The first increases b to a critical value, and the second increases a to infinity (thus, as discussed in Remark 12.31, the family of cones approaches the line of positive constant curvature as $s \to \infty$).

Lemma 13.3. *For $s \in [0, \frac{1}{2}]$, let*

$$a = \frac{(n-2)s^2 + 2s}{2 + 2(n-2)s^2}, \qquad b = s, \qquad p = \frac{(n-2)s^2}{1 + (n-2)s^2}.$$

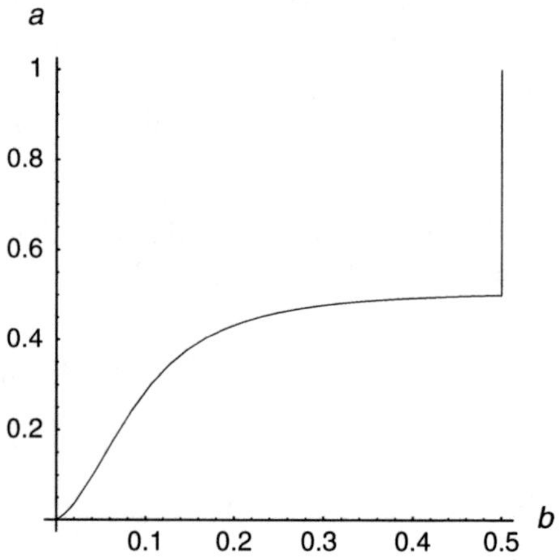

Fig. 13.1 The trajectory of the cone parameterised by s in the (a, b)-plane when $n = 89$

Then the vector field Q points strictly into the cone

$$C(s) = l_{a,b}\left(\left\{R \in \mathrm{Curv} : R \in C(0), \mathrm{Ric} \geq p\frac{\mathrm{tr}\,\mathrm{Ric}}{n}\right\}\right)$$

for $0 < s \leq \frac{1}{2}$.

Proof. By Lemma 12.32, we must prove that the vector field $X_{a,b}$ points strictly into the untransformed cone $C(0) \cap C_p$ at any non-zero boundary point R, where

$$C_p = \left\{R \in \mathrm{Curv} : \mathrm{Ric} \geq p\frac{\mathrm{tr}\,\mathrm{Ric}}{n}\right\}.$$

That is, we must check that $X_{a,b}(R)$ is in the interior of the tangent cone $\mathcal{T}_R(C(0) \cap C_p)$. By Theorem B.7, it suffices to check that $X_{a,b}(R) \in \mathcal{T}_R C(0)$ for $R \in \partial C(0) \cap C_p$, and that $X_{a,b}(R) \in \mathcal{T}_R C_p$ for $R \in \partial C_p \cap C(0)$. We consider these two cases in turn:

1. The boundary of $C(0)$: Suppose $R \in \partial C(0) \cap C_p$. By assumption $C(0)$ contains the cone C_+ of positive curvature operators, and hence $R + C_+ \subset C(0)$, and so

$$C_+ \subset \mathcal{T}_R C(0) = \bigcup_{h>0} \frac{C(0) - R}{h}.$$

Since we know by assumption that Q points into $C(0)$, it suffices to prove that $D_{a,b}$ strictly points into $C(0)$, and for this it suffices to prove that $D_{a,b}$ is in C_+. We can do this by checking the positivity of the eigenvalues of $D_{a,b}$ (given by Corollary 12.35).

As $R \in C_p$, we have the following estimate for the eigenvalues of Ric_0:

$$\lambda_i \geq -(1-p)\bar{\lambda}.$$

Next observe with our parametrisation that

$$2(a-b) = \frac{1-2b}{1+(n-2)b^2}(n-2)b^2 \qquad \text{and} \qquad (n-2)b^2 + 2b = \frac{2a}{1-p}.$$

In which case Corollary 12.35 implies that

$$\begin{aligned} d_{ij} = &\left(\frac{2a}{1-p} - 2a\right)\lambda_i\lambda_j + 2a(\lambda_i + \bar{\lambda})(\lambda_j + \bar{\lambda}) + b^2(\lambda_i^2 + \lambda_j^2) \\ &+ \frac{(1-2b)\sigma}{1+2(n-1)a}\left(nb^2 - \frac{(n-2)b^2}{1+(n-2)b^2}(1-2b+nb^2)\right) \end{aligned}$$

$$
\begin{aligned}
&= 2a\Big(\frac{1}{1-p}\lambda_i\lambda_j + (1-p)\bar{\lambda}^2 + (\lambda_i+\lambda_j)\bar{\lambda}\Big) + 2ap\bar{\lambda}^2 + b^2(\lambda_i + \lambda_j^2) \\
&\qquad + \frac{2(1-2b)b^2\sigma}{1+2(n-1)a}\frac{1-(n-2)b}{1-(n-2)b^2} \\
&> \frac{2a}{1-p}(\lambda_i + (1-p)\bar{\lambda})(\lambda_j + (1-p)\bar{\lambda}) \\
&\geq 0
\end{aligned}
$$

since $0 \leq b \leq 1/2$ and $\lambda_i + \bar{\lambda} \geq p\bar{\lambda}$.

2. The boundary of C_p: The cone C_p may be given as an intersection of half-spaces in the following way (compare the general construction in Sect. 12.1.2):

$$
C_p = \bigcap_{\mathbb{O}\in O(n)} \{R \in \mathrm{Curv} : \; \ell(R^{\mathbb{O}}) \leq 0\}
$$

where $\ell(R) = p\frac{\mathrm{scal}(R)}{n} - \mathrm{Rc}(R)(e_1, e_1)$. By symmetry we can assume that we are working at a boundary point R with $\ell(R) = 0$ and $R \in C(0)$, and by Theorem B.7 it is sufficient to prove that $\ell(X_{a,b}(R)) < 0$ (for $R \neq 0$).

Since $\ell(R^{\mathbb{O}}) \leq 0$ for all $\mathbb{O}$ we have $\lambda_i + \bar{\lambda} \geq p\bar{\lambda}$, so that $\lambda_i \geq -(1-p)\bar{\lambda}$, with equality holding for $i = 1$ (corresponding to the e_1 direction). We have to show that

$$
\mathrm{Rc}(X_{a,b}(R))_{11} > p\frac{\mathrm{scal}(X_{a,b}(R))}{n}.
$$

Firstly, by Proposition 12.24 we have

$$
\mathrm{scal}(Q(R)) = \sum_k |\mathrm{Ric}_{kk}|^2 = \sum_k (\lambda_k + \bar{\lambda})^2 = n(\sigma + \bar{\lambda}^2). \tag{13.1}
$$

As $\sum \lambda_i = 0$ and $\sum \lambda_i^2 = n\sigma$, we find from Corollary 12.35 that

$$
\begin{aligned}
\mathrm{scal}(D_{a,b}(R)) &= \sum_i r_i \\
&= -2bn\sigma + 2an(n-1)\bar{\lambda}^2 \\
&\qquad + \frac{n\sigma}{1+2(n-1)a}\big(n^2b^2 - 2(n-1)(a-b)(1-2b)\big) \\
&= 2n(n-1)a\bar{\lambda}^2 - n\sigma + \frac{n(1+(n-2)b)^2}{1+2(n-1)a}\sigma.
\end{aligned}
$$

Combining this result with (13.1) gives

$$
\frac{\mathrm{scal}(X_{a,b}(R))}{n} = (1+2(n-1)a)\bar{\lambda}^2 + \frac{(1+(n-2)b)^2}{1+2(n-1)a}\sigma. \tag{13.2}
$$

Note that this equation holds for any $a \neq -\frac{1}{2(n-1)}$ and is *independent* of the choice of parametrisation.

Now by Proposition 12.24 with $\mathrm{Ric}_{kk} \geq p\bar{\lambda}$ we get

$$\mathrm{Rc}(R^2 + R^\#)_{ii} = \sum_k \mathrm{Ric}_{kk} R_{ikik} \geq p\bar{\lambda}\mathrm{Ric}_{ii} \geq p^2\bar{\lambda}^2.$$

Also, from the previous step our given parametrisation implies that

$$\begin{aligned}
d_{ij} &= \frac{2a}{1-p}(\lambda_i + (1-p)\bar{\lambda})(\lambda_j + (1-p)\bar{\lambda}) + 2ap\bar{\lambda}^2 + b^2(\lambda_i^2 + \lambda_j^2) \\
&\quad + \frac{\sigma}{1+2(n-1)a}\big(nb^2(1-2b) - 2(a-b)(1-2b+nb^2)\big) \\
&\geq 2ap\bar{\lambda}^2 + b^2(\lambda_i^2 + \lambda_j^2) \\
&\quad + \frac{\sigma}{1+2(n-1)a}\big(nb^2(1-2b) - 2(a-b)(1-2b+nb^2)\big).
\end{aligned}$$

In which case we find that

$$\begin{aligned}
\mathrm{Rc}(X_{a,b})_{ii} &\geq p^2\bar{\lambda}^2 + \sum_{j\neq i} d_{ij} \\
&\geq p^2\bar{\lambda}^2 + 2a(n-1)p\bar{\lambda}^2 + (n-2)b^2\lambda_i^2 + nb^2\sigma \\
&\qquad + \frac{(n-1)\sigma}{1+2(n-1)a}\big(nb^2(1-2b) - 2(a-b)(1-2b+nb^2)\big) \\
&= p^2\bar{\lambda}^2 + 2a(n-1)p\bar{\lambda}^2 + (n-2)b^2\lambda_i^2 \\
&\qquad + \Big(\frac{(1+(n-2)b)^2}{1+2(n-1)a} + 2b - 1\Big)\sigma.
\end{aligned}$$

By combining this result with (13.2) we find that

$$\begin{aligned}
\mathrm{Rc}(X_{a,b})_{ii} - p\frac{\mathrm{scal}(X_{a,b})}{n} &\geq p(p-1)\bar{\lambda}^2 + (n-2)b^2\lambda_i^2 \\
&\quad + \Big((1-p)\frac{(1+(n-2)b)^2}{1+2(n-1)a} + 2b - 1\Big)\sigma.
\end{aligned}$$

From the stated parametrisation of $p = p(b)$ and $a = a(b)$ it is straightforward to check that

$$\begin{aligned}
p(p-1)\bar{\lambda}^2 + (n-2)b^2\lambda_i^2 &= (n-2)b^2(\lambda_i^2 - (1-p)^2\bar{\lambda}^2) \\
(1-p)\frac{(1+(n-2)b)^2}{1+2(n-1)a} + 2b - 1 &= \frac{2nb^2}{nb+1}.
\end{aligned}$$

Hence we get

$$\begin{aligned}\operatorname{Rc}(X_{a,b})_{ii} - p\frac{\operatorname{scal}(X_{a,b})}{n} &\geq (n-2)b^2(\lambda_i^2 - (1-p)^2\bar{\lambda}^2) + \frac{2nb^2}{nb+1} \\ &\geq \frac{2nb^2}{nb+1} > 0\end{aligned}$$

since $\lambda_i = -(1-p)\bar{\lambda}$. □

Lemma 13.4. *For $s \in [1/2, \infty)$, let*

$$a = \frac{1+2s}{4}, \qquad b = \frac{1}{2}, \qquad p = 1 - \frac{4}{n+4s}.$$

Then the vector field $Q(R)$ strictly points into the cone $C(s) = l_{a,b}\left(C(0) \cap C_p\right)$ at points $R \in \partial C(s) \setminus \{0\}$.

Remark 13.5. Notice that $\lim_{s\to\infty} l_{a,b}(R) = 2(n-1)R_I$. Consequently the cones of the lemma converge to $\mathbb{R}^+ I$ for $s \to \infty$.

Proof. The proof is very similar to the previous one: We have two cases to check, the first for $R \in (\partial C(0)) \cap C_p$, and the other for $R \in C(0) \cap \partial C_p$. For convenience we define $u = s - \frac{1}{2} \geq 0$.

1. The boundary of $C(0)$: As before it is sufficient to prove that the eigenvalues of $D_{a,b}$ are positive. Using Corollary 12.35 with $a = (1+u)/2$ and $b = 1/2$, we find that

$$d_{ij} = \Big(\frac{n-2}{4} - u\Big)\lambda_i\lambda_j + (1+u)(\bar{\lambda}+\lambda_i)(\bar{\lambda}+\lambda_j) + \frac{1}{4}(\lambda_i^2+\lambda_j^2) - \frac{\sigma n u}{4n+4(n-1)u}.$$

We observe that:

Claim 13.6 The quantity $\sigma = \frac{1}{n}\sum\lambda_i^2$ is bounded above in terms of $\bar{\lambda}$:

$$\sigma \leq \frac{16(n-1)\bar{\lambda}^2}{(n+2+4u)^2}. \tag{13.3}$$

Proof of Claim. Since $\lambda_i + \bar{\lambda} \geq p\bar{\lambda}$ for all i, consider the optimisation problem for $\sigma = \sigma(\lambda_1, \ldots, \lambda_n)$ under the constraint $\sum\lambda_i = 0$. As the function σ has a strictly positive Hessian, it achieves its maximum on the boundary of the set

$$\Big\{(\lambda_1, \ldots, \lambda_n) : \lambda_i + \bar{\lambda} \geq p\bar{\lambda} \text{ for all } i, \text{ and} \sum\lambda_i = 0\Big\}$$

and furthermore at the vertices. Therefore the extremal points of σ are at $(-(1-p)\bar{\lambda}, \ldots, (n-1)(1-p)\bar{\lambda})$ and permutations thereof. Hence $\sigma \leq (n-1)(1-p)^2\bar{\lambda}^2$. □

In addition, as $\lambda_i + \bar{\lambda} \geq p\bar{\lambda}$ and $1 - p = \frac{4}{n+2+4u}$, we also have

$$\lambda_i + \frac{4\bar{\lambda}}{n+2} \geq \lambda_i + \frac{4\bar{\lambda}}{n+2+4u} \geq 0. \tag{13.4}$$

From these inequalities we find that

$$\begin{aligned} d_{ij} &= \frac{n+2}{4}\lambda_i\lambda_j + (1+u)\bar{\lambda}^2 + (1+u)(\lambda_i+\lambda_j)\bar{\lambda} \\ &\quad + \frac{1}{4}(\lambda_i^2+\lambda_j^2) - \frac{\sigma n u}{4n+4(n-1)u} \\ &= \frac{n+2}{4}\Big(\lambda_i + \frac{4\bar{\lambda}}{n+2}\Big)\Big(\lambda_j + \frac{4\bar{\lambda}}{n+2}\Big) + \frac{1}{4}(\lambda_i^2+\lambda_j^2) \\ &\quad + \bar{\lambda}^2\Big(\frac{n-2}{n+2} + u\Big) + u(\lambda_i+\lambda_j)\bar{\lambda} - \frac{\sigma n u}{4n+4(n-1)u} \\ &\geq \Big(\frac{n-2}{n+2} + u - \frac{8u}{n+2+4u} - \frac{4(n-1)nu}{(n+(n-1)u)(n+2+4u)^2}\Big)\bar{\lambda}^2 \\ &\geq \Big(\frac{n-2}{n+2} + u\frac{n-6+4u}{n+2+4u} - \frac{4(n-1)u}{(n+2)(n+2+4u)}\Big)\bar{\lambda}^2 \\ &= \Big(4u^2 + \big(n-6-\frac{4}{n+2}\big)u + n - 2\Big)\frac{\bar{\lambda}^2}{n+2+4u}. \end{aligned}$$

To get the first inequality we discarded the first two terms, and used the inequality (13.4) for the fourth term and (13.3) for the last. The second inequality is due to $\frac{n}{(n+(n-1)u)(n+2+4u)} \leq \frac{1}{n+2}$ for $n \geq 1$ and $u \geq 0$. Moreover, when $n = 3$ we have $4u^2 + (n-6-\frac{4}{n+2})u + n - 2 = 4u^2 - \frac{19}{5}u + 1 > 0$. Also for fixed $u \geq 0$, the function $4u^2 + (n-6-\frac{4}{n+2})u + n - 2$ is monotone increasing in n. Therefore $d_{ij} > 0$ for all $n \geq 3$ and $u \geq 0$.

2. The boundary of C_p: By the same argument as in the previous case, it suffices to show that:

$$\mathrm{Rc}(X_{a,b})_{ii} = \mathrm{Rc}(D_{a,b})_{ii} + \mathrm{Rc}(R^2 + R^{\#})_{ii} > p\frac{\mathrm{scal}(X_{a,b})}{n}.$$

By Corollary 12.35, with $a = (1+u)/2$ and $b = 1/2$, we find that

$$r_i = -\lambda_i^2 + (u+1)\bar{\lambda}(n-2)\lambda_i + (u+1)(n-1)\bar{\lambda}^2 + \frac{\sigma n^2}{4n+4(n-1)u}.$$

Also, Proposition 12.24 with $\mathrm{Ric}_{kk} \geq p\bar{\lambda}$ implies that $\mathrm{Rc}(R^2 + R^{\#})_{ii} = \sum_k \mathrm{Ric}_{kk} R_{ikik} \geq p\bar{\lambda}\,\mathrm{Ric}_{ii} \geq p^2\bar{\lambda}^2$ and by (13.2) we also have

$$\frac{\mathrm{scal}(X_{a,b})}{n} = (n+(n-1)u)\bar{\lambda}^2 + \frac{n^2\sigma}{4n+4(n-1)u}.$$

In which case we may suppose that $\lambda_i = -(1-p)\bar{\lambda}$, so it suffices to show

$$r_i + p^2\bar{\lambda}^2 > p\Big((n+(n-1)u)\bar{\lambda}^2 + \frac{n^2\sigma}{4n+4(n-1)u}\Big)$$

or equivalently

$$\begin{aligned}0 \le p^2\bar{\lambda}^2 &- (1-p)^2\bar{\lambda}^2 - (u+1)\bar{\lambda}^2(n-2)(1-p) + (u+1)(n-1)\bar{\lambda}^2\\ &+ \frac{\sigma n^2}{4n+4(n-1)u} - p\Big((n+(n-1)u)\bar{\lambda}^2 + \frac{n^2}{4n+4(n-1)u}\sigma\Big).\end{aligned}$$

Since $\sigma \ge 0$ we can neglect the terms containing σ. Dividing by $\bar{\lambda}^2$ gives

$$p^2 - (1-p)^2 + (u+1) + (u+1)p(n-2) - p(n+(n-1)u) = u(1-p)$$

which is clearly positive. □

13.2 Generalised Pinching Sets

Hamilton [Ham86, p. 163] introduced the notion of a pinching set, which he used to prove that solutions of the Ricci flow on four-manifolds with positive curvature operator converge to space-forms. His definition is as follows:

Definition 13.7 (Pinching Set). A subset $Z \subset \mathrm{Curv}$ is a *pinching set* if

1. Z is closed and convex.
2. Z is $O(n)$-invariant.
3. Z is preserved by the ODE $\frac{d}{dt}R = R^2 + R^{\#}$.
4. There exist $\delta > 0$ and $K < \infty$ such that

$$|\widetilde{R}| \le K|R|^{1-\delta}$$

for all $R \in Z$, where $\widetilde{R} = R - \frac{1}{N}\mathrm{tr}R$ is the trace-free part of R.

This definition is closely analogous to the conditions he used in his paper on three-manifolds [Ham82b], where he proved that sets of the form

$$\{R : \mathrm{Ric} \ge \varepsilon \mathrm{Scal}\, g\} \cap \{R : \ |\mathrm{Ric}_0|^2 \le C\mathrm{Scal}^{2-\gamma}\}$$

are preserved by the Ricci flow for suitable γ depending on ε (see Sect. 7.5.3 for a different construction of pinching sets for this situation). The important point is that if the scalar curvature becomes large then the traceless part becomes relatively small, so that the curvature is close to that of a constant sectional curvature space.

13.2.1 Generalised Pinching Set Existence Theorem

Böhm and Wilking [BW08, Theorem 4.1] observed that a weaker notion of pinching set suffices for applications. The theorem below (which is similar to that of Böhm and Wilking but modified following ideas of Brendle and Schoen [BS09a, Sect. 3]) still provides a set which guarantees pinching to constant sectional curvatures where the scalar curvature is large. The result is a useful tool, because it proves the existence of such a pinching set simply from the existence of a suitable family of cones. We will apply it in the final argument (see Sect. 15.1) to the family of cones constructed above.

Theorem 13.8. *Let $\{C(s)\}_{s\in[0,\infty)}$ be a continuous family of closed convex $O(n)$-invariant cones in* Curv *of maximal dimension, contained in the half-space of curvature operators with positive scalar curvature. Suppose that for any $s>0$ and any $R\in\partial C(s)\setminus\{0\}$, the vector $Q(R)=R^2+R^{\#}$ is contained in the interior of the tangent cone of $C(s)$ at R. Then if K is any compact set contained in the interior of $C(0)$, there exists a closed convex $O(n)$-invariant set $F\subset C(0)$ with the following properties:*

1. *$Q(R)$ is in the tangent cone of F for every $R\in\partial F$.*
2. *$K\subset F$.*
3. *For each $s>0$ there exists $\rho(s)$ such that $F+\rho(s)I\subset C(s)$.*

Proof. By the continuity of the family $\{C(s)\}$, given any compact K in the interior of $C(0)$ we have $K\subset C(s_0)$ for some $s_0>0$. Adapting an idea of Brendle and Schoen, we will construct a set F of the form

$$F=C(s_0)\cap\bigcap_{i=1}^{\infty}\{R:\ R+2^ihI\in C(s_i)\},$$

where $h>0$ and (s_i) is an increasing sequence approaching infinity. Thus F is produced by a sequence of intersections with translated copies of the cones $C(s_i)$. The idea is to choose (s_i) in such a way that each intersection only changes F where the scalar curvature is large, and the vector field $Q(R)$ points into the new part of the boundary at each stage (that is, into the boundary of the translated cone at points of large scalar curvature).

The set F is manifestly convex, and condition (3) of the theorem holds automatically. Also, since each set $C(s)$ is $O(n)$-invariant, so is F. We first choose $s_0>0$ and $h>0$ such that

$$K\subset C(s_0)\cap\{\mathrm{scal}(R)\leq h\}.$$

Lemma 13.9. *For any $\bar{s}\geq s_0$ there exists $N(\bar{s})\geq 1$ (non-decreasing in $\bar{s}$) such that if $s\in[s_0,\bar{s}]$ and $R\in\partial C(s)$ with $\mathrm{scal}(R)\geq N(\bar{s})$, then $Q(S)$ is in the tangent cone to $C(s)$ at R for every S with $|S-R|\leq 2|I|$.*

Proof of Lemma. The set $Z = \{(s,R):\ s \in [s_0,\bar{s}],\ R \in \partial C(s),\ \mathrm{scal}(R) = 1\}$ is compact. We claim:

Claim. There exists $r > 0$ such that $(s,R) \in Z$, $|S - R| \le r$ implies $Q(S)$ is in the tangent cone to $C(s)$ at R.

Proof of Claim. If not, there exists a sequence (s_i, R_i, S_i) with $(s_i, R_i) \in Z$ and $|S_i - R_i| \to 0$, but $Q(S_i)$ not in $\mathcal{T}_{R_i}C(s_i)$. Thus for each i there exists a unit norm linear function ℓ_i with $\ell_i(R_i) = \sup_{C(s_i)} \ell_i$ so that ℓ is in the normal cone $\mathcal{N}_{R_i}C(s_i)$, but $\ell_i(Q(S_i)) > 0$ (see Definition B.4 in Appendix B). By compactness, after passing to a subsequence we have $s_i \to s \in [s_0, \bar{s}]$, and (by continuity of the family $\{C(s)\}$) $R_i \to R \in \partial C(s) \cap \{\mathrm{scal}(R) = 1\}$. By compactness of the set of unit norm linear functions we also have $\ell_i \to \ell$, and $\ell(R) = \sup_{C(s)} \ell$ so that ℓ is in the normal cone to $C(s)$ at R. Since $|S_i - R_i| \to 0$ we also have $S_i \to R$, and by continuity of Q we have $\ell(Q(R)) \ge 0$. But this is a contradiction to the assumption that $Q(R)$ is in the interior of $\mathcal{T}_R C(s)$ (equivalently $\ell(Q(R)) < 0$ for every $\ell \in \mathcal{N}_R C(s)$). □

We claim the lemma holds with $N(\bar{s}) = 2|I|/r$. For if $R \in \partial C(s)$ with $\mathrm{scal}(R) \ge N(\bar{s})$, then $(s, R/\mathrm{scal}(R)) \in Z$ and $|S - R| \le 2|I|$ gives $|S/\mathrm{scal}(R) - R/\mathrm{scal}(R)| \le 2|I|/\mathrm{scal}(R) \le 2|I|/N(\bar{s}) = r$. Hence $Q(S/\mathrm{scal}(R))$ is in the tangent cone to $C(s)$ at $R/\mathrm{scal}(R)$. The result follows since $Q(S) = \mathrm{scal}(R)^2 Q\left(S/\mathrm{scal}(R)\right)$ and the tangent cone to $C(s)$ at R is the same as the tangent cone to $C(s)$ at $R/\mathrm{scal}(R)$. □

Lemma 13.10. *There exists a non-increasing function* $\delta : [s_0, \infty) \to \mathbb{R}_+$ *such that whenever* $s \in [s_0, \bar{s}]$ *and* $R + I \in C(s)$ *with* $\mathrm{scal}(R) \le N(\bar{s})$, *then* $R + 2I \in C(s + \delta(\bar{s}))$.

Proof of Lemma. The set $Z = \{(s, R+2I):\ s \in [s_0,\bar{s}],\ \mathrm{scal}(R) \le N(\bar{s}),\ R + I \in C(s)\}$ is a compact set in the interior of $\{(s,A):\ s \in [s_0,\bar{s}],\ A \in C(s)\}$. By continuity of the family $\{C(s)\}$, there exists $\delta > 0$ such that $Z \subset \{(s,A):\ s \in [s_0,\bar{s}],\ A \in C(s+\delta)\}$. □

We now construct the sequence s_i inductively by taking $s_{i+1} = s_i + \delta(s_i)$ for each $i \ge 1$. Note that since δ is a non-increasing positive function, $\lim_{i\to\infty} s(i) = \infty$ (otherwise we would have $\delta(s) = 0$ for $s > \lim_{i\to\infty} s_i$).

Let

$$F_j = C(s_0) \cap \bigcap_{i=1}^{j} \{R:\ R + 2^i h I \in C(s_i)\}.$$

We prove that for each j,

$$F_{j+1} \cap \{\mathrm{scal}(R) \le 2^j N(s_j) h\} = F_j \cap \{\mathrm{scal}(R) \le 2^j N(s_j) h\}. \tag{13.5}$$

To see this we must show that the set on the right is contained in that on the left. If $R \in F_j \cap \{\mathrm{scal}(R) \le 2^j N(s_j)h\}$ then $R + 2^j hI \in C(s_j)$ and $\mathrm{scal}(R) \le 2^j N(s_j)h$. Therefore $R/(2^j h) + I \in C(s_j)$ and $\mathrm{scal}(R/(2^j h)) \le N(s_j)$, so by Lemma 13.10,

$$R/(2^j h) + 2I \in C(s_j + \delta(s_j)) = C(s_{j+1})$$

and $R + 2^{j+1}hI \in C(s_{j+1})$. Thus $R \in F_{j+1}$ as required.

Now (13.5) says that as j increases above an index i, the part of F_j with $\mathrm{scal}(R) \leq 2^i N(s_i)h$ does not change. It follows that $F \cap \{\mathrm{scal}(R) \leq 2^i N(s_i)h\} = F_i \cap \{\mathrm{scal}(R) \leq 2^i N(s_i)h\}$ for each i. In particular F is locally the intersection of finitely many closed sets, so is closed. Also, we have

$$K \subset C(s_0) \cap \{\mathrm{scal}(R) \leq h\} = F \cap \{\mathrm{scal}(R) \leq h\} \subset F.$$

It remains to prove that $Q(R)$ is in the tangent cone of F for every $R \in \partial F$. In such a case R is in $\partial C(s_0)$ or in $\partial(C(s_i) - 2^i hI)$ for finitely many values of i, where necessarily $\mathrm{scal}(R) \geq 2^{i-1}N(s_i)h$ by the inclusion (13.5). We must show that $Q(R)$ lies in $\mathcal{T}_R\left(C(s_i) - 2^i hI\right) = \mathcal{T}_{R+2^i hI}C(s_i)$ for each such i. This can be done by letting

$$R' = 2^{1-i}h^{-1}(R + 2^i hI) \qquad \text{and} \qquad S = 2^{i-1}h^{-1}R,$$

so that $\mathrm{scal}(R') \geq N(s_i)$ and $|S - R'| = 2|I|$. Now by Lemma 13.9, $Q(S)$ is in the tangent cone to $C(s_i)$ at R', and hence $Q(R) \in \mathcal{T}_R\left(C(s_i) - 2^i hI\right)$ as required. □

Chapter 14
Preserving Positive Isotropic Curvature

The condition of positive curvature on totally isotropic 2-planes was first introduced by Micallef and Moore [MM88]. They were able to prove the following sphere theorem:

Theorem 14.1. *Let M be a compact simply connected n-dimensional Riemannian manifold which has positive curvature on totally isotropic two-planes, where $n \geq 4$. Then M is homeomorphic to a sphere.*

One nice feature of their condition is that it is implied by strict *pointwise* 1/4-pinching; so in particular this theorem implies that compact simply connected manifolds with pointwise 1/4-pinched sectional curvatures are homeomorphic to spheres, thus refining the classical Berger–Klingenberg–Rauch result.

Recently Brendle and Schoen [BS09a] used this condition, with the machinery of Böhm and Wilking (as seen in Chaps. 12 and 13), to prove the differentiable pointwise 1/4-pinching sphere theorem. One of the nontrivial aspects of their proof involves showing that positive isotropic curvature is preserved by the Ricci flow in all dimensions $n \geq 4$. In regard to this, Brendle and Schoen [BS09a, p. 289] remarked that:

> 'This is a very intricate calculation which exploits special identities and inequalities for the curvature tensor arising from the first and second variations applied to a set of four orthonormal vectors which minimise the isotropic curvature. After this paper was written, we learned that H. Nguyen [Ngu08] has independently proved that positive isotropic curvature is preserved under the Ricci flow.'

In this chapter we will prove that positive isotropic curvature and positive complex sectional curvature are preserved by the Ricci flow. We also show in Corollary 14.13 that positive isotropic curvature on $M \times \mathbb{R}^2$ is sufficient for pointwise 1/4-pinching.

B. Andrews and C. Hopper, *The Ricci Flow in Riemannian Geometry*, Lecture Notes in Mathematics 2011, DOI 10.1007/978-3-642-16286-2_14,

14.1 Positive Isotropic Curvature

Given a real vector space V, we consider its complexification $V_{\mathbb{C}}$ which extends scalar multiplication to include multiplication by complex numbers. Formally this is achieved by letting $V_{\mathbb{C}} = V \otimes_{\mathbb{R}} \mathbb{C}$ be the tensor product of V with the complex numbers $\mathbb{C}$. To make $V_{\mathbb{C}}$ into a complex vector space, we define complex scalar multiplication by $\lambda(v \otimes \mu) := v \otimes (\lambda\mu)$ for all $\lambda, \mu \in \mathbb{C}$ and $v \in V$. We define the complex conjugate of $v \otimes \lambda$ to be $v \otimes \bar{\lambda}$. By the nature of the tensor product, every vector $v \in V_{\mathbb{C}}$ can be written uniquely in the form $v = v_1 \otimes 1 + v_2 \otimes i$, where $v_1, v_2 \in V$. Furthermore, we can regard $V_{\mathbb{C}}$ as the direct sum of two copies of V, that is $V_{\mathbb{C}} \simeq V \oplus iV$. Note that the complex dimension $\dim_{\mathbb{C}}(V_{\mathbb{C}}) = \dim_{\mathbb{R}}(V)$ is equal to the real dimension of V. It is a common practice to drop the tensor product symbol and just write $v = v_1 + iv_2$. The archetypical example is the complexification of $\mathbb{R}^n$, which to no surprise is simply $\mathbb{C}^n$.

Given a real inner product space $V = (V, \langle \cdot, \cdot \rangle)$, one can naturally extend the real inner product to be complex-linear in both arguments (not Hermitian) by defining

$$(x + iy) \cdot (u + iv) := \langle x, u \rangle - \langle y, v \rangle + i \langle y, u \rangle + i \langle x, v \rangle . \tag{14.1}$$

On the other hand there is also a natural extension of the inner product on V to a Hermitian inner product $\langle\langle \cdot, \cdot \rangle\rangle$ by defining

$$\langle\langle x + iy, u + iv \rangle\rangle = (x + iy, u - iv) = \langle x, u \rangle + \langle y, v \rangle + i \langle y, u \rangle - i \langle x, v \rangle .$$

Note that $\langle\langle U, V \rangle\rangle = U \cdot \bar{V}$.

Other tensors which act on V can also be extended in various ways, with each argument extending to be either linear or conjugate linear. In particular, if R is an algebraic curvature operator and $V = \mathbb{R}^n$, then it is natural to extend R as a Hermitian bilinear form acting on $\bigwedge^2 V_{\mathbb{C}}$. In particular, for any two-dimensional complex subspace Π of $V_{\mathbb{C}}$, R defines a *complex sectional curvature* of Π, as follows:

$$K_{\mathbb{C}}(\Pi) = R(Z, W, \bar{Z}, \bar{W}),$$

where W and Z are an orthonormal basis for Π with respect to the Hermitian inner product $\langle\langle \cdot, \cdot \rangle\rangle$. Note that the symmetries of R imply that $K_{\mathbb{C}}(\Pi)$ is real and independent of the choice of orthonormal basis. Writing $Z = X + iY$ and $W = U + iV$, and using the first Bianchi identity, we find that

$$\begin{aligned} K_{\mathbb{C}}(\Pi) = R(X, U, X, U) + R(X, V, X, V) + R(Y, U, Y, U) \\ + R(Y, V, Y, V) - 2R(X, Y, U, V). \end{aligned} \tag{14.2}$$

A complex subspace U of $V_{\mathbb{C}}$ is said to be *isotropic*, with respect to the complex form defined in (14.1), if there is at least one non-zero vector $u \in U$ such that $u \cdot v = 0$ for all $v \in U$. The subspace is called *totally isotropic* if this is true for every vector u (equivalently, the restriction of the complex form (14.1) to U vanishes). In particular, we say a complex vector $z \neq 0$ is *isotropic* if $z \cdot z = 0$. Note that if $z = x + iy$ then $z \cdot z = \|x\|^2 - \|y\|^2 + 2i \langle x, y \rangle$, so z is isotropic if and only if $x \perp y$ and $\|x\| = \|y\|$. A subspace is totally isotropic precisely when it is composed entirely of isotropic vectors.

We are interested in *totally isotropic 2-planes*, i.e. two-dimensional complex subspaces, of $\mathbb{R}^n_{\mathbb{C}} = \mathbb{C}^n$. Every such plane can be spanned by two vectors $Z = X + iY$ and $W = U + iV$ where X, Y, U, V are all orthonormal (so there are no such planes for $n \leq 3$).

In this chapter we are interested in the following curvature conditions:

Definition 14.2 (PCSC Condition). We say $R \in \mathrm{Curv}$ has *positive complex sectional curvature* (PCSC) if all complex sectional curvatures of R are positive. This defines an invariant closed convex cone in Curv:

$$\begin{aligned} C_{\mathrm{PCSC}} &= \left\{R \in \mathrm{Curv} : \ R(X, Y, \bar{X}, \bar{Y}) \geq 0 \text{ for all } X, Y \in \mathbb{R}^n_{\mathbb{C}}\right\} \\ &= \bigcap_{X,Y \in \mathbb{R}^n_{\mathbb{C}}} \{R \in \mathrm{Curv} : \ \ell_{X,Y}(R) \geq 0\} \end{aligned}$$

where $\ell_{X,Y}(R) := R(X, Y, \bar{X}, \bar{Y})$. We say a Riemannian manifold M has positive complex sectional curvature if $R \in \mathrm{Int}\,(C_{\mathrm{PCSC}}(T_pM))$ for all points $p \in M$.

Definition 14.3 (PIC Condition). We say $R \in \mathrm{Curv}$ has *positive isotropic curvature* (PIC) if the complex sectional curvatures of all totally isotropic 2-planes in $\mathbb{R}^n_{\mathbb{C}}$ are positive. This again defines an invariant closed convex cone:

$$C_{\mathrm{PIC}} = \{R \in \mathrm{Curv} : \ \ell_{X,Y}(R) \geq 0 \text{ if } X \cdot X = X \cdot Y = Y \cdot Y = 0\} \quad (14.3)$$

$$= \bigcap_{\mathbb{O} \in O(n)} \{R \in \mathrm{Curv} : \ \tilde{\ell}(R^{\mathbb{O}}) \geq 0\} \quad (14.4)$$

where $\tilde{\ell} := \ell_{e_1+ie_2, e_3+ie_4}$, so by (14.2) we have

$$\tilde{\ell}(R) = R_{1313} + R_{2424} + R_{1414} + R_{2323} - 2R_{1234}. \quad (14.5)$$

We say that a Riemannian manifold M has positive isotropic curvature if $R \in \mathrm{Int}\,(C_{\mathrm{PIC}}(T_pM))$ for all points $p \in M$.

Remark 14.4. By applying the vector bundle maximum principle, we will show that both of these conditions are preserved by the Ricci flow, by proving that the vector field Q on Curv points into the corresponding invariant cones. Moreover – with an eye to the differentiable sphere theorem – we will relate the positive isotropic curvature condition to the pointwise 1/4-pinching condition as well.

14.2 The 1/4-Pinching Condition and PIC

In this section we relate the positive isotropic curvature condition with the 1/4-pinching condition needed for the differentiable sphere theorem. This is achieved indirectly by working with the PIC condition on $M \times \mathbb{R}^2$ and the related cone $\widehat{C}_{\mathrm{PIC}_2}$.

14.2.1 The Cone $\widehat{C}_{\mathrm{PIC}_k}$

To avoid certain unpleasant analytical aspects, Brendle and Schoen [BS09a, Sect. 3] consider the PIC condition on $M \times \mathbb{R}^2$ rather than that on M.[1]

The curvature operator $\widehat{R}_k$ of $M \times \mathbb{R}^k$ is an element of $\mathrm{Curv}(TM \times \mathbb{R}^k)$ given by the natural injection $\pi_k^* : \mathrm{Curv}(\mathbb{R}^n) \hookrightarrow \mathrm{Curv}(\mathbb{R}^{n+k})$ defined by

$$\widehat{R}_k(u, v, w, z) := (\pi_k^* R)(u, v, w, z) = R(\pi u, \pi v, \pi w, \pi z)$$

for all $u, v, w, z \in \mathbb{R}^{n+k}$, where $\pi_k : \mathbb{R}^{n+k} \to \mathbb{R}^n$ is the orthogonal projection. We define $\widehat{C}_{\mathrm{PIC}_k}$ to be the cone of all curvature operators R which produce positive isotropic curvature on $M \times \mathbb{R}^k$, i.e.

$$\widehat{C}_{\mathrm{PIC}_k} := \left\{R \in \mathrm{Curv}(\mathbb{R}^n) : \widehat{R}_k \in C_{\mathrm{PIC}}(\mathbb{R}^{n+k})\right\} = (\pi_k^*)^{-1}\left(C_{\mathrm{PIC}}(\mathbb{R}^{n+k})\right).$$

Clearly, $\widehat{C}_{\mathrm{PIC}_k}$ is a closed, convex, $O(n)$-invariant cone in Curv.

Remark 14.5. The same construction with PIC replaced by PCSC is rather dull: Since $\widehat{R}_k(X, Y, \bar{X}, \bar{Y}) = R(\pi_k X, \pi_k Y, \pi_k \bar{X}, \pi_k \bar{Y})$, we have that $\widehat{R}_k \in C_{\mathrm{PCSC}}(\mathbb{R}^{n+k})$ if and only if $R \in C_{\mathrm{PCSC}}(\mathbb{R}^n)$. However with the PIC condition, the cone does change:

$$\begin{aligned}\widehat{C}_{\mathrm{PIC}_k} &= \left\{R \in \mathrm{Curv} :\ \ell_{\pi_k X, \pi_k Y}(R) \geq 0 \text{ if } X \cdot X = X \cdot Y = Y \cdot Y = 0\right\}\\ &= \bigcap_{(X,Y) \in A_k} \{R \in \mathrm{Curv} :\ \ell_{X,Y}(R) \geq 0\}\end{aligned}$$

[1] At the time of writing, nobody knows how to handle Ricci flow on PIC manifolds in dimensions higher than 4. One reason for the difficulty is that the curvature of the manifold $S^{n-1} \times \mathbb{R}$ lies strictly inside C_{PIC}, so that singularities can form and surgery arguments are required (see [Ham97, CZ06, CTZ08] for the case $n = 4$).

where A_k is the set of all $(X,Y) \in \mathbb{R}^n_{\mathbb{C}} \times \mathbb{R}^n_{\mathbb{C}}$ such that there exists $(\widehat{X},\widehat{Y}) \in \mathbb{R}^{n+k}_{\mathbb{C}} \times \mathbb{R}^{n+k}_{\mathbb{C}}$ with $\pi_k(\widehat{X}) = X$, $\pi_k(\widehat{Y}) = Y$ and $\widehat{X}\cdot\widehat{X} = \widehat{X}\cdot\widehat{Y} = \widehat{Y}\cdot\widehat{Y} = 0$. Note that A_k increases with k, so $\widehat{C}_{\mathrm{PIC}_k}$ decreases with k (i.e. $\widehat{C}_{\mathrm{PIC}_k} \supset \widehat{C}_{\mathrm{PIC}_\ell}$, for $k < \ell$). We will obtain a clearer understanding of these cones in Sect. 14.4.3.

In order to apply the maximum principle with these curvature conditions, we need to prove that the Ricci flow reaction vector field Q points into $\widehat{C}_{\mathrm{PIC}_k}$ (i.e. that $\widehat{C}_{\mathrm{PIC}_k}$ is a preserved cone). One would expect that this condition should relate closely to whether the cone C_{PIC} itself is preserved: If (M, g_0) is a Riemannian manifold for which $M \times \mathbb{R}^k$ has PIC, and $g(t)$ is a solution of Ricci flow on M with initial data g_0, then the product metrics on $M \times \mathbb{R}^k$ – or on $M \times (S^1)^k$ – also evolve by Ricci flow (cf. Sect. 4.1.1.3). Thus if we know that C_{PIC} is a preserved cone, the metric on $M \times (S^1)^k$ remains PIC, and so R remains in $\widehat{C}_{\mathrm{PIC}_k}$. This strongly suggests that it should be sufficient to prove that PIC is preserved. The following simple observations confirm this:

Lemma 14.6. *For $R \in \mathrm{Curv}$, $\pi_k^*(Q(R)) = Q(\pi_k^*(R))$.*

Proof. Since Q is $O(n+k)$-invariant, we can choose to compute it in a basis with $e_1, \dots e_n$ an orthonormal basis for $\mathbb{R}^n$ and $e_{n+1}, \dots, e_{n+k}$ an orthonormal basis for $\mathbb{R}^k$. Then for $1 \le a,b,c,d \le n+k$ we have that

$$
\begin{aligned}
B(\pi_k^* R)_{abcd} &= \sum_{1\le p,q\le n+k} \widehat{R}_{apbq}\widehat{R}_{cpdq} \\
&= \sum_{1\le p,q\le n+k} R(\pi_k e_a, \pi_k e_p, \pi_k e_b, \pi_k e_q) R(\pi_k e_c, \pi_k e_p, \pi_k e_d, \pi_k e_q) \\
&= \sum_{1\le p,q\le n} R(\pi_k e_a, e_p, \pi_k e_b, e_q) R(\pi_k e_c, e_p, \pi_k e_d, e_q) \\
&= \left(\pi_k^* B(R)\right)_{abcd}.
\end{aligned}
$$

Since Q is a sum of such terms, the result follows. □

Lemma 14.7. *For any $R \in \partial\widehat{C}_{\mathrm{PIC}_k}$,*

$$\mathcal{T}_R \widehat{C}_{\mathrm{PIC}_k} = (\pi_k^*)^{-1}\left(\mathcal{T}_{\widehat{R}_k} C_{\mathrm{PIC}}\right).$$

Proof. The map π_k^* is a linear isomorphism from $\mathrm{Curv}(\mathbb{R}^n)$ to the subspace $L = \bigcap_{v\in\{0\}\times\mathbb{R}^k} \left\{R \in \mathrm{Curv}(\mathbb{R}^{n+k}) : \ R(v,\cdot,\cdot,\cdot) = 0\right\}$. In particular, π_k^* maps $\widehat{C}_{\mathrm{PIC}_k}$ to $L \cap C_{\mathrm{PIC}}$. Therefore

$$\pi_k^*(\mathcal{T}_R \widehat{C}_{\mathrm{PIC}_k}) = \pi_k^*\Big(\bigcup_{h>0} \frac{\widehat{C}_{\mathrm{PIC}_k} - R}{h}\Big)$$

$$= \bigcup_{h>0} \frac{L \cap C_{\mathrm{PIC}} - \pi_k^* R}{h}$$

$$= L \cap \Big(\bigcup_{h>0} \frac{C_{\mathrm{PIC}} - \widehat{R}_k}{h} \Big) = L \cap \mathcal{T}_{\widehat{R}_k} C_{\mathrm{PIC}}. \qquad \square$$

Theorem 14.8. *If $C_{\mathrm{PIC}}(\mathbb{R}^{n+k})$ is a preserved cone, then so is $\widehat{C}_{\mathrm{PIC}_k}(\mathbb{R}^n)$.*

Proof. For $R \in \partial \widehat{C}_{\mathrm{PIC}_k}$, we have

$$\pi_k^* \left(Q(R)\right) = Q(\widehat{R}) \in L \cap \mathcal{T}_{\widehat{R}} C_{\mathrm{PIC}} = \pi_k^* (\mathcal{T}_R \widehat{C}_{\mathrm{PIC}_k}),$$

so $Q(R) \in \mathcal{T}_R \widehat{C}_{\mathrm{PIC}_k}$ since π_k^* is injective. $\square$

For present purposes our interest is mainly in the cone $\widehat{C}_{\mathrm{PIC}_2}$. It has the following elementary properties:

Proposition 14.9. *The cone $\widehat{C}_{\mathrm{PIC}_k}$ contains the cone of nonnegative curvature operators $\{R : R \geq 0\}$ for any k, and every $R \in \widehat{C}_{\mathrm{PIC}_2}$ has nonnegative sectional curvature.*

Proof. We first show that the cone of nonnegative curvature operators in contained in C_{PIC}:

Suppose $R \in \mathrm{Curv}$ is a nonnegative curvature operator, and let $\{e_1, e_2, e_3, e_4\}$ be an orthonormal 4-frame in $\mathbb{R}^n$. By setting $\varphi, \psi \in \bigwedge^2 \mathbb{R}^n$ to be

$$\varphi = e_1 \wedge e_3 + e_4 \wedge e_2$$
$$\psi = e_1 \wedge e_4 + e_2 \wedge e_3$$

we see that

$$0 \leq R(\varphi, \varphi) + R(\psi, \psi)$$
$$= R_{1313} + R_{1414} + R_{2323} + R_{2424} - 2R_{1234}$$

and so $R \in C_{\mathrm{PIC}}$. The inclusion of the non-negative curvature operators in $\widehat{C}_{\mathrm{PIC}_k}$ now follows, since ι_k maps non-negative curvature operators to non-negative curvature operators.

Now we show that $\widehat{C}_{\mathrm{PIC}_2}$ is contained in the cone of positive sectional curvature operators: If $R \in \widehat{C}_{\mathrm{PIC}_2}$, let $\{e_1, e_2\} \subset \mathbb{R}^n$ be an orthonormal 2-frame, and define

$$\widehat{e}_1 = (e_1, 0, 0) \qquad \widehat{e}_2 = (0, 0, 1)$$
$$\widehat{e}_3 = (e_2, 0, 0) \qquad \widehat{e}_4 = (0, 1, 0)$$

which is an orthonormal 4-frame for $\mathbb{R}^n \times \mathbb{R}^2$. As $\widehat{R} \in C_{\mathrm{PIC}}$,

$$\begin{aligned} 0 &\le \widehat{R}_{1313} + \widehat{R}_{1414} + \widehat{R}_{2323} + \widehat{R}_{2424} - 2\widehat{R}_{1234} \\ &= R_{1212} \end{aligned}$$

since $\widehat{R}(\widehat{e}_1, \widehat{e}_3, \widehat{e}_1, \widehat{e}_3) = R(e_1, e_2, e_1, e_2)$. Therefore R has nonnegative sectional curvatures. □

14.2.2 An Algebraic Characterisation of the Cone $\widehat{C}_{\mathrm{PIC}_2}$

By following [BS09a, Sect. 4], we characterise the cone $\widehat{C}_{\mathrm{PIC}_2}$ by providing a necessary and sufficient condition for $\widehat{R}$ to have PIC. We use this to show that all curvature operators R which are 1/4-pinched must lie in the cone $\widehat{C}_{\mathrm{PIC}_2}$, thus relating the 1/4-pinching condition to the PIC condition on $M \times \mathbb{R}^2$.

To begin we quote the following linear algebra result:

Lemma 14.10. (**[Che91, Lemma 3.1], [BS09a. p. 303]**) *Suppose $\varphi, \psi \in \bigwedge^2 \mathbb{R}^4$ are bivectors that satisfy $\varphi \wedge \psi = 0$, $\varphi \wedge \varphi = \psi \wedge \psi$, and $\langle \varphi, \psi \rangle = 0$. Then there exists an orthonormal basis $\{e_1, e_2, e_3, e_4\}$ for $\mathbb{R}^4$ such that*

$$\varphi = a_1\, e_1 \wedge e_3 + a_2\, e_4 \wedge e_2$$

$$\psi = b_1\, e_1 \wedge e_4 + b_2\, e_2 \wedge e_3$$

where $a_1 a_2 = b_1 b_2$.

Working from this lemma, we can prove the following 'rotated version':

Lemma 14.11. *Suppose $\varphi, \psi \in \bigwedge^2 \mathbb{R}^4$ satisfy $\varphi \wedge \psi = 0$ and $\varphi \wedge \varphi = \psi \wedge \psi$. Then there exists an orthonormal frame $\{e_1, e_2, e_3, e_4\}$ and $\theta \in \mathbb{R}$ such that*

$$\cos\theta\, \varphi + \sin\theta\, \psi = a_1\, e_1 \wedge e_3 + a_2\, e_4 \wedge e_2$$

$$-\sin\theta\, \varphi + \cos\theta\, \psi = b_1\, e_1 \wedge e_4 + b_2\, e_2 \wedge e_3$$

where $a_1 a_2 = b_1 b_2$.

Proof. Firstly, define θ to be such that

$$\frac{1}{2} \sin 2\theta \,(|\varphi|^2 - |\psi|^2) = \cos 2\theta\, \langle \varphi, \psi \rangle\,.$$

Now set $\varphi' = \cos\theta\,\varphi + \sin\theta\,\psi$ and $\psi' = -\sin\theta\,\varphi + \cos\theta\,\psi$. By hypothesis, the following quantities both vanish:

$$\begin{aligned}\varphi'\wedge\varphi' - \psi'\wedge\psi' &= \cos 2\theta\,(\varphi\wedge\varphi - \psi\wedge\psi) + 2\sin 2\theta\,\varphi\wedge\psi = 0\\ \varphi'\wedge\psi' &= \frac{1}{2}\sin 2\theta\,(\varphi\wedge\varphi - \psi\wedge\psi) + \cos 2\theta\,\varphi\wedge\psi = 0.\end{aligned}$$

Moreover, by the definition of θ, we also have that

$$\langle\varphi',\psi'\rangle = \frac{1}{2}\sin 2\theta\,(|\psi|^2 - |\varphi|^2) + \cos 2\theta\,\langle\varphi,\psi\rangle = 0,$$

from which the assertion follows by Lemma 14.10. □

Proposition 14.12 (Characterisation of $\widehat{C}_{\mathrm{PIC}_2}$). *$R \in \widehat{C}_{\mathrm{PIC}_2}$ if and only if*

$$R_{1313} + \lambda^2 R_{1414} + \mu^2 R_{2323} + \lambda^2\mu^2 R_{2424} - 2\lambda\mu R_{1234} \geq 0,$$

for all orthonormal 4-frames $\{e_1, e_2, e_3, e_4\}$ and all $\lambda, \mu \in [-1, 1]$.

Proof. Suppose $\widehat{R} \in C_{\mathrm{PIC}}$. Let $\{e_1, e_2, e_3, e_4\} \subset \mathbb{R}^n$ be an orthonormal 4-frame, and let $\lambda, \mu \in [-1, 1]$. Define

$$\begin{aligned}\widehat{e}_1 &= (e_1, 0, 0) & \widehat{e}_2 &= (\mu e_2, 0, \sqrt{1-\mu^2})\\ \widehat{e}_3 &= (e_3, 0, 0) & \widehat{e}_4 &= (\lambda e_4, \sqrt{1-\lambda^2}, 0)\end{aligned}$$

to be such that the vectors $\{\widehat{e}_1, \widehat{e}_2, \widehat{e}_3, \widehat{e}_4\}$ form an orthonormal 4-frame for $\mathbb{R}^n \times \mathbb{R}^2$. With this, it follows that

$$\begin{aligned}0 &\leq \widehat{R}_{1313} + \widehat{R}_{1414} + \widehat{R}_{2323} + \widehat{R}_{2424} - 2\widehat{R}_{1234}\\ &= R_{1313} + \lambda^2 R_{1414} + \mu^2 R_{2323} + \lambda^2\mu^2 R_{2424} - 2\lambda\mu R_{1234}.\end{aligned}$$

To show the reverse implication, suppose $\{\widehat{e}_1, \widehat{e}_2, \widehat{e}_3, \widehat{e}_4\}$ is an orthonormal 4-frame for $\mathbb{R}^n \times \mathbb{R}^2$. By definition each vector $\widehat{e}_j$ is of the form $\widehat{e}_j = (v_j, x_j)$, where $v_j \in \mathbb{R}^n$ and $x_j \in \mathbb{R}^2$. Let V be a 4-dimensional subspace containing $\{v_1, v_2, v_3, v_4\}$, and define $\varphi, \psi \in \bigwedge^2 V$ by

$$\begin{aligned}\varphi &= v_1 \wedge v_3 + v_4 \wedge v_2\\ \psi &= v_1 \wedge v_4 + v_2 \wedge v_3.\end{aligned}$$

It is clear that $\varphi\wedge\varphi = \psi\wedge\psi$, $\varphi\wedge\psi = 0$ and $V \simeq \mathbb{R}^4$. So by Lemma 14.11 there exists an orthonormal basis $\{e_1, e_2, e_3, e_4\}$ for V such that

$$\begin{aligned}\varphi' &= \cos\theta\,\varphi + \sin\theta\,\psi = a_1\, e_1 \wedge e_3 + a_2\, e_4 \wedge e_2\\ \psi' &= -\sin\theta\,\varphi + \cos\theta\,\psi = b_1\, e_1 \wedge e_4 + b_2\, e_2 \wedge e_3\end{aligned}$$

where $a_1a_2 = b_1b_2$. Using the first Bianchi identity we find that

$$\begin{aligned} R(\varphi,\varphi) + R(\psi,\psi) &= R(\varphi',\varphi') + R(\psi',\psi') \\ &= a_1^2 R_{1313} + b_1^2 R_{1414} + b_2^2 R_{2323} + a_2^2 R_{2424} - 2a_1a_2 R_{1234}. \end{aligned}$$

By setting $\lambda = b_1/a_1$ and $\mu = b_2/a_1$, our hypothesis implies that the right-hand side is nonnegative. So it follows that

$$\begin{aligned} 0 &\le R(\varphi,\varphi) + R(\psi,\psi) \\ &= R(v_1,v_3,v_1,v_3) + R(v_1,v_4,v_1,v_4) + R(v_2,v_3,v_2,v_3) \\ &\quad + R(v_2,v_4,v_2,v_4) - 2R(v_1,v_2,v_3,v_4) \\ &= \widehat{R}_{1313} + \widehat{R}_{1414} + \widehat{R}_{2323} + \widehat{R}_{2424} - 2\widehat{R}_{1234}. \end{aligned}$$

□

Corollary 14.13. *Let* $R \in \mathrm{Curv}$. *If the sectional curvatures of* R *are* 1/4-*pinched, then* $\widehat{R} \in C_{\mathrm{PIC}}$ *and so* $R \in \widehat{C}_{\mathrm{PIC}_2}$.

Proof. Scale the metric (if need be) so that sectional curvatures of R lie in the interval $(1,4]$. Let $\{e_1, e_2, e_3, e_4\}$ be an orthonormal 4-frame in $\mathbb{R}^n$ and let $\lambda, \mu \in [-1,1]$. By Berger's Lemma (see Sect. 2.7.7) we have that $|R(e_1,e_2,e_3,e_4)| \le 2$. In which case the result now follows by Proposition 14.12, since

$$\begin{aligned} R_{1313} + \lambda^2 R_{1414} + \mu^2 R_{2323} + \lambda^2\mu^2 R_{2424} - 2\lambda\mu R_{1234} & \\ \ge 1 + \lambda^2 + \mu^2 + \lambda^2\mu^2 - 4|\lambda\mu| & \\ = (1 - |\lambda\mu|)^2 + (|\lambda| - |\mu|)^2 & \\ \ge 0. & \end{aligned}$$

□

14.3 PIC is Preserved by the Ricci Flow

It was first shown by Hamilton [Ham97] that positive isotropic curvature on a 4-manifold is preserved by the Ricci flow. The result for $n \ge 4$ was settled independently by Brendle and Schoen [BS09a], and by Nguyen [Ngu08].

Theorem 14.14 (Brendle, Schoen, Nguyen). *Let* M *be a compact manifold of dimension* $n \ge 4$ *with a family of metrics* $\{g(t)\}_{t\in[0,T)}$ *evolving under Ricci flow. If* $g(0)$ *has positive isotropic curvature, then* $g(t)$ *has positive isotropic curvature for all* $t \in [0,T)$.

In this section we prove this theorem using ideas from [BS09a, Ngu08, Ngu10, AN07]. In the remaining sections we will look at alternative arguments and simplifications.

By the previous considerations the proof requires just one step: We must prove that C_{PIC} is preserved, in the sense that the vector field Q on Curv is in the tangent cone to C_{PIC} at any boundary point. By Definition 14.3,

$$C_{\mathrm{PIC}} = \bigcap_{\mathbb{O}\in O(n)} \{R \in \mathrm{Curv} : \tilde{\ell}(R^{\mathbb{O}}) \geq 0\}$$

where $\tilde{\ell}(R) = R_{1313} + R_{2424} + R_{1414} + R_{2323} - 2R_{1234}$. Since C_{PIC} is an $O(n)$-invariant cone explicitly presented as an intersection of half-spaces, Theorem B.7 of Appendix B implies that it is enough to check that

$$\tilde{\ell}(Q(R)) \geq 0$$

for any $R \in \partial C_{\mathrm{PIC}}$ for which $\tilde{\ell}(R) = 0$ and $\tilde{\ell}(R^{\mathbb{O}}) \geq 0$ for all $\mathbb{O} \in O(n)$.

Lemma 14.15 (Brendle–Schoen Decomposition). *For any* $R \in \mathrm{Curv}$,

$$\begin{aligned}\tilde{\ell}(Q(R)) = \frac{1}{2}\Big((R_{13pq} - R_{24pq})^2 + (R_{14pq} + R_{23pq})^2\Big) \\ + \Big((R_{1p1q} + R_{2p2q})(R_{3p3q} + R_{4p4q}) - R_{12pq}R_{34pq} \\ - (R_{1p3q} + R_{2p4q})(R_{3p1q} + R_{4p2q}) \\ - (R_{1p4q} - R_{2p3q})(R_{4p1q} - R_{3p2q})\Big).\end{aligned}$$

Proof. By Remark 12.12, $Q = Q(R)$ satisfies the first Bianchi identity (although R^2 and $R^\#$ do not). In which case

$$\begin{aligned}\tilde{\ell}(Q(R)) &= Q_{1313} + Q_{1414} + Q_{2323} + Q_{2424} - 2Q_{1234} \\ &= Q_{1313} + Q_{1414} + Q_{2323} + Q_{2424} + 2Q_{1342} + 2Q_{1423}.\end{aligned}$$

So by Lemma 12.17 and the first Bianchi identity we find that

$$\begin{aligned}R^2_{1313} + R^2_{1414} + R^2_{2323} + R^2_{2424} + 2R^2_{1342} + 2R^2_{1423} \\ = \frac{1}{2}\Big((R_{13pq} - R_{24pq})^2 + (R_{14pq} + R_{23pq})^2\Big)\end{aligned}$$

and

$$\begin{aligned}R^\#_{1313} + R^\#_{1414} + R^\#_{2323} + R^\#_{2424} + 2R^\#_{1342} + 2R^\#_{1423} \\ = (R_{1p1q} + R_{2p2q})(R_{3p3q} + R_{4p4q}) \\ - (R_{1p3q} + R_{2p4q})(R_{3p1q} + R_{4p2q}) \\ - (R_{1p4q} - R_{2p3q})(R_{4p1q} - R_{3p2q}) \\ - 2R_{1p2q}R_{3p4q} + 2R_{1p2q}R_{4p3q}\end{aligned}$$

$$
\begin{aligned}
&= (R_{1p1q} + R_{2p2q})(R_{3p3q} + R_{4p4q}) \\
&\quad - (R_{1p3q} + R_{2p4q})(R_{3p1q} + R_{4p2q}) \\
&\quad - (R_{1p4q} - R_{2p3q})(R_{4p1q} - R_{3p2q}) \\
&\quad - R_{12pq}R_{34pq}.
\end{aligned}
$$

□

From this lemma it is clear that

$$
\frac{1}{2}\sum_{p,q=1}^{n}\left((R_{13pq} - R_{24pq})^2 + (R_{14pq} + R_{23pq})^2\right) \geq 0,
$$

so in order to prove Theorem 14.14 all that is needed is to verify that:

Claim 14.16. If $R \in C_{\mathrm{PIC}}$ with $\tilde{\ell}(R) = 0$ then

$$
\begin{aligned}
\sum_{p,q=1}^{n}\Big((R_{1p1q} + R_{2p2q})(R_{3p3q} + R_{4p4q}) - R_{12pq}R_{34pq} \\
- (R_{1p3q} + R_{2p4q})(R_{3p1q} + R_{4p2q}) \\
- (R_{1p4q} - R_{2p3q})(R_{4p1q} - R_{3p2q})\Big) \geq 0.
\end{aligned}
\tag{14.6}
$$

Remark 14.17. We observe there is some redundancy in our description of C_{PIC}, since not all of the half-spaces given by $\{\tilde{\ell}(R^{\mathbb{O}}) \geq 0\}$ are distinct: $\tilde{\ell}(R)$ computes the complex sectional curvature of R in the plane generated by $e_1 + ie_2$ and $e_3 + ie_4$, but this is unchanged if we choose a different basis for the same complex 2-plane. In particular we get the same result if we replace e_1+ie_2 by $e_2-ie_1 = -i(e_1+ie_2)$, or replace e_3+ie_4 by e_4-ie_3, or interchange $e_1 + ie_2$ with $e_3 + ie_4$. Thus for any inequalities we prove for $\tilde{\ell}(R)$ there are corresponding inequalities for $\tilde{\ell}(R^{\mathbb{O}})$, where $\mathbb{O}$ is an $O(n)$ matrix which has the top-most 4×4 block given by any of

$$
\begin{bmatrix} 0 & 1 & 0 & 0 \\ -1 & 0 & 0 & 0 \\ 0 & 0 & 1 & 0 \\ 0 & 0 & 0 & 1 \end{bmatrix}, \quad
\begin{bmatrix} 1 & 0 & 0 & 0 \\ 0 & 1 & 0 & 0 \\ 0 & 0 & 0 & 1 \\ 0 & 0 & -1 & 0 \end{bmatrix}, \quad \text{or} \quad
\begin{bmatrix} 0 & 0 & 1 & 0 \\ 0 & 0 & 0 & 1 \\ 1 & 0 & 0 & 0 \\ 0 & 1 & 0 & 0 \end{bmatrix}.
$$

14.3.1 Inequalities from the Second Derivative Test

In this section we prove Claim 14.16 by applying the second derivative test to the function Z on $O(n)$, defined by

$$
Z(\mathbb{O}) = \tilde{\ell}(R^{\mathbb{O}}),
$$

along integral curves in the Lie group $O(n)$ through the identity.

Integral curves in $O(n)$ can be conveniently computed by following the flow of left-invariant vector fields: Given $\Lambda \in \mathfrak{so}(n)$, the corresponding left-invariant vector field $X \in \mathrm{Lie}(O(n))$ is given by $X_{\mathbb{O}} = \mathbb{O} \circ \Lambda$.[2] The integral curve through the identity in direction Λ is given by solving the ODE[3]

$$\begin{aligned} \frac{d}{ds}\mathbb{O}(s) &= \mathbb{O}(s) \circ \Lambda \\ \mathbb{O}(0) &= I_n \end{aligned}$$

Equivalently, if we write $v_i(s) = (\mathbb{O}(s))e_i$, then this is equivalent to the system

$$\frac{d}{ds}v_i = \frac{d}{ds}(\mathbb{O}e_i) = \mathbb{O}(\Lambda_i^j e_j) = \Lambda_{ij}v_j, \tag{14.7}$$

with $v_i(0) = e_i$.

We observe that $Z(\mathbb{O}(0)) = \tilde{\ell}(R) = 0$ by construction, while $Z(\mathbb{O}(s)) = \tilde{\ell}(R^{\mathbb{O}(s)}) \geq 0$ for every s. So the first and second derivative tests imply that

$$\begin{aligned} \frac{d}{ds}Z(\mathbb{O}(s))\Big|_{s=0} &= 0 \\ \frac{d^2}{ds^2}Z(\mathbb{O}(s))\Big|_{s=0} &\geq 0. \end{aligned}$$

By explicitly evaluating these first and second derivative conditions, we show in Sect. 14.3.1.1 that the sums over indices $p \leq 4 < q$, $q \leq 4 < p$ and over $1 \leq p, q \leq 4$ of (14.6) all vanish. In Sect. 14.3.1.2 we show that the sum over $p, q \geq 5$ of (14.6) is nonnegative, thus proving Claim 14.16.

14.3.1.1 First Order Terms

We write

$$\begin{aligned} Z(\mathbb{O}(s)) &= \tilde{\ell}(R^{\mathbb{O}(s)}) \\ &= R(v_1, v_3, v_1, v_3) + R(v_2, v_4, v_2, v_4) + R(v_1, v_4, v_1, v_4) \\ &\quad + R(v_2, v_3, v_2, v_3) - 2R(v_1, v_2, v_3, v_4), \end{aligned}$$

---

[2] Recall that we are identifying $O(n)$ with a subset of $\mathrm{End}(\mathbb{R}^n) \simeq (\mathbb{R}^n)^* \otimes \mathbb{R}^n$, namely the set of linear transformations of $\mathbb{R}^n$ which are isometries. The tangent space $T_{I_n}O(n)$ to the identity is then the subspace of anti-self-adjoint transformations $\mathfrak{so}(n) = \{\Lambda : \Lambda^T + \Lambda = 0\} \simeq \bigwedge^2(\mathbb{R}^n)$ (see [Lee02, Example 8.39]).

[3] Note that the flow Ψ of the left-invariant vector field X is given by $\Psi_s = R_{\exp s\Lambda}$ (i.e. right multiplication by $\exp s\Lambda$), so that the path $\mathbb{O}(s) = \Psi_s(I_n) = \exp s\Lambda$ (see [Lee02, Proposition 20.8(g)]).

and differentiate directly using (14.7). The first derivative gives

$$\begin{aligned}\frac{1}{2}\frac{d}{ds}Z\Big|_{s=0} = &(R_{p313} + R_{p414} - R_{p234})\Lambda_{1p} \\ &+ (R_{p323} + R_{p424} - R_{1p34})\Lambda_{2p} \\ &+ (R_{1p13} + R_{2p23} - R_{12p4})\Lambda_{3p} \\ &+ (R_{1p14} + R_{2p24} - R_{123p})\Lambda_{4p}\end{aligned} \tag{14.8}$$

where we have used

$$\frac{d}{ds}R(v_a, v_b, v_c, v_d)\Big|_{s=0} = \Lambda_{aj}R_{j234} + \Lambda_{bj}R_{1j34} + \Lambda_{cj}R_{12j4} + \Lambda_{dj}R_{123j}.$$

From the first derivative condition, the right-hand side of (14.8) vanishes for any choice of antisymmetric Λ. Choosing $\Lambda = e_p \wedge e_q$ with $p \leq 4 < q$ gives

$$p = 1: \quad 0 = R_{133q} + R_{144q} + R_{432q} \tag{14.9a}$$

$$p = 2: \quad 0 = R_{233q} + R_{244q} + R_{341q} \tag{14.9b}$$

$$p = 3: \quad 0 = R_{131q} + R_{232q} + R_{124q} \tag{14.9c}$$

$$p = 4: \quad 0 = R_{141q} + R_{242q} + R_{213q} \tag{14.9d}$$

The remaining cases are $\Lambda = e_p \wedge e_q$ with $1 \leq p < q \leq 4$. By Remark 14.17, the path $\mathbb{O}(s) = \exp s(e_1 \wedge e_2)$ corresponds to multiplying $e_1 + ie_2$ by e^{is}. Similarly, the choice $\mathbb{O}(s) = \exp, s(e_3 \wedge e_4)$ multiplies $e_3 + ie_4$ by e^{is}. Finally, $\Lambda = e_1 \wedge e_3 + e_2 \wedge e_4$ and $\Lambda = e_1 \wedge e_4 - e_2 \wedge e_3$ give changes of basis in the complex 2-plane. All of these keep $Z = \ell(R^{\exp s\Lambda})$ fixed. Thus the only nontrivial identities are given by choosing Λ equal to $e_1 \wedge e_3$ and $e_1 \wedge e_4$:

$$e_1 \wedge e_3: \qquad R_{3414} + R_{3423} + R_{1223} + R_{1214} = 0 \tag{14.10a}$$

$$e_1 \wedge e_4: \qquad R_{3424} + R_{3134} + R_{2113} + R_{1224} = 0. \tag{14.10b}$$

Lemma 14.18 ([BS09a, Proposition 5]).

$$\begin{aligned}\sum_{p,q=1}^{4} \Big(&(R_{1p1q} + R_{2p2q})(R_{3p3q} + R_{4p4q}) - R_{12pq}R_{34pq} \\ &- (R_{1p3q} + R_{2p4q})(R_{3p1q} + R_{4p2q}) \\ &- (R_{1p4q} - R_{2p3q})(R_{4p1q} - R_{3p2q})\Big) = 0.\end{aligned}$$

Proof. By direct computation we get

$$\begin{aligned}
\sum_{p,q=1}^{4} \Big(&(R_{1p1q} + R_{2p2q})(R_{3p3q} + R_{4p4q}) - R_{12pq}R_{34pq} \\
&\quad - (R_{1p3q} + R_{2p4q})(R_{3p1q} + R_{4p2q}) \\
&\quad - (R_{1p4q} - R_{2p3q})(R_{4p1q} - R_{3p2q}) \Big) \\
= &(R_{1212} + R_{3434})(R_{1313} + R_{1414} + R_{2323} + R_{2424} - 2R_{1234}) \\
&+ 2R_{1234}(R_{1313} + R_{1414} + R_{2323} + R_{2424} + 2R_{1342} + 2R_{1423}) \\
&- (R_{1213} + R_{1242} + R_{3413} + R_{3442})^2 \\
&- (R_{1214} + R_{1223} + R_{3414} + R_{3423})^2 \\
= &(R_{1212} + R_{3434} + 2R_{1234})(R_{1313} + R_{1414} + R_{2323} + R_{2424} - 2R_{1234}) \\
&- (R_{1213} + R_{1242} + R_{3413} + R_{3442})^2 \\
&- (R_{1214} + R_{1223} + R_{3414} + R_{3423})^2.
\end{aligned}$$

The expression now vanishes by (14.10a) and (14.10b). □

Lemma 14.19 ([BS09a, Proposition 7])**.** *For fixed $q \geq 5$, we have*

$$\begin{aligned}
\sum_{p=1}^{4} \Big(&(R_{1p1q} + R_{2p2q})(R_{3p3q} + R_{4p4q}) - R_{12pq}R_{34pq} \\
&\quad - (R_{1p3q} + R_{2p4q})(R_{3p1q} + R_{4p2q}) \\
&\quad - (R_{1p4q} - R_{2p3q})(R_{4p1q} - R_{3p2q}) \Big) = 0.
\end{aligned}$$

Proof. Using (14.9a) and (14.9b) a direct computation shows that

$$\begin{aligned}
\sum_{p=1}^{2} \Big(&(R_{1p1q} + R_{2p2q})(R_{3p3q} + R_{4p4q}) - R_{12pq}R_{34pq} \Big) \\
&= R_{212q}(R_{313q} + R_{414q}) + R_{121q}(R_{323q} + R_{424q}) \\
&\quad - R_{121q}R_{341q} - R_{122q}R_{342q} \\
&= R_{212q}(R_{313q} + R_{414q} + R_{342q}) \\
&\quad + R_{121q}(R_{323q} + R_{424q} - R_{341q}) \\
&= 0
\end{aligned}$$

and

$$\begin{aligned}
\sum_{p=3}^{4} \Big(&(R_{1p3q} + R_{2p4q})(R_{3p1q} + R_{4p2q}) - (R_{1p4q} - R_{2p3q})(R_{4p1q} - R_{3p2q}) \Big) \\
= &\ (R_{133q} + R_{234q})R_{432q} + (R_{143q} + R_{244q})R_{341q} \\
&+ (R_{134q} - R_{233q})R_{431q} - (R_{144q} + R_{243q})R_{342q}
\end{aligned}$$

$$
\begin{aligned}
&= (R_{133q} + R_{234q} + R_{144q} - R_{243q})R_{432q} \\
&\quad + (R_{143q} + R_{244q} - R_{134q} + R_{233q})R_{341q} \\
&= (R_{133q} + R_{144q} + R_{432q})R_{432q} \\
&\quad + (R_{341q} + R_{244q} + R_{233q})R_{341q} \\
&= 0.
\end{aligned}
$$

Replacing $\{e_1, e_2, e_3, e_4\}$ by $\{e_3, e_4, e_1, e_2\}$ (see Remark 14.17) yields

$$
\sum_{p=1}^{2} \Big((R_{1p3q} + R_{2p4q})(R_{3p1q} + R_{4p2q}) + (R_{1p4q} - R_{2p3q})(R_{4p1q} - R_{3p2q}) \Big) = 0
$$

$$
\sum_{p=3}^{4} \Big((R_{1p1q} + R_{2p2q})(R_{3p3q} + R_{4p4q}) - R_{12pq}R_{34pq} \Big) = 0.
$$

The result now follows by putting these sums together. □

14.3.1.2 Second Order Terms

We now calculate the second derivative of Z and establish the nonnegativity of the final final sum over indices $p, q \geq 5$ of (14.6). By grouping similar terms we find that the second order derivative of Z, calculated from (14.8), is equal to

$$
\begin{aligned}
\frac{1}{2}\frac{d^2 Z}{ds^2} &= (R_{j3k3} + R_{j4k4})\Lambda_{1j}\Lambda_{1k} && + (R_{j3k3} + R_{j4k4})\Lambda_{2j}\Lambda_{2k} \\
&\quad + (R_{1j1k} + R_{2j2k})\Lambda_{3j}\Lambda_{3k} && + (R_{1j1k} + R_{2j2k})\Lambda_{4j}\Lambda_{4k} \\
&\quad + 2(R_{jk13} + R_{j31k} - R_{j2k4})\Lambda_{1j}\Lambda_{3k} && + 2(R_{jk14} + R_{j41k} - R_{j23k})\Lambda_{1j}\Lambda_{4k} \\
&\quad + 2(R_{jk23} + R_{j32k} - R_{1jk4})\Lambda_{2j}\Lambda_{3k} && + 2(R_{jk24} + R_{j42k} - R_{1j3k})\Lambda_{2j}\Lambda_{4k} \\
&\quad + (R_{k313} + R_{k414} - R_{k234})\Lambda_{1j}\Lambda_{jk} && + (R_{k323} + R_{k424} - R_{1k34})\Lambda_{2j}\Lambda_{jk} \\
&\quad + (R_{1k13} + R_{2k23} - R_{12k4})\Lambda_{3j}\Lambda_{jk} && + (R_{1k14} + R_{2k24} - R_{123k})\Lambda_{4j}\Lambda_{jk} \\
&\quad - 2R_{jk34}\Lambda_{1j}\Lambda_{2k} && - 2R_{12jk}\Lambda_{3j}\Lambda_{4k}.
\end{aligned}
$$

By identifying the following coefficients:

$$
\begin{aligned}
a_{jk} &= R_{1j1k} + R_{2j2k} & b_{jk} &= R_{3j3k} + R_{4j4k} \\
c_{jk} &= R_{3j1k} + R_{4j2k} & d_{jk} &= R_{4j1k} - R_{3j2k} \\
e_{jk} &= R_{12jk} & f_{jk} &= R_{34jk}
\end{aligned}
$$

we can rewrite the second derivative as

$$
\begin{aligned}
\frac{1}{2}\frac{d^2}{ds^2}Z(\mathbb{O}(s))\bigg|_{s=0} &= 2([R_{3k1j} - R_{4k2j}] - 2R_{3j1k})\Lambda_{1j}\Lambda_{3k} \\
&+ 2([R_{4k2j} - R_{3k1j}] - 2R_{4j2k})\Lambda_{2j}\Lambda_{4k} \\
&+ 2([R_{4k1j} + R_{3k2j}] - 2R_{4j1k})\Lambda_{1j}\Lambda_{4k} \\
&+ 2([R_{3k2j} + R_{4k1j}] - 2R_{3j2k})\Lambda_{2j}\Lambda_{3k} \\
&+ b_{jk}\Lambda_{1j}\Lambda_{1k} + b_{jk}\Lambda_{2j}\Lambda_{2k} + a_{jk}\Lambda_{3j}\Lambda_{3k} + a_{jk}\Lambda_{4j}\Lambda_{4k} \\
&+ 0\Lambda_{1j}\Lambda_{jk} + 0\Lambda_{2j}\Lambda_{jk} + 0\Lambda_{3j}\Lambda_{jk} + 0\Lambda_{4j}\Lambda_{jk} \\
&- 2f_{jk}\Lambda_{1j}\Lambda_{2k} - 2e_{jk}\Lambda_{3j}\Lambda_{4k},
\end{aligned}
$$

where the zero coefficients are the result of applying (14.9a)–(14.9d). Observe that by making the following frame switch[4]

$$\{e_1, e_2, e_3, e_4\} \mapsto \{e_2, -e_1, e_4, -e_3\},$$

the terms $a_{jk}, b_{jk}, e_{jk}, f_{jk}$ remain invariant. In which case, if we denote the new orthonormal frame by $\mathbb{O}'$, we find that

$$
\begin{aligned}
\frac{1}{2}\frac{d^2}{ds^2}Z(\mathbb{O}'(s))\bigg|_{s=0} &= 2([\ \ R_{4k2j} - R_{3k1j}] - 2R_{4j2k})\Lambda_{1j}\Lambda_{3k} \\
&+ 2([\ \ R_{3k1j} - R_{4k2j}] - 2R_{3j1k})\Lambda_{2j}\Lambda_{4k} \\
&+ 2([-R_{3k2j} - R_{4k1j}] + 2R_{3j2k})\Lambda_{1j}\Lambda_{4k} \\
&+ 2([-R_{4k1j} - R_{3k2j}] + 2R_{4j1k})\Lambda_{2j}\Lambda_{3k} \\
&+ b_{jk}\Lambda_{1j}\Lambda_{1k} + b_{jk}\Lambda_{2j}\Lambda_{2k} + a_{jk}\Lambda_{3j}\Lambda_{3k} + a_{jk}\Lambda_{4j}\Lambda_{4k} \\
&+ 0\Lambda_{1j}\Lambda_{jk} + 0\Lambda_{2j}\Lambda_{jk} + 0\Lambda_{3j}\Lambda_{jk} + 0\Lambda_{4j}\Lambda_{jk} \\
&- 2f_{jk}\Lambda_{1j}\Lambda_{2k} - 2e_{jk}\Lambda_{3j}\Lambda_{4k}.
\end{aligned}
$$

We take the sum of the second derivatives in these two frames to get

$$
\begin{aligned}
\frac{1}{2}\left(\frac{d^2}{ds^2}Z(\mathbb{O}(s)) + \frac{d^2}{ds^2}Z(\mathbb{O}'(s))\right)\bigg|_{s=0} &= -4c_{jk}\Lambda_{1j}\Lambda_{3k} - 4c_{jk}\Lambda_{2j}\Lambda_{4k} \\
&- 4d_{jk}\Lambda_{1j}\Lambda_{4k} + 4d_{jk}\Lambda_{2j}\Lambda_{3k} \\
&+ 2b_{jk}\Lambda_{1j}\Lambda_{1k} + 2b_{jk}\Lambda_{2j}\Lambda_{2k} \\
&+ 2a_{jk}\Lambda_{3j}\Lambda_{3k} + 2a_{jk}\Lambda_{4j}\Lambda_{4k} \\
&- 4f_{jk}\Lambda_{1j}\Lambda_{2k} - 4e_{jk}\Lambda_{3j}\Lambda_{4k}
\end{aligned}
$$

[4] Note that this the same switch used in [BS09a, Proposition 8] (see also Remark 14.17).

which is a considerably simpler looking expression involving only a_{jk}, b_{jk}, e_{jk} and f_{jk} terms. Moreover, since $Z(\mathbb{O}(0)) = Z(\mathbb{O}'(0)) = 0$ and $Z(\mathbb{O}) \geq 0$ for all $\mathbb{O}$, the sum

$$\frac{d^2}{ds^2} Z(\mathbb{O}(s))\Big|_{s=0} + \frac{d^2}{ds^2} Z(\mathbb{O}'(s))\Big|_{s=0}$$

is positive semi-definite by construction. Thus the corresponding matrix

$$L = 2 \begin{pmatrix} B & -F & -C & -D \\ F & B & D & -C \\ -C^T & D^T & A & -E \\ -D^T & -C^T & E & A \end{pmatrix}$$

is also positive semi-definite (note $E^T = -E$ and $F^T = -F$). Defining

$$U = \begin{pmatrix} 0 & 0 & I & 0 \\ 0 & 0 & 0 & -I \\ -I & 0 & 0 & 0 \\ 0 & I & 0 & 0 \end{pmatrix},$$

the conjugation of L by U is

$$L^{\vee} = ULU^T = 2 \begin{pmatrix} A & E & C^T & D^T \\ -E & A & -D^T & C^T \\ C & -D & B & F \\ D & C & -F & B \end{pmatrix}.$$

In which case positive semi-definiteness implies that

$$\begin{aligned} 0 &\leq \frac{1}{4} \operatorname{tr} LL^{\vee} \\ &= \operatorname{tr} AB + \operatorname{tr} EF - \operatorname{tr} C^2 - \operatorname{tr} D^2 \\ &= \sum_{j,k=5}^{n} a_{jk} b_{jk} - \sum_{j,k=5}^{n} \Big(e_{jk} f_{jk} + c_{jk} c_{kj} + d_{jk} d_{kj} \Big). \end{aligned}$$

Therefore

$$\sum_{j,k=5}^{n} \Big(a_{jk} b_{jk} - e_{jk} f_{jk} - c_{jk} c_{kj} - d_{jk} d_{kj} \Big) \geq 0.$$

This proves Claim 14.16. Theorem 14.14 follows by the maximum principle.

14.4 PCSC is Preserved by the Ricci Flow

In this section we show directly that nonnegative complex sectional curvature is preserved under the Ricci flow. This provides an alternative approach to showing that $\widehat{C}_{\mathrm{PIC}_2}$ is preserved without discussing the PIC condition. We also use the argument as motivation for the next section where we adapt it to give a notationally simpler proof that PIC is preserved.

14.4.1 The Mok Lemma

The main technical tool in our proof will be the following simple case of a result originally proved by Mok [Mok88].

Lemma 14.20. *Let* A *be a non-negative Hermitian form on* $\mathbb{C}^{2n}$ *given by*

$$\mathsf{A} = \begin{pmatrix} A & B \\ B^* & C \end{pmatrix}.$$

Then

$$\operatorname{tr} A\bar{C} \geq \operatorname{tr} B\bar{B}.$$

Proof. Firstly, note that if A is non-negative definite, then so is the matrix

$$\mathsf{B} = \begin{pmatrix} \bar{C} & -\bar{B}^* \\ -\bar{B} & \bar{A} \end{pmatrix}$$

since

$$\begin{pmatrix} x & y \end{pmatrix} \mathsf{B} \begin{pmatrix} x^* \\ y^* \end{pmatrix} = \begin{pmatrix} y & -x \end{pmatrix} \bar{\mathsf{A}} \begin{pmatrix} y^* \\ -x^* \end{pmatrix} \geq 0.$$

We observe that $\operatorname{tr} \mathsf{AB} \geq 0$, since If $\mathbb{U} \in U(2n)$ diagonalises A, then

$$\operatorname{tr} \mathsf{AB} = \operatorname{tr} (\mathbb{U}^* \mathsf{A} \mathbb{U})(\mathbb{U}^* \mathsf{B} \mathbb{U}) = \sum_i a_{ii} b_{ii} \geq 0$$

where $a_{ii} = (\mathbb{U}e_i)^* \mathsf{A} (\mathbb{U}e_i) \geq 0$ and $b_{ii} = (\mathbb{U}e_i)^* \mathsf{B} (\mathbb{U}e_i) \geq 0$ for each i. Now observe

$$\operatorname{tr} \mathsf{AB} = \operatorname{tr} (A\bar{C} - B\bar{B}) + \operatorname{tr} (C\bar{A} - B^* \bar{B}^*).$$

Since $\operatorname{tr} C\bar{A} = \operatorname{tr} A^* C^T = \operatorname{tr} A\bar{C}$ and $\operatorname{tr} B^* \bar{B}^* = \operatorname{tr} B\bar{B}$ we have that

$$\operatorname{tr} \mathsf{AB} = 2\operatorname{tr} (A\bar{C} - B\bar{B})$$

and so the lemma follows. □

14.4.2 Preservation of PCSC Proof

It is now straightforward to prove that positive complex sectional curvature is preserved under the Ricci flow.[5] To prove that the vector field Q points into C_{PCSC} it suffices to consider (by the characterisation of the cone C_{PCSC} in Definition 14.2) $R \in C_{\mathrm{PCSC}}$ and $X, Y \in \mathbb{R}^n_{\mathbb{C}}$ for which $\ell_{X,Y}(R) = 0$, and prove that $\ell_{X,Y}(Q(R)) \geq 0$. Noting that $\ell_{X,Y}(R) = X^i Y^i \bar{X}^k \bar{Y}^\ell R_{ijkl}$, this amounts to the following:

Claim 14.21. If $R(X, Y, \bar{X}, \bar{Y}) = X^i Y^i \bar{X}^k \bar{Y}^\ell R_{ijk\ell} = 0$ and all complex sectional curvatures are nonnegative, then

$$X^i Y^j \bar{X}^k \bar{Y}^\ell Q_{ijk\ell} \geq 0.$$

Proof. To prove this claim, consider $X(t) = X + tW$ and $Y(t) = Y + tZ$. Also, define

$$h(t) := X^i(t) Y^j(t) \bar{X}^k(t) \bar{Y}^\ell(t) R_{ijk\ell}.$$

So we have $h(0) = 0$ and $h(t) \geq 0$ for all t. Hence $h'(0) = 0$ and $h''(0) \geq 0$.

Computing $\frac{1}{2}h''(t)$ directly, we find

$$\Big(W^i Z^j \bar{X}^k \bar{Y}^\ell + W^i Y^j \bar{W}^k \bar{Y}^\ell + W^i Y^j \bar{X}^k \bar{Z}^\ell + X^i Z^j \bar{W}^k \bar{Y}^\ell + X^i Z^j \bar{X}^k \bar{Z}^\ell + X^i Y^j \bar{W}^k \bar{Z}^\ell\Big) R_{ijk\ell} \geq 0.$$

The inequality holds for arbitrary W and Z in $\mathbb{R}^n_{\mathbb{C}}$. In particular, the same inequality with W replaced by iW and Z replaced by iZ gives the following:

$$\Big(-W^i Z^j \bar{X}^k \bar{Y}^\ell + W^i Y^j \bar{W}^k \bar{Y}^\ell + W^i Y^j \bar{X}^k \bar{Z}^\ell + X^i Z^j \bar{W}^k \bar{Y}^\ell + X^i Z^j \bar{X}^k \bar{Z}^\ell - X^i Y^j \bar{W}^k \bar{Z}^\ell\Big) R_{ijk\ell} \geq 0.$$

Adding these two inequalities gives:

$$\Big(W^i Y^j \bar{W}^k \bar{Y}^\ell + X^i Z^j \bar{X}^k \bar{Z}^\ell + W^i Y^j \bar{X}^k \bar{Z}^\ell + X^i Z^j \bar{W}^k \bar{Y}^\ell\Big) R_{ijk\ell} \geq 0. \tag{14.11}$$

Writing this as the positivity of a Hermitian form on $\mathbb{R}^n_{\mathbb{C}} \times \mathbb{R}^n_{\mathbb{C}}$ we get

$$(W\ Z) \begin{pmatrix} A & B \\ B^* & C \end{pmatrix} \begin{pmatrix} W^* \\ Z^* \end{pmatrix} \geq 0$$

[5] We learnt this argument from unpublished work of Ni and Wolfson.

where

$$A_{pq} = Y^j \bar{Y}^\ell R_{jp\ell q}$$
$$B_{pq} = -Y^j \bar{X}^k R_{pjqk}$$
$$C_{pq} = X^i \bar{X}^k R_{ipkq}.$$

Then Mok's lemma implies

$$0 \le \mathrm{tr}\,(A\bar{C} - B\bar{B}) = X^i Y^j \bar{X}^k \bar{Y}^\ell R_{ipkq} R_{jp\ell q} - X^i Y^j \bar{X}^k \bar{Y}^\ell R_{ip\ell q} R_{jpqk}$$

which is precisely the $R^\#$ term in $X^i Y^j \bar{X}^k \bar{Y}^\ell Q_{ijk\ell}$. The remanding squared term is trivially nonnegative, since

$$\frac{1}{2} X^i Y^j \bar{X}^k \bar{Y}^\ell R_{ijpq} R_{pqk\ell} = \frac{1}{2} \sum_{p,q} |X^i Y^j R_{ijpq}|^2 \ge 0.$$

This completes the proof. □

14.4.3 Relating PCSC to PIC

We can relate the complex sectional curvature condition to the cone construction discussed in Sect. 14.2.

Proposition 14.22. *The cone* $C_{\mathrm{PCSC}} = \widehat{C}_{\mathrm{PIC}_k}$ *for* $k \ge 2$.[6]

Proof. Certainly $C_{\mathrm{PCSC}} \subset \widehat{C}_{\mathrm{PIC}_k}$, since if $R \in C_{\mathrm{PCSC}}$ then the complex sectional curvature of any 2-plane in $\mathbb{C}^{n+k}$ is proportional to the complex curvature of its projection onto $\mathbb{C}^n$, which is non-negative.

In order to see the converse, it is enough to show that given $Z, W \in \mathbb{C}^n$ linearly independent, there exist extensions $\tilde{Z}, \tilde{W}$ on $\mathbb{C}^n \times \mathbb{C}^2$ such that the complex 2-plane they generate is totally isotropic and

$$K_M(Z, W, \bar{Z}, \bar{W}) = K_{M \times \mathbb{R}^2}(\tilde{Z}, \tilde{W}) \tag{14.12}$$

To show this, let $\tilde{Z} = Z + ue_1 + ve_2$ and $\tilde{W} = W + xe_1 + ye_2$ be extensions of Z, W which span a totally isotropic plane. Here $\{e_1, e_2\}$ is an orthonormal basis for the factor $\mathbb{R}^2$. Clearly (14.12) is satisfied since the extra $\mathbb{R}^2$ part is flat. To show the isotropic condition, one needs $\tilde{Z} \cdot \tilde{Z} = \tilde{W} \cdot \tilde{W} = \tilde{Z} \cdot \tilde{W} = 0$. By expanding the inner product, this is equivalent to

[6] This observation is due to Nolan Wallach. A similar argument also identifies $\widehat{C}_{\mathrm{PIC}_1}$ the cone of operators which has positive complex sectional curvature on isotropic (not necessarily *totally isotropic*) 2-planes (cf. Sect. 14.1).

$$Z \cdot Z + u^2 + v^2 = 0$$

$$W \cdot W + x^2 + y^2 = 0$$

$$Z \cdot W + ux + vy = 0.$$

These can be written in the matrix form $XX^T = A$, with

$$X = \begin{pmatrix} u & v \\ x & y \end{pmatrix} \qquad A = \begin{pmatrix} a & c \\ c & b \end{pmatrix}$$

where $a = -Z \cdot Z$, $b = -W \cdot W$, and $c = -Z \cdot W$. The matrix equation can be solved since A can be diagonalised by a $\mathrm{GL}(2, \mathbb{C})$ transformation. □

14.5 Preserving PIC Using the Complexification

In this section we give a proof that PIC is preserved, by working directly with the description (14.3) of PIC in terms of the complex sectional curvatures, instead of the expression (14.4). While the argument is fundamentally the same, working with the complexification results in considerable notational simplification, and perhaps elucidates the particular choices and combinations which were made in the previous approach.[7] We write

$$C_{\mathrm{PIC}} = \{R \in \mathrm{Curv} : \ \ell(R) \geq 0, \ \text{for all } \ell \in B\}$$

where $B = \{\ell_{u,v} : u, v \in \mathbb{R}^n_{\mathbb{C}}, \ u \cdot u = u \cdot v = v \cdot v = 0\}$, and $\ell_{u,v}(R) = R(u, v, \bar{u}, \bar{v})$. By the maximum principle and the results of Sect. B.5 in Appendix B, it suffices to show that $\ell_{(u,v)}(Q(R)) \geq 0$ whenever R is an element of C_{PIC} with $\ell_{u,v}(R) = 0$ and $\ell_{u,v} \in B$. Noting that $Q(R) = R^2 + R^\#$, and that $\ell_{u,v}(R^2) = R^2(u, v, \bar{u}, \bar{v}) \geq 0$, it suffices to prove that $\ell_{u,v}(R^\#) \geq 0$.

In order to prove this, we first observe that ℓ is really defined on a quotient of B: $\ell_{au+bv, cu+dv} = |ad - bc|^2 \ell_{u,v}$, so we have an action of $\mathrm{GL}(2, \mathbb{C})$ which only changes ℓ by a positive factor. Also note that $(au + bv, cu + dv)$ still satisfies the totally isotropic conditions. Therefore we can choose u and v so that $u \cdot \bar{u} = v \cdot \bar{v} = 1$ and $u \cdot \bar{v} = 0$. It follows that we can choose an orthonormal basis for $\mathbb{R}^n_{\mathbb{C}}$ consisting of $e_1 = u$, $e_2 = \bar{u}$, $e_3 = v$, $e_4 = \bar{v}$, and $(n - 4)$ other vectors e_p, $p > 4$, which can be chosen to be real ($\bar{e}_p = e_p$).

[7] We recently learned of an independent argument of Wilking which has a related idea but applies in greater generality, proving that a large family of cones are preserved by the Ricci flow, including C_{PCSC}, $\widehat{C}_{\mathrm{PIC}_1}$ and $\widehat{C}_{\mathrm{PIC}_2}$.

Let X and Y be arbitrary vectors in the span of $\{e_p : p > 4\}$. We will consider the smooth family $(U(t), V(t))$ defined by

$$U(t) = u + tX - \frac{t^2 X \cdot X}{2}\bar{u} - \frac{t^2 X \cdot Y}{2}\bar{v}$$
$$V(t) = v + tY - \frac{t^2 X \cdot Y}{2}\bar{u} - \frac{t^2 Y \cdot Y}{2}\bar{v}.$$

A direct computation (using the fact that $X \cdot u = X \cdot \bar{u} = X \cdot v = X \cdot \bar{v} = 0$, and similarly for Y) shows that $(U(t), V(t))$ is totally isotropic for every t. Since R is in C_{PIC}, we have $\ell_{U(t),V(t)}(R) \geq 0$ for all t, but $\ell_{U(0),V(0)}(R) = 0$, and hence the first derivative vanishes and the second derivative is non-negative when $t = 0$. We compute directly:

$$\frac{d}{dt}\ell_{(U(t),V(t))}(R)$$
$$= R(\dot{U}, V, \bar{U}, \bar{V}) + R(U, \dot{V}, \bar{U}, \bar{V}) + R(U, V, \dot{\bar{U}}, \bar{V}) + R(U, V, \bar{U}, \dot{\bar{V}}) \quad (14.13)$$
$$= 2\Re\big(R(\dot{U}, V, \bar{U}, \bar{V}) + R(U, \dot{V}, \bar{U}, \bar{V})\big). \quad (14.14)$$

Evaluating at $t = 0$ gives

$$0 = \Re\big(R(X, v, \bar{u}, \bar{v}) + R(u, Y, \bar{u}, \bar{v})\big).$$

The same equation with X and Y replaced by iX and iY gives

$$0 = \Im\big(R(X, v, \bar{u}, \bar{v}) + R(u, Y, \bar{u}, \bar{v})\big).$$

Since X and Y are arbitrary vectors in the span of $\{e_p : p > 4\}$, this implies

$$R(X, v, \bar{u}, \bar{v}) = R(u, Y, \bar{u}, \bar{v}) = 0 \quad (14.15)$$

for any $X, Y \in \text{span}\{e_p : p > 4\}$.

Now we differentiate (14.13) again in t:

$$\frac{d^2}{dt^2}\ell_{(U(t),V(t))}(R) = R(\ddot{U}, V, \bar{U}, \bar{V}) + R(U, \ddot{V}, \bar{U}, \bar{V})$$
$$+ R(U, V, \ddot{\bar{U}}, \bar{V}) + R(U, V, \bar{U}, \ddot{\bar{V}})$$
$$+ 2R(\dot{U}, \dot{V}, \bar{U}, \bar{V}) + R(\dot{U}, V, \dot{\bar{U}}, \bar{V}) + 2R(\dot{U}, V, \bar{U}, \dot{\bar{V}})$$
$$+ 2R(U, \dot{V}, \dot{\bar{U}}, \bar{V}) + 2R(U, \dot{V}, \bar{U}, \dot{\bar{V}}) + R(U, V, \dot{\bar{U}}, \dot{\bar{V}}).$$

Now we take the same with X and Y replaced by iX and iY respectively, and add. Note that this reverses the signs of $\ddot{U}$ and $\ddot{V}$, so all the terms involving these second derivatives cancel. The terms involving $R(\dot{U},\dot{V},\bar{\dot{U}},\bar{\dot{V}})$ and its conjugate also cancel. Evaluating at $t=0$ (so that $\dot{U}=X$ and $\dot{V}=Y$) we obtain the following:

$$\begin{aligned}R(X,v,\bar{X},\bar{v}) &+ R(u,Y,\bar{u},\bar{Y})\\ &+ R(X,v,\bar{u},\bar{Y}) + R(u,Y,\bar{X},\bar{v}) \geq 0\end{aligned} \tag{14.16}$$

for all $X,Y \in \operatorname{span}\{e_p:\ p>4\}$. Note the similarity to (14.11). It follows by the same argument as used there that

$$\sum_{p,q>4}\left(R(u,e_p,\bar{u},e_q)R(v,e_p,\bar{v},e_q) - R(u,e_p,\bar{v},e_q)R(v,e_p,\bar{u},e_q)\right) \geq 0.$$

This is almost the same as $R^{\#}(u,v,\bar{u},\bar{v})$, except that the sum is not over all values of p. To fix this problem we look in more detail at the terms which are missing: These are

$$\begin{aligned}\sum_{1\leq p,q\leq 4}&\left(R(u,e_p,\bar{u},e_q)R(v,\bar{e}_p,\bar{v},\bar{e}_q) - R(u,e_p,\bar{v},e_q)R(v,\bar{e}_p,\bar{u},\bar{e}_q)\right)\\ &+\sum_{p\leq 4<q}\left(R(u,e_p,\bar{u},e_q)R(v,\bar{e}_p,\bar{v},\bar{e}_q) - R(u,e_p,\bar{v},e_q)R(v,\bar{e}_p,\bar{u},\bar{e}_q)\right)\\ &+\sum_{q\leq 4<p}\left(R(u,e_p,\bar{u},e_q)R(v,\bar{e}_p,\bar{v},\bar{e}_q) - R(u,e_p,\bar{v},e_q)R(v,\bar{e}_p,\bar{u},\bar{e}_q)\right).\end{aligned}$$

The second and third sums above are zero by the first derivative condition (14.15): The second sum expands to give (for each $q>4$)

$$\begin{aligned}R(u,\bar{u},\bar{u},e_q)R(v,u,\bar{u},e_q) &+ R(u,v,\bar{u},e_q)R(v,\bar{v},\bar{v},e_q)\\ &- R(u,\bar{u},\bar{v},e_q)R(v,u,\bar{u},e_q) - R(u,v,\bar{v},e_q)R(v,\bar{v},\bar{u},e_q).\end{aligned}$$

The second factor in the first and third terms, and the first factor in the second and fourth terms, all vanish by (14.15). The third sum expands similarly. This leaves only the terms where both p and q are less than or equal to 4, which expand as follows:

$$\begin{aligned}R(u,\bar{u},\bar{u},\bar{v})R(v,u,\bar{v},\bar{u}) &+ R(u,\bar{u},\bar{u},\bar{v})R(v,u,\bar{v},v)\\ &+ R(u,v,\bar{u},u)R(v,\bar{v},\bar{v},\bar{u}) + R(u,v,\bar{u},\bar{v})R(v,\bar{v},\bar{v},v)\\ &- R(u,\bar{u},\bar{v},\bar{u})R(v,u,\bar{u},u) - R(u,\bar{u},\bar{v},v)R(v,u,\bar{u},\bar{v})\\ &- R(u,v,\bar{v},\bar{u})R(v,\bar{v},\bar{u},u) - R(u,v,\bar{v},v)R(v,\bar{v},\bar{u},\bar{v}).\end{aligned}$$

Here the second factor in the first and sixth terms, and the first factor in the fourth and seventh terms, vanish since $R(u, v, \bar{u}, \bar{v}) = \ell_{(u,v)}R = 0$. The remaining terms factor to give

$$\big(R(u, \bar{u}, \bar{u}, \bar{v}) + R(v, \bar{v}, \bar{u}, \bar{v})\big)\big(R(u, v, v, \bar{v}) + R(u, v, u, \bar{u})\big).$$

For the punchline we need to consider two more smooth curves in B: First let $U(t) = u + t\bar{v}$ and $V(t) = v - t\bar{u}$. Then $\ell_{U(t),V(t)} \in B$ for every t, so

$$\begin{aligned} 0 = \frac{d}{dt}\ell_{(U(t),V(t))}R\Big|_{t=0} &= R(\bar{v}, v, \bar{u}, \bar{v}) - R(u, \bar{u}, \bar{u}, \bar{v}) \\ &\quad + R(u, v, v, \bar{v}) - R(u, v, \bar{u}, u). \end{aligned}$$

Now the same computation with $U(t) = u + it\bar{v}$ and $V(t) = v - it\bar{u}$ gives

$$0 = iR(\bar{v}, v, \bar{u}, \bar{v}) - iR(u, \bar{u}, \bar{u}, \bar{v}) - iR(u, v, v, \bar{v}) + iR(u, v, \bar{u}, u).$$

These two identities imply

$$0 = R(u, v, v, \bar{v}) - R(u, v, \bar{u}, u) = R(\bar{v}, v, \bar{u}, \bar{v}) - R(u, \bar{u}, \bar{u}, \bar{v}),$$

so that all of the terms arising from the sum with $1 \le p, q \le 4$ vanish. We have proved that

$$\begin{aligned} \ell_{(u,v)}(R^{\#}) = \sum_{p,q>4} \Big(&R(u, e_p, \bar{u}, e_q)R(v, e_p, \bar{v}, e_q) \\ &- R(u, e_p, \bar{v}, e_q)R(v, e_p, \bar{u}, e_q)\Big) \ge 0, \qquad (14.17) \end{aligned}$$

so by the maximum principle PIC is preserved by Ricci flow.

Chapter 15
The Final Argument

15.1 Proof of the Sphere Theorem

We now have all the ingredients in place to prove the following:

Theorem 15.1 (Differentiable 1/4-Pinched Sphere Theorem). *A compact, pointwise 1/4-pinched Riemannian manifold of dimension $n \geq 4$ is diffeomorphic to a spherical space form.*

Remark 15.2. In fact we prove the stronger result that any compact Riemannian manifold (M, g) such that $M \times \mathbb{R}^2$ has positive curvature on totally isotropic 2-planes (equivalently, any compact Riemannian manifold with positive complex sectional curvatures) is diffeomorphic to a spherical space-form. This implies the pointwise differentiable 1/4-pinching sphere theorem by Corollary 14.13.

Proof. Suppose (M, g_0) is a compact Riemannian manifold with positive complex sectional curvatures, so that the curvature operator R is in the interior of $C_{\mathrm{PCSC}} = \widehat{C}_{\mathbf{PIC}_2}$ for every $p \in M$ (cf. Sect. 14.4.3). Let $(M, g(t))$ be the solution of Ricci flow with initial metric g_0 on a maximal time interval $[0, T)$ (the existence of which is guaranteed by the results of Chap. 4). Also note that $T < \infty$ since the maximum principle gives a lower bound on scalar curvature which approaches infinity in finite time (see Sect. 7.2.1). The curvature operators of (M, g_0) lie in a compact set K in the interior of the cone C_{PCSC}. By Theorem 14.8 together with Theorem 14.14 (or alternatively the results of Sect. 14.4.2) the cone C_{PCSC} is a closed convex cone preserved by Ricci flow, which is contained in the cone of positive sectional curvature and contains the positive curvature operator cone (by Proposition 14.9). Therefore by Theorem 13.2 there exists a pinching family of cones $C(s)$, $0 \leq s < 1$ with $C(0) = C_{\mathrm{PCSC}}$, and by Theorem 13.8 there exists a preserved closed convex set F containing the compact set K, and numbers $\rho(s)$ such that $F + \rho(s) I \subset C(s)$ for each $s > 0$. By the maximum principle of Theorem 7.15, the curvature operators of $g(t)$ lie in F for all $t \in [0, T)$.

Now we perform the blow-up procedure described in Sect. 9.5. As indicated in Theorem 11.22, by the compactness theorem of Chap. 9 together

B. Andrews and C. Hopper, *The Ricci Flow in Riemannian Geometry*, Lecture Notes in Mathematics 2011, DOI 10.1007/978-3-642-16286-2_15,

with the injectivity radius lower bound from Chap. 11 (and the regularity estimates from Chap. 8) there exists a limit $(M_\infty, g_\infty, O_\infty)$ of a sequence of blow-up metrics $g_i(t) = Q_i g(t_i + Q_i^{-1} t)$ about points O_i with $Q_i = |R|(O_i, t_i) = \sup_{M\times[0,t_i]} |R| \to \infty$, and the limit has $|R|(x,t) \le 1$ for $t \le 0$. Therefore the curvatures of g_i are given by Q_i^{-1} times the curvatures of g, and so lie in the set $Q_i^{-1}F$. On a neighbourhood of O_∞, the pullback metrics $\Phi_i^* g_i$ have scalar curvature bounded away from zero, and so $R(g_\infty)$ lies in the line $\mathbb{R}_+ I$ by condition (3) of Theorem 13.8. Schur's Lemma states that a smooth manifold of dimension at least 3 for which $R(x) = \alpha(x)I$ has constant curvature, i.e. $\alpha(x)$ is constant (see Sect. 12.3.1). Therefore (M_∞, g_∞) is a complete Riemannian manifold with constant positive curvature; hence a spherical space-form by the theorem of Hopf mentioned in the introduction chapter. Since M_∞ is compact, the convergence is in the C^∞ sense, rather than only C^∞ on compact subsets. In particular M is diffeomorphic to a spherical space form. his completes the proof. □

15.2 Refined Convergence Result

In this section we discuss more recent work of Brendle [Bre08] in which he proved the following theorem:

Theorem 15.3. *Let (M, g_0) be a compact Riemannian manifold of dimension $n \ge 3$ such that $M \times \mathbb{R}$ has positive isotropic curvature. Then M carries a metric of constant positive curvature, and is diffeomorphic to a spherical space form.*

As usual, the last part of the statement is given by Hopf's classification of spaces with constant positive curvature; the idea of proof is to use Ricci flow to deform the metric g_0 to a metric of constant positive curvature.

Remark 15.4. As shown in Sect. 14.2.1, the cone $\widehat{C}_{\mathbf{PIC}_1}$ contains the cone $\widehat{C}_{\mathbf{PIC}_2} = C_{\mathrm{PCSC}}$, so the assumptions of Theorem 15.3 are weaker than in the result proved above. We shall see that the cone $\widehat{C}_{\mathbf{PIC}_2}$ contains the cone of 2-positive curvature operators (see Lemma 15.6), so Theorem 15.3 also generalises the main results of Chen [Che91] and of Böhm and Wilking [BW08].

We have seen that the cone $\widehat{C}_{\mathbf{PIC}_1}$ is a closed convex cone in the space Curv of algebraic curvature operators, which is preserved by the Ricci flow. The difficulty is that $\widehat{C}_{\mathbf{PIC}_1}$ is no longer contained in the cone of positive sectional curvature, so we can no longer apply Theorem 13.2 to find a pinching family of cones. The work to be done is to find a replacement for this missing step, so that the rest of the argument can be applied as before.

For $n > 3$ Brendle [Bre08, Proposition 4] gives a characterization similar to the result of Proposition 14.12:

Proposition 15.5 (Characterisation of $\widehat{C}_{\mathbf{PIC}_1}$). *Let $R \in \mathrm{Curv}(\mathbb{R}^n)$, for $n \geq 4$. Then $R \in \widehat{C}_{\mathbf{PIC}_1}$ if and only if*

$$R_{1313} + \lambda^2 R_{1414} + R_{2323} + \lambda^2 R_{2424} - 2\lambda R_{1234} \geq 0$$

for all orthonormal frames $\{e_1, e_2, e_3, e_4\}$ and all $\lambda \in [-1, 1]$.

This implies that PIC on $M \times \mathbb{R}$ is weaker than 2-positive curvature:[1]

Lemma 15.6. *If $R \in \mathrm{Curv}(\mathbb{R}^n)$, for $n \geq 4$, is 2-positive, then $R \in \widehat{C}_{\mathbf{PIC}_1}$.*

Proof. If R is a 2-positive curvature operator, then for any 4-frame

$$\begin{aligned} 0 &\leq R(e_1 \wedge e_3 - \lambda e_2 \wedge e_4, e_1 \wedge e_3 - \lambda e_2 \wedge e_4) \\ &\qquad + R(e_2 \wedge e_3 + \lambda e_1 \wedge e_4, e_2 \wedge e_3 + \lambda e_1 \wedge e_4) \\ &= R_{1313} + \lambda^2 R_{1414} + R_{2323} + \lambda^2 R_{2424} - 2\lambda R_{1234}, \end{aligned}$$

where $|e_1 \wedge e_3 - \lambda e_2 \wedge e_4|^2 = |e_2 \wedge e_3 + \lambda e_1 \wedge e_4|^2 = 1 + \lambda^2$ and

$$\langle e_1 \wedge e_3 - \lambda e_2 \wedge e_4, e_2 \wedge e_3 + \lambda e_1 \wedge e_4 \rangle = 0.$$

Therefore by Proposition 15.5, $R \in \widehat{C}_{\mathbf{PIC}_1}$. □

Another reason why the result of Theorem 15.3 seems very satisfying is because of its meaning in the three-dimensional case:

Lemma 15.7. *On a 3-manifold, the cones*

$$\begin{aligned} \widehat{C}_{\mathbf{PIC}_1}(\mathbb{R}^3) &= \{R \in \mathrm{Curv}(\mathbb{R}^3) : \ \mathrm{Ric}(R) \geq 0\} \\ \widehat{C}_{\mathbf{PIC}_2}(\mathbb{R}^3) &= \{R \in \mathrm{Curv}(\mathbb{R}^3) : \ R \geq 0\}. \end{aligned}$$

That is, the result of Theorem 15.3 for $n=3$ recovers Hamilton's result [Ham82b] for manifolds with positive Ricci curvature, while $R \in \widehat{C}_{\mathbf{PIC}_2}$ amounts to the assumption of positive sectional curvatures.

Proof. The cone $\widehat{C}_{\mathbf{PIC}_2}$ is easy to identify: From Proposition 14.9, C_{PCSC} is contained in the cone of positive sectional curvatures, and contains the cone of positive curvature operators. However in three dimensions these coincide, since the curvature tensor, when $n=3$, can be rewritten as a symmetric bilinear form Λ given by (7.5). So if the sectional curvatures are all positive, $\Lambda(v,v)$ is also positive since it is $\|v\|^2$ times the sectional curvature of the 2-plane normal to v.

---

[1] A self-adjoint bilinear form is 2-*positive* if it has positive trace on 2-dimensional subspaces (equivalently, the sum of the smallest two eigenvalues is positive).

Now consider the cone $\widehat{C}_{\mathbf{PIC}_1}$. First, we show that this is contained in the cone of positive Ricci curvature: Let $R \in \widehat{C}_{\mathbf{PIC}_1}(\mathbb{R}^3)$ and $v \in \mathbb{R}^3$ with $\|v\| = 1$. Choose an orthonormal basis $\{e_1, e_2, e_3 = v\}$ for $\mathbb{R}^3 \times \{0\}$, and let $e_4 = (0, 1)$. Then by (14.2) we have

$$\begin{aligned} 0 < K_{\mathbb{C}}(e_1 + ie_2, e_3 + ie_4) &= \widehat{R}_{1313} + \widehat{R}_{2424} + \widehat{R}_{1414} + \widehat{R}_{2323} - 2\widehat{R}_{1234} \\ &= R_{1313} + R_{2323} \\ &= R_{j3j3} = \mathrm{Ric}(R)_{33} = \mathrm{Ric}(R)(v, v). \end{aligned}$$

Since v is arbitrary, $\mathrm{Ric}(R) > 0$. For the converse, we must suppose that $\mathrm{Ric}(R) > 0$, and show that all complex sectional curvatures of totally isotropic 2-planes in $\mathbb{R}^4_{\mathbb{C}}$ are positive. Let Π be any such 2-plane, and let X be a vector in Π orthogonal to e_4. As X is isotropic, we can write $X = e_1 + ie_2$ for some orthonormal pair e_1, e_2 in $\mathbb{R}^3 \times \{0\}$. Moreover Π is totally isotropic, so Π is spanned by $e_1 + ie_2$ and $(\cos\theta e_3 + \sin\theta e_4) + i(-\sin\theta e_3 + \cos\theta e_4)$ for some θ. Now we can compute the complex sectional curvature directly:

$$\begin{aligned} K_{\mathbb{C}}(\Pi) &= K(\widehat{R})_{\mathbb{C}}(e_1 + ie_2, (\cos\theta e_3 + \sin\theta e_4) + i(-\sin\theta e_3 + \cos\theta e_4)) \\ &= K(R)_{\mathbb{C}}(e_1 + ie_2, (\cos\theta - i\sin\theta)(e_3 + ie_4)) \\ &= K(R)_{\mathbb{C}}(e_1 + ie_2, e_3 + ie_4) \\ &= R_{1313} + R_{2323} = \mathrm{Ric}(R)_{33} > 0. \end{aligned}$$

□

15.2.1 A Preserved Set Between $\widehat{C}_{\mathbf{PIC}_1}$ and $\widehat{C}_{\mathbf{PIC}_2}$

The main ingredient in the proof of Theorem 15.3 is the construction of a new preserved set E. Given $R \in \mathrm{Curv}(\mathbb{R}^n)$, define $S \in \mathrm{Curv}(\mathbb{R}^{n+2})$ by

$$S(\widehat{v}_1, \widehat{v}_2, \widehat{v}_3, \widehat{v}_4) = R(v_1, v_2, v_3, v_4) + \langle x_1 \wedge x_2, x_3 \wedge x_4 \rangle$$

for all $\widehat{v}_j = (v_j, x_j) \in \mathbb{R}^{n+2} \simeq \mathbb{R}^n \times \mathbb{R}^2$. Then E is defined as follows:[2]

$$E = \{R \in \mathrm{Curv}(\mathbb{R}^n) : S \in C_{\mathrm{PIC}}(\mathbb{R}^{n+2})\}.$$

This is a closed, convex $O(n)$-invariant set.

[2] Note that S is the curvature operator of $M \times S^2$, so E corresponds to PIC on $M \times S^2$.

Proposition 15.8 (Characterisation of E). *Let $R \in \mathrm{Curv}(\mathbb{R}^n)$, for $n \geq 4$. Then $R \in E$ if and only if*

$$R_{1313} + \lambda^2 R_{1414} + \mu^2 R_{2323} + \lambda^2\mu^2 R_{2424} - 2\lambda\mu R_{1234} + (1-\lambda^2)(1-\mu^2) \geq 0$$

for all orthonormal frames $\{e_1, e_2, e_3, e_4\}$ and all $\lambda, \mu \in [-1, 1]$.

We refer the reader to [Bre08, Proposition 7] for a proof. From Propositions 14.12, 15.5 and 15.8 we can relate E to the cones $\widehat{C}_{\mathbf{PIC}_1}$ and $\widehat{C}_{\mathbf{PIC}_2}$:

Corollary 15.9. *The set E lies between $\widehat{C}_{\mathbf{PIC}_1}$ and $\widehat{C}_{\mathbf{PIC}_2}$ in the sense that $\widehat{C}_{\mathbf{PIC}_2} \subset E \subset \widehat{C}_{\mathbf{PIC}_1}$. Furthermore we have*

$$\overline{\bigcup_{c>0} (cE)} = \widehat{C}_{\mathbf{PIC}_1} \qquad \text{and} \qquad \bigcap_{c>0} (cE) = \widehat{C}_{\mathbf{PIC}_2}.$$

In particular, if K is any compact subset in the interior of $\widehat{C}_{\mathbf{PIC}_1}$, then $K \subset \lambda E$ for some $\lambda > 0$.

Proof. If $R \in \widehat{C}_{\mathbf{PIC}_2}$ then

$$R_{1313} + \lambda^2 R_{1414} + \mu^2 R_{2323} + \lambda^2\mu^2 R_{2424} - 2\lambda\mu R_{1234} \geq 0,$$

so for any $c > 0$,

$$\begin{aligned} R_{1313} + \lambda^2 R_{1414} + \mu^2 R_{2323} + \lambda^2\mu^2 R_{2424} - 2\lambda\mu R_{1234} & \\ + c(1-\lambda^2)(1-\mu^2) &\geq 0, \end{aligned}$$

and $R \in cE$ for any $c > 0$.

If $R \in cE$, then choosing $\mu = 1$ in the characterization of Proposition 15.8 gives

$$R_{1313} + \lambda^2 R_{1414} + R_{2323} + \lambda^2 R_{2424} - 2\lambda R_{1234} \geq 0$$

for every 4-frame and every $|\lambda| \leq 1$, so $R \in \widehat{C}_{\mathbf{PIC}_1}$.

Now we show $\bigcup_{c>0}(cE) = \widehat{C}_{\mathbf{PIC}_1}$. The forward inclusion is clear, so we prove the reverse: Suppose R is not in cE for any $c > 0$. That is, for any c there exists a frame $\mathbb{O}_c$ and numbers $\lambda_c, \mu_c \in [-1, 1]$ such that

$$R^{\mathbb{O}_c}_{1313} + \lambda_c^2 R^{\mathbb{O}_c}_{1414} + \mu_c^2 R^{\mathbb{O}_c}_{2323} + \lambda_c^2\mu_c^2 R^{\mathbb{O}_c}_{2424} - 2\lambda_c\mu_c R^{\mathbb{O}_c}_{1234} + c(1-\lambda_c^2)(1-\mu_c^2) < 0.$$

By compactness we can find a subsequence $c_j \to \infty$ such that $\mathbb{O}_{c_j} \to \mathbb{O}$, $\lambda_{c_j} \to \lambda$, and $\mu_{c_j} \to \mu$. If $(1-\lambda^2)(1-\mu^2) > 0$ then we have a contradiction since the last term approaches infinity, so we can assume without loss of generality that $\mu = 1$. But then $R^{\mathbb{O}}_{1313} + \lambda^2 R^{\mathbb{O}}_{1414} + R^{\mathbb{O}}_{2323} + \lambda^2 R^{\mathbb{O}}_{2424} - 2\lambda R^{\mathbb{O}}_{1234} \leq 0$, contradicting the fact that R is in the interior of $\widehat{C}_{\mathbf{PIC}_1}$. This proves that $\bigcup_{c>0} cE = \widehat{C}_{\mathbf{PIC}_1}$.

The characterization of the intersection is clear from Propositions 15.8 and 14.12. □

The first step in proving that E is preserved by the Ricci flow is the following lemma:

Lemma 15.10. *Let $R \in \mathrm{Curv}(\mathbb{R}^n)$, and let S be the induced operator in $\mathrm{Curv}(\mathbb{R}^{n+2})$. Then*

$$S^{\#}(\widehat{v}_a, \widehat{v}_b, \widehat{v}_c, \widehat{v}_d) = R^{\#}(v_a, v_b, v_c, v_d)$$

for all vectors $\widehat{v}_i = (v_i, x_i) \in \mathbb{R}^n \times \mathbb{R}^2$.

Proof. Let $\{e_1, \dots, e_n\}$ be a basis for $\mathbb{R}^n$. Suppose also that $\{\widehat{e}_1, \dots, \widehat{e}_{n+2}\}$ is an orthonormal basis of $\mathbb{R}^n \times \mathbb{R}^2$ such that $\widehat{e}_k = (e_k, (0,0))$ for $k = 1, \dots, n$. Note that

$$\begin{aligned}\sum_{p,q=1}^{n+2} S_{apcq} S_{bpdq} &= \sum_{p,q=1}^{n} S_{apcq} S_{bpdq} + \sum_{p,q=n+1}^{n+2} S_{apcq} S_{bpdq} \\ &= \sum_{p,q=1}^{n} R_{apcq} R_{bpdq} + \langle x_a, x_b \rangle \langle x_c, x_d \rangle .\end{aligned}$$

By interchanging $\widehat{v}_c$ and $\widehat{v}_d$ we obtain

$$\sum_{p,q=1}^{n+2} S_{apdq} S_{bpcq} = \sum_{p,q=1}^{n} R_{apdq} R_{bpcq} + \langle x_a, x_b \rangle \langle x_d, x_c \rangle .$$

So by subtracting the first equation from this we get $S^{\#}_{abcd} = R^{\#}_{abcd}$. □

Proposition 15.11. *Let $\{e_1, e_2, e_3, e_4\}$ be an orthonormal 4-frame, let $\lambda, \mu \in [-1, 1]$ and suppose that the curvature operator $R \in E$. If*

$$\begin{aligned}R_{1313} + \lambda^2 R_{1414} + \mu^2 R_{2323} + \lambda^2 \mu^2 R_{2424} - 2\lambda\mu R_{1234} \\ + (1-\lambda^2)(1-\mu^2) = 0,\end{aligned} \tag{15.1}$$

then we have

$$Q_{1313} + \lambda^2 Q_{1414} + \mu^2 Q_{2323} + \lambda^2 \mu^2 Q_{2424} - 2\lambda\mu Q_{1234} \geq 0.$$

Proof. Define the orthonormal 4-frame $\{\widehat{e}_1, \widehat{e}_2, \widehat{e}_3, \widehat{e}_4\}$ for $\mathbb{R}^n \times \mathbb{R}^2$ by

$$\begin{aligned}\widehat{e}_1 &= (e_1, 0, 0) \qquad & \widehat{e}_2 &= (\mu e_2, 0, \sqrt{1-\mu^2}) \\ \widehat{e}_3 &= (e_3, 0, 0) \qquad & \widehat{e}_4 &= (\lambda e_4, \sqrt{1-\lambda^2}, 0).\end{aligned}$$

By hypothesis S has nonnegative isotropic curvature, and so (15.1) implies that

$$S_{1313} + S_{1414} + S_{2323} + S_{2424} - 2S_{1234} = 0.$$

From the argument of Sect. 14.3 (in particular Claim 14.16), or alternatively that of Sect. 14.5 (in particular (14.17)), we also have

$$S^{\#}_{1313} + S^{\#}_{1414} + S^{\#}_{2323} + S^{\#}_{2424} + 2S^{\#}_{1342} + 2S^{\#}_{1423} \geq 0.$$

Using Lemma 15.10 we obtain

$$R^{\#}_{1313} + \lambda^2 R^{\#}_{1414} + \mu^2 R^{\#}_{2323} + \lambda^2\mu^2 R^{\#}_{2424} + 2\lambda\mu R^{\#}_{1342} + 2\lambda\mu R^{\#}_{1423} \geq 0. \quad (15.2)$$

Moreover,

$$\begin{aligned} &R^2_{1313} + \lambda^2 R^2_{1414} + \mu^2 R^2_{2323} + \lambda^2\mu^2 R^2_{2424} + 2\lambda\mu R^2_{1342} + 2\lambda\mu R^2_{1423} \\ &\quad = (R_{13pq} - \lambda\mu R_{24pq})^2 + (\lambda R_{14pq} + \mu R_{23pq})^2 \geq 0. \end{aligned} \quad (15.3)$$

The result now follows by adding (15.2) and (15.3), together with the fact that $Q = Q(R)$ satisfies the first Bianchi identity by Remark 12.12. □

Now by the same argument used in Sect. 14.3, we see that Proposition 15.11 implies:

Proposition 15.12. *The set E is preserved by the Ricci flow.*

15.2.2 A Pinching Set Argument

We are now in a position to give the outline of the proof of Theorem 15.3, of which the main ingredient is Proposition 15.12. The picture looks very nice: We already know that the cone $\widehat{C}_{\mathbf{PIC}_2}$ has a pinching family linking it to the constant positive curvature ray. Now we have the region between $\widehat{C}_{\mathbf{PIC}_1}$ and $\widehat{C}_{\mathbf{PIC}_2}$ filled out by a family of preserved sets, each of which is asymptotic to the cone $\widehat{C}_{\mathbf{PIC}_2}$ near infinity. It seems highly plausible that with this information we could find a pinching set to finish the proof.

One might hope to modify the argument in the proof of Theorem 13.2 to produce a suitable pinching set. The fact that the sets λE are not cones does not pose any great difficulty here, however there is a more serious difficulty: The vector field Q does not point *strictly* into them. For example these sets all contain the radial line through the curvature operator of $S^{n-1} \times \mathbb{R}$, and the vector Q at such points is also radial, so is tangent to the boundary of E. Thus our first step is to modify E, using an idea from [BW08, Proposition 3.2], to fix this problem. We use the notation of Sect. 12.4.

Lemma 15.13. *Let G be a closed $O(n)$-invariant preserved convex set, such that*

1. *$G \setminus \{0\}$ is in the open half-space of positive scalar curvature.*
2. *G is contained in the cone of non-negative Ricci curvature.*
3. *G contains the cone of positive curvature operators.*

Suppose that

$$b \in \left(0, \frac{\sqrt{2n(n-2)+4}-2}{n(n-2)}\right), \qquad \text{and} \qquad a = b + \frac{n-2}{2}b^2.$$

Then the set $l_{a,b}(G)$ is strictly preserved, in the sense that $Q(R)$ is in the interior of the tangent cone of $l_{a,b}(G)$ at any non-zero boundary point R.

Proof. By Lemma 12.32 it suffices to show that the vector field $X_{a,b} = D_{a,b}$ $+Q$ points strictly into G, and since we know Q points into G it remains to show $D_{a,b}$ points strictly into G. Using Theorem 12.33, $D_{a,b}(R)$ is given by

$$\begin{aligned} D_{a,b}(R) = &\big((n-2)b^2 - 2(a-b)\big)\overset{\circ}{\mathrm{Ric}} \wedge \overset{\circ}{\mathrm{Ric}} \\ &+ 2a \,\mathrm{Ric} \wedge \mathrm{Ric} + 2b^2 \,\overset{\circ}{\mathrm{Ric}}{}^2 \wedge \mathrm{id} \\ &+ \frac{\mathrm{tr}(\overset{\circ}{\mathrm{Ric}}{}^2)}{n+2n(n-1)a}\big(nb^2(1-2b) - 2(a-b)(1-2b+nb^2)\big)I. \end{aligned}$$

By our choice of a and b, the first term vanishes; by assumption (2) the second term is non-negative; and the third is manifestly non-negative. The last term is non-negative, since the term in the bracket can be rewritten as $b^2(2-4b-(n-2)nb^2)$ by the choice of a and b, i.e. $2(a-b) = (n-2)b^2$. By examining the roots of the quadratic equation $2-4b-n(n-2)b^2$ one see that the desired quantity is strictly positive for the chosen range of b. Therefore $D_{a,b}(R) \geq 0$ for $R \in G$. Furthermore, the inequality is strict: In this range the last term is strictly positive unless $\sigma = 0$, and in that case we have $\mathrm{Ric} = \frac{\mathrm{Scal}}{n}I$, so the first term is strictly positive (since $\mathrm{Scal} > 0$ unless $R = 0$ by assumption (1)).

Finally, by assumption (3), we have $S/\varepsilon \in G$ for any non-negative curvature operator S and any $\varepsilon \in (0,1)$. So by convexity $(1-\varepsilon)R + S = (1-\varepsilon)R + \varepsilon(S/\varepsilon) \in G$. As G is closed, $R + S \in G$. But then $S \in G - R \subset \overline{\bigcup_{h>0} \frac{G-R}{h}} = \mathcal{T}_R G$. Thus the non-negative curvature operators are in $\mathcal{T}_R G$, and the positive curvature operators are in the interior of $\mathcal{T}_R G$, so in particular $D_{a,b}(R)$ points strictly into G at R. □

Corollary 15.14. *For a and b as given in Lemma 15.13, the sets $l_{a,b}(E)$ and $l_{a,b}(\widehat{C}_{\mathbf{PIC}_2})$ are strictly preserved.*

Proof. We verify the assumptions of Lemma 15.13 in these cases: The sets E contain $\widehat{C}_{\mathbf{PIC}_2}$, so assumption (3) of the lemma follows from Proposition 14.9. Assumptions (1) and (2) hold for $n = 3$ by Lemma 15.7. For $n \geq 4$, Proposition 15.5 with $\lambda = 0$ implies $R_{1313} + R_{2323} \geq 0$ for all orthonormal frames and all $R \in \widehat{C}_{\mathbf{PIC}_1}$. For any k, $\mathrm{Ric}_{kk} = \frac{1}{2(n-2)} \sum_{i \neq j,\ i,j \neq k} (R_{ikik} + R_{jkjk}) \geq 0$, and $\mathrm{Scal} = \frac{1}{2(n-2)} \sum_k \sum_{i \neq j\ i,j \neq k} (R_{ikik} + R_{jkjk}) \geq 0$. If $\mathrm{Scal} = 0$ then each of the non-negative summands must be zero, so $R_{1313} + R_{1212} = 0$, and subtracting gives $R_{2323} - R_{1212} = 0$. Since this is true for all frames, we have $R_{1212} = 0$ for all frames, and hence $R = 0$. □

Now things look much better: Consider the nested family of preserved sets $A(s)$ for $0 < s < \infty$ defined by

$$A(s) = \begin{cases} \frac{1-s}{s} l_{a(s),b(s)}(E) & \text{for } 0 < s < 1 \\ l_{a(1),b(1)}(\widehat{C}_{\mathbf{PIC}_2}) & \text{for } s = 1 \\ C(s-1) \bigcap l_{a(1),b(1)}(\widehat{C}_{\mathbf{PIC}_2}) & \text{for } s > 1 \end{cases}$$

where $C(s)$ is the pinching family of preserved cones constructed by applying Theorem 13.2 with $C(0) = \widehat{C}_{\mathbf{PIC}_2}$, and we choose

$$a(s) = b(s) + \frac{n-2}{2} b(s)^2 \qquad \text{and} \qquad b(s) = \frac{s}{2} \frac{\sqrt{2n(n-2)+4} - 2}{n(n-2)}$$

for $0 < s \leq 1$. By Lemma 15.13, $A(s)$ is strictly preserved for $0 < s \leq 1$, and $A(s)$ is also strictly preserved for $s > 1$ since it is an intersection of two strictly preserved sets. Since $C(s)$ approaches the ray of constant positive curvature operators as $s \to \infty$, so does $A(s)$. We also have that $A(s)$ approaches $\widehat{C}_{\mathbf{PIC}_1}$ as $s \to 0$ by Proposition 15.9, since $\frac{1-s}{s} E$ does and $l_{a(s),b(s)}$ approaches the identity transformation.

It is now straightforward to modify the argument of Theorem 13.8 to find a pinching set: The Theorem below is only a slight modification, and the structure of the proof is the same.

Theorem 15.15. *Let $\{A(s)\}_{s \in [0,\infty)}$ be a continuous nested family of closed convex $O(n)$-invariant preserved sets in* Curv *of maximal dimension, contained in the half-space of curvature operators with positive scalar curvature. Assume that for each $s > 0$ and for each $\lambda \in (0,1)$ we have $\lambda A(s) \subseteq A(s)$ and $\tilde{A}(s) = \bigcap_{\lambda > 0} (\lambda A(s))$ is a strictly preserved cone, in the sense that for all $R \in \partial \tilde{A}(s) \setminus \{0\}$, $Q(R)$ is in the interior of the tangent cone to $\tilde{A}(s)$ at R. Assume further that the family of cones $\{\tilde{A}(s)\}$ is continuous. Then if K is any compact set contained in the interior of $A(0)$, there exists a closed convex $O(n)$-invariant set $F \subset A(0)$ with the following properties:*

1. *$Q(R) \in \mathcal{T}_R F$ for every $R \in \partial F$.*
2. *$K \subset F$.*
3. *For each $s > 0$ there exists $\rho(s)$ such that $F + \rho(s) I \subset A(s)$.*

Proof. The family $\{A(s)\}$ is continuous in s, and $A(s)$ approaches $\widehat{C}_{\mathbf{PIC}_1}$ as $s \to 0$, so given any compact K in the interior of $A(0)$ we have $K \subset A(s_0)$ for some $s_0 > 0$. We will construct a set F of the form

$$F = A(s_0) \cap \bigcap_{i=1}^{\infty} \left(A(s_i) - 2^i hI \right),$$

where $h > 0$ and (s_i) is an increasing sequence approaching infinity.

Such a set is manifestly convex. This construction also gives condition (3) of the theorem automatically. Also, since each set $A(s)$ is $O(n)$-invariant, so is the set F. We first choose $s_0 > 0$ and $h \geq 1$ such that

$$K \subset A(s_0) \cap \{\mathrm{scal}(R) \leq h\}.$$

Lemma 15.16. *For any $\bar{s} \geq s_0$ there exists $N(\bar{s}) \geq 1$ (non-decreasing in $\bar{s}$) such that if $s \in [s_0, \bar{s}]$, $T \in \partial A(s)$ with $\mathrm{scal}(T) \geq N(\bar{s})$, then $Q(S)$ is in the tangent cone to $A(s)$ at T for every S with $|S - T| \leq 2|I|\mathrm{scal}(T)/N(\bar{s})$.*

Proof of Lemma. If the Lemma fails, then there exist sequences $s_i \in [s_0, \bar{s}]$, $T_i \in \partial A(s_i)$ and $S_i \in \mathrm{Curv}$ with $\mathrm{scal}(T_i) \geq i$ and $|S_i - T_i| \leq 2\mathrm{scal}(T_i)/i$, such that $Q(S_i)$ is not in the tangent cone to $A(s_i)$ at T_i.

Passing to a subsequence we can assume that $s_i \to s \in [s_0, \bar{s}]$. For each $N \in \mathbb{N}$, and for each $i \geq N$ we have $T_i/\mathrm{scal}(T_i) \in A(s_i)/\mathrm{scal}(T_i) \cap \{\mathrm{scal}(R) = 1\} \subseteq A(s_i)/N \cap \{\mathrm{scal}(R) = 1\}$, which is compact. Therefore there is a subsequence converging in $A(s)/N \cap \{\mathrm{scal}(R) = 1\}$. Taking a diagonal subsequence gives $T_i/\mathrm{scal}(T_i) \to T \in \bigcap_{N \geq 1} (A(s)/N) \cap \{\mathrm{scal}(R) = 1\} = \tilde{A}(s) \cap \{\mathrm{scal}(R) = 1\}$.

Since $Q(S_i)$ is not in the tangent cone of $A(s_i)$ at T_i, there exists a linear function ℓ_i with $\|\ell_i\| = 1$ such that $\ell_i(T_i) = \sup_{A(s_i)} \ell_i$ and $\ell_i(Q(S_i)) > 0$. Passing to a subsequence we have $\ell_i \to \ell$, and $\ell(T) = \sup_{\tilde{A}(s)} \ell$, so $T \in \partial \tilde{A}(s)$ and $\ell \in \mathcal{N}_T \tilde{A}(s)$. Also we have $\ell_i(Q(S_i/\mathrm{scal}(T_i))) = \ell_i(Q(S_i))/(\mathrm{scal}(T_i))^2 \geq 0$, and $|S_i/\mathrm{scal}(T_i) - T_i/\mathrm{scal}(T_i)| \leq 2/i \to 0$. So $S_i/\mathrm{scal}(T_i) \to T$, and by continuity of Q, $Q(S_i/\mathrm{scal}(T_i)^2) \to Q(T)$ while $\ell(Q(T)) = \lim_{i \to \infty} \ell_i(Q(S_i/\mathrm{scal}(T_i))) \geq 0$. But this is a contradiction to the assumption that $\tilde{A}(s)$ is strictly preserved, since it implies that $\ell(Q(T)) < 0$ for every $\ell \in \mathcal{N}_T \tilde{A}(s)$. □

Lemma 15.17. *There exists a non-increasing function $\delta : [s_0, \infty) \to \mathbb{R}_+$ such that whenever $s \in [s_0, \bar{s}]$ and $R + cI \in A(s)$ with $\mathrm{scal}(R) \leq cN(\bar{s})$ for $c \geq 1$, then $R + 2cI \in A(s + \delta(\bar{s}))$.*

Proof of Lemma. Otherwise there exist sequences $s_i \in [s_0, \bar{s}]$, $c_i \geq 1$ and $R_i \in \mathrm{Curv}$ with $R_i + c_i I \in A(s_i)$ and $\mathrm{scal}(R_i) \leq c_i N(\bar{s})$, but $R_i + 2c_i I \notin A(s_i + 1/i)$. If there exists a subsequence with c_i bounded, then we can find a subsequence with $c_i \to c$, $s_i \to s$, $R_i \to R$ with $R + cI \in A(s)$, $\mathrm{scal}(R) \leq cN(\bar{s})$, but $R + 2cI \notin A(s')$ for all $s' > s$. But this is impossible, since $R + cI \in A(s)$ implies that $R + 2cI$ is in the interior of $A(s)$, so $R + 2cI \in A(s + \delta)$ for some $\delta > 0$ by the continuity of the family $\{A(s)\}$.

The remaining possibility is that $c_i \to \infty$. In this case we have $R_i/c_i + I \in A(s_i)/c_i \subset A(s)/k$ for $i \geq k$, and $\mathrm{scal}(R_i/c_i) \leq N(\bar{s})$. So for a subsequence we have $R_i/c_i \to \tilde{R}$ with $\tilde{R} + I \in \tilde{A}(s)$ for some $s \in [s_0, \bar{s}]$. However, we have $R_i + 2c_i I \notin A(s + 1/i)$, so $\tilde{R} + 2I \notin \tilde{A}(s + \delta)$ for all $\delta > 0$. But this is a contradiction since $\tilde{R} + 2I$ is in the interior of $\tilde{A}(s)$, and hence is contained in $\tilde{A}(s + \delta)$ for some $\delta > 0$, by continuity of the family $\{\tilde{A}(s)\}$. □

We now construct the sequence s_i inductively by taking $s_{i+1} = s_i + \delta(s_i)$ for each $i \geq 1$. Note that since δ is a non-increasing positive function, $\lim_{i\to\infty} s(i) = \infty$. Let

$$F_j = A(s_0) \cap \bigcap_{i=1}^{j} \left(A(s_i) - 2^i hI\right).$$

We prove that for each j,

$$F_{j+1} \cap \{\mathrm{scal}(R) \leq 2^j N(s_j)h\} = F_j \cap \{\mathrm{scal}(R) \leq 2^j N(s_j)h\}. \tag{15.4}$$

To see this we must show that the set on the right is contained in that on the left. If $R \in F_j \cap \{\mathrm{scal}(R) \leq 2^j N(s_j)h\}$ then $R + 2^j hI \in A(s_j)$ and $\mathrm{scal}(R) \leq 2^j N(s_j)h$. Therefore $R + 2^{j+1}hI \in C(s_{j+1})$ by Lemma 15.17 (with $c = 2^j h$).

It follows that $F \cap \{\mathrm{scal}(R) \leq 2^j h\} = F_j \cap \{\mathrm{scal}(R) \leq 2^j h\}$, so F is closed. Also, we have $K \subset A(s_0) \cap \{\mathrm{scal}(R) \leq h\} = F \cap \{\mathrm{scal}(R) \leq h\} \subset F$.

It remains to show that $Q(R)$ is in $\mathcal{T}_R F$ for every $R \in \partial F$. To prove this it is enough to prove, by the inclusion (15.4), that $Q(R)$ is in $\mathcal{T}_R(A(s_j) - 2^j hI) = \mathcal{T}_{R+2^j hI} A(s_j)$ for every $R \in \partial(A(s_j) - 2^j hI)$ with $\mathrm{scal}(R) \geq 2^{j-1}N(s_j)h$. Let $T = R + 2^j hI$ and $S = R$, so that $\mathrm{scal}(T) \geq N(s_j)$ and $|S - T| \leq 2|I|\mathrm{scal}(T)/N(s_j)$. By Lemma 15.16 $Q(S)$ is in the tangent cone to $C(s_j)$ at T, hence $Q(R) \in \mathcal{T}_{R+2^j hI} A(s_j)$ as required. □

Remark 15.18. Having constructed the pinching set, the remainder of the proof of Theorem 15.3 is exactly the same as that given in Sect. 15.1.

Appendix A
Gâteaux and Fréchet Differentiability

Following [LP03] there are two basic notions of differentiability for functions $f : X \to Y$ between Banach spaces X and Y.

Definition A.1. A function f is said to be *Gâteaux differentiable* at x if there exists a bounded linear[1] operator $T_x \in \mathcal{B}(X, Y)$ such that $\forall\, v \in X$,

$$\lim_{t \to 0} \frac{f(x + tv) - f(x)}{t} = T_x v.$$

The operator T_x is called the *Gâteaux derivative* of f at x.

If for some fixed v the limits

$$\delta_v f(x) := \frac{d}{dt}\bigg|_{t=0} f(x + tv) = \lim_{t \to 0} \frac{f(x + tv) - f(x)}{t}$$

exists, we say f has a directional derivative at x in the direction v. Hence f is Gâteaux differentiable at x if and only if all the directional derivatives $\delta_v f(x)$ exist and form a bounded linear operator $Df(x) : v \mapsto \delta_v f(x)$.

If the limit (in the sense of the Gâteaux derivative) exists *uniformly* in v on the unit sphere of X, we say f is *Fréchet differentiable* at x and T_x is the *Fréchet derivative* of f at x. Equivalently, if we set $y = tv$ then $t \to 0$ if and only if $y \to 0$. Thus f is Fréchet differentiable at x if for all y,

$$f(x + y) - f(x) - T_x(y) = o(\|y\|)$$

and we call $T_x = Df(x)$ the derivative of f at x.

Note that the distinction between the two notion of differentiability is made by how the limit is taken. The importance being that the limit in the Fréchet case only depends on the norm of y.[2]

[1] Some authors drop the requirement for linearity here.

[2] In terms of ε-δ notation the differences can expressed as follows. Gâteaux: $\forall \varepsilon > 0$ and $\forall v \neq 0$, $\exists\, \delta = \delta(\varepsilon, v) > 0$ such that, $\|f(x + tv) - f(x) - tTv\| \leq \varepsilon|t|$ whenever $|t| < \delta$. Fréchet: $\forall \varepsilon > 0$, $\exists\, \delta = \delta(\varepsilon) > 0$ such that $\|f(x + v) - f(x) - Tv\| \leq \varepsilon\|v\|$ whenever $\|v\| < \delta$.

B. Andrews and C. Hopper, *The Ricci Flow in Riemannian Geometry*, Lecture Notes in Mathematics 2011, DOI 10.1007/978-3-642-16286-2,

A.1 Properties of the Gâteaux Derivative

If the Gâteaux derivative exists it unique, since the limit in the definition is unique if it exists.

A function which is Fréchet differentiable at a point is continuous there, but this is not the case for Gâteaux differentiable functions (even in the finite dimensional case). For example, the function $f : \mathbb{R}^2 \to \mathbb{R}$ defined by $f(0,0) = 0$ and $f(x,y) = x^4 y/(x^6 + y^3)$ for $x^2 + y^2 > 0$ has 0 as its Gâteaux derivative at the origin, but fails to be continuous there. This also provides an example of a function which is Gâteaux differentiable but not Fréchet differentiable. Another example is the following: If X is a Banach space, and $\varphi \in X'$ a discontinuous linear functional, then the function $f(x) = \|x\|\varphi(x)$ is Gâteaux differentiable at $x = 0$ with derivative 0, but $f(x)$ is not Fréchet differentiable since φ does not have limit zero at $x = 0$.

Proposition A.2 (Mean Value Formula). *If f is Gâteaux differentiable then*

$$\|f(y) - f(x)\| \leq \|x - y\| \sup_{0 \leq \theta \leq 1} \|Df(\theta x + (1-\theta)y)\|.$$

Proof. Choose $u^* \in X$ such that $\|u^*\| = 1$ and $\|f(y) - f(x)\| = \langle u^*, f(y) - f(x)\rangle$. By applying the mean value theorem to $h(t) = \langle u^*, f(x + t(y-x))\rangle$ we find that $|\langle u^*, f(y)\rangle - \langle u^*, f(x)\rangle| = \|h(1) - h(0)\| \leq \sup_{0 \leq t \leq 1} \|h'(t)\|$ and

$$\begin{aligned} h'(t) &= \left\langle u^*, \frac{d}{dt} f(x + t(y-x)) \right\rangle \\ &= \left\langle u^*, \lim_{s \to 0} \frac{f(x + (t+s)(y-x)) - f(x + t(y-x))}{s} \right\rangle \\ &= \langle u^*, Df(x + t(y-x))\,(y-x)\rangle \,. \end{aligned}$$

So $|h'(t)| \leq \|Df(x + t(y-x))\,(y-x)\| \leq \|Df(x + t(y-x))\|\,\|y - x\|$. □

If the Gateaux derivative exist and is continuous in the following sense, then the two notions coincide.

Proposition A.3. *If f is Gâteaux differentiable on an open neighbourhood U of x and $Df(x)$ is continuous,[3] then f is also Fréchet differentiable at x.*

---

[3] In the sense that $Df : U \to \mathcal{B}(X,Y)$ is continuous at x so that $\lim_{\tilde{x} \to x} \|Df(x) - Df(\tilde{x})\| = 0$. In words, the derivative depends continuous on the point x.

Proof. Fix v and let $g(t) = f(x+tv) - f(x) - t\,Df(x)v$, so $g(0) = 0$. By continuity of the Gâteaux derivative with the mean value theorem we find that

$$\begin{aligned} \|f(x+tv) - f(x) - t\,Df(x)v\| &= \|g(1)\| \\ &\leq \|v\| \sup_{0\leq t\leq 1} \|Df(x+tv) - Df(x)\| \\ &= o(\|v\|) \end{aligned}$$

□

The notion of Gâteaux differentiability and Fréchet differentiability also coincide if f is Lipschitz and $\dim(X) < \infty$, that is:

Proposition A.4. *Suppose $f : X \to Y$ is a Lipschitz function from a finite-dimensional Banach space X to a (possibly infinite-dimensional) Banach space Y. If f is Gâteaux differentiable at some point x, then it is also Fréchet differentiable at that point.*

Proof. As the unit sphere S_X of X is compact, it is totally bounded. So given $\varepsilon > 0$ there exists a finite set $F = F(\varepsilon) \subset X$ such that $S_X = \bigcup_{u_j \in F} B_\varepsilon(u_j)$. Thus for all $u \in S_X$ there is an index j such that $\|u - u_j\| < \varepsilon$.

By hypothesis choose $\delta > 0$ such that

$$\|f(x+tu_j) - f(x) - t\,Df(x)u_j\| < \varepsilon\,|t|$$

for $|t| < \delta$ and any index j. It follows that for any $u \in S_X$,

$$\begin{aligned} \|f(x+tu) - f(x) - tDf(x)u\| &\leq \|f(x+tu) - f(x+tu_j)\| \\ &\quad + \|f(x+tu_j) - f(x) - t\,Df(x)u_j\| \\ &\quad + \|t\,Df(x)(u_j - u)\| \\ &\leq (C + \|Df(x)\| + 1)\varepsilon\,|t| \end{aligned}$$

for $|t| < \delta$, where C is the Lipschitz constant of f. Hence δ is independent of u and so f is also Fréchet differentiable at x. □

In the infinite dimensional case the story is very different. Broadly speaking in such a situation there are reasonably satisfactory results on the existence of Gâteaux derivatives of Lipschitz functions, while results on existence of Fréchet derivatives are rare and usually very hard to prove. On the other hand, in many applications it is important to have Fréchet derivatives of f, since they provide genuine local linear approximation to f, unlike the much weaker Gâteaux derivatives.

Appendix B
Cones, Convex Sets and Support Functions

The geometric concept of tangency is one of the most important tools in analysis. Tangent lines to curves and tangent planes to surfaces are defined classically in terms of differentiation. In convex analysis, the opposite approach is exploited. A generalised tangency is defined geometrically in terms of separation; it is expressed by supporting hyperplanes and half-spaces. Here we look at convex sets (particularly when they are defined by a set of linear inequalities) and the characterisation of their tangent and normal cones. This will be needed in the proof of the maximum principle for vector bundles discussed in Sect. 7.4.

B.1 Convex Sets

Let E be a (finite-dimensional) inner product space, and E^* its dual space. A subset $A \subset E$ is *convex set* if for every $v, w \in A$, $\theta v + (1-\theta)w \in A$ for all $\theta \in [0,1]$. A set $\Gamma \subset E$ is a *cone* with vertex $u \in E$ if for every $v \in \Gamma$ we have $u + \theta(v-u) \in \Gamma$ for all $\theta \geq 0$. A *half-space* is a set of the form $\{x \in E : \ell(x) \leq c\}$ where ℓ is a non-trivial linear function on E, i.e. $\ell \in E^* \setminus \{0\}$. In such a case we normalise so that ℓ is an element of $S^* = \{\omega \in E^* : \|\omega\| = 1\}$.

A *supporting half-space* to a closed convex set A is a half-space which contains A and has points of A arbitrarily close to its boundary. A *supporting hyperplane* to A is a hyperplane which is the boundary of a supporting half-space to A. That is, supporting hyperplanes to A take the form $\{x : \ell(x) = c\}$ where $\ell \in E^* \setminus \{0\}$ and $c = \sup\{\ell(v) : v \in A\}$.

B.2 Support Functions

If A is a closed convex set in E, the *support function* of A is a function $s = s_A : E^* \to \mathbb{R} \cup \{\infty\}$ defined by

$$s(\ell) = \sup\{\ell(x) : x \in A\}$$

for each $\ell \in E^* \setminus \{0\}$. Here s is a homogeneous degree one convex function on E^*. For each ℓ with $s(\ell) < \infty$, the half-space $\{x : \ell(x) \leq s(\ell)\}$ is the unique supporting half-space of A which is parallel to $\{x : \ell(x) \leq 0\}$.

Theorem B.1. *The convex set A is the intersection of its supporting half-spaces:*

$$A = \bigcap_{\ell \in S^*} \{x \in E : \ell(x) \leq s(\ell)\}.$$

Proof. Firstly, the set A is contained in this intersection since it is contained in each of the half-spaces. To prove the reverse inclusion it suffices to show for any $y \notin A$ there exists $\ell \in S^*$ such that $\ell(y) > s(\ell)$.

Let x be the closest point to y in A, and define $\ell \in E^*$ by $\ell(z) = \langle z, y-x \rangle$. Suppose $\ell(w) > \ell(x)$ for some $w \in A$. Then $x + t(w-x) \in A$ for $0 \leq t \leq 1$, and

$$\frac{d}{dt} \|y - (x + t(w-x))\|^2 \Big|_{t=0} = -2\langle y-x, w-x \rangle = -2(\ell(w) - \ell(x)) < 0,$$

contradicting the fact that x is the closed point to y in A. Therefore $\ell(z) \leq \ell(x)$ for all $z \in A$, so $s(\ell) = \sup_A \ell = \ell(x) < \ell(y)$. The same holds for $\tilde{\ell} = \ell / \|\ell\| \in S^*$. □

B.3 The Distance From a Convex Set

For a closed convex set A in E, the function $d_A : E \to \mathbb{R}$ given by

$$d_A(x) = \inf\{\|x - y\| : y \in A\}$$

is Lipschitz continuous, with Lipschitz constant 1, and is strictly positive on $E \setminus A$. We call this the distance to A. It has the following characterisation in terms of the support function of A.

Theorem B.2. *For any $y \notin A$,*

$$d_A(y) = \sup\{\ell(y) - s(\ell) : \ell \in S^*\}.$$

Proof. Let x be the closest point to y in A. So for any $\ell \in S^*$ we have

$$\begin{aligned} \ell(y) - s(\ell) = \ell(y) - \sup_A \ell \leq \ell(y) - \ell(x) = \ell(y - x) \\ \leq \|\ell\| \, \|y - x\| = \|y - x\| = d_A(y), \end{aligned}$$

while the particular choice of $\ell(\,\cdot\,) = \langle y - x, \cdot \rangle / \|y - x\|$ gives equality throughout. □

B.4 Tangent and Normal Cones

A convex set may have non-smooth boundary, so there will not in general be a well-defined normal vector or tangent plane. Nevertheless we can make sense of a *set* of normal vectors, as follows:

Definition B.3. Let A be a closed bounded convex set in E, and let $x \in \partial A$. The *normal cone* to A at x is defined by

$$\mathcal{N}_x A = \{\ell \in E^* : \ \ell(x) = s(\ell)\}.$$

In other words, $\mathcal{N}_x A$ is the set of linear functions which achieve their maximum over A at the point x (so that the corresponding supporting half-spaces have x in their boundary). The set $\mathcal{N}_x A$ is a convex cone in E^* with vertex at the origin.

Complementary to this is the following definition:

Definition B.4. The *tangent cone* $\mathcal{T}_x A$ to A at x is the set

$$\mathcal{T}_x A = \bigcap_{\ell \in \mathcal{N}_x A} \{z \in E : \ \ell(z) \le 0\}.$$

That is, $x + \mathcal{T}_x A$ is the intersection of the supporting half-spaces of A with x on their boundary. It follows that $A - x \subset \mathcal{T}_x A$. Indeed $\mathcal{T}_x A$ may alternatively be characterised as the closure of $\bigcup\{\frac{1}{h}(A - x) : h > 0\}$. The tangent cone $\mathcal{T}_x A$ is a closed convex cone in E with vertex at the origin (in fact it is the smallest such cone containing $A - x$).

B.5 Convex Sets Defined by Inequalities

In many cases the convex set A of interest is explicitly presented as an intersection of half-spaces, in the form

$$A = \bigcap_{\ell \in B} \{x \in E : \ \ell(x) \le \phi(\ell)\} \tag{B.1}$$

where B is a given closed subset of $E^* \setminus \{0\}$ and $\phi : B \to \mathbb{R}$ is given. If B does not intersect every ray from the origin, this definition will involve only a subset of the supporting half-spaces of A. In this situation we have the following characterisation of the support function of A:

Theorem B.5. *Let E be of dimension n, and suppose A is defined by* (B.1). *For any $\ell \in E^*$ with $s(\ell) < \infty$ there exist $\ell_1, \ldots, \ell_{n+1} \in B$ and $\lambda_1, \ldots, \lambda_{n+1} \geq 0$ such that*

$$\ell = \sum_{i=1}^{n+1} \lambda_i \ell_i \quad \textit{and} \quad s(\ell) = \sum_{i=1}^{n+1} \lambda_i \phi(\ell_i).$$

It follows that the support function s of A on all of E^* can be recovered from the given function ϕ on B.

Proof. Firstly, for $\ell \in B$ note that if $\ell(x) = \phi(\ell)$ for some $x \in A$, then $\ell(x) = \sup\{\ell(y) : y \in A\} = s(\ell)$. Now define

$$\widetilde{B} = \mathbb{R}^+\{\ell \in B : \exists x \in A \text{ with } \ell(x) = \phi(\ell)\}.$$

That is, $\widetilde{B}$ consists of positive scalar multiples of those ℓ in B for which equality holds in equation (B.1). Note that $\widetilde{B}$ is closed. Also let

$$\widetilde{\phi}(\vartheta) = \begin{cases} c\phi(\ell) & \text{if } \vartheta = c\ell \text{ where } c \geq 0, \ell \in \widetilde{B} \\ +\infty & \text{otherwise} \end{cases}$$

Thus we have that $\widetilde{\phi}(\vartheta) = s(\vartheta)$ for $\vartheta \in \widetilde{B}$. From (B.1) and by the construction of $\widetilde{\phi}$ we have

$$A = \bigcap_{\vartheta \in E^*} \{x \in E : \vartheta(x) \leq \widetilde{\phi}(\vartheta)\}.$$

In which case we see that

$$\begin{aligned} s(\ell) &= \sup\{\ell(x) : x \in A\} \\ &= \sup\{\ell(x) : x \in E, \vartheta(x) \leq \widetilde{\phi}(\vartheta), \forall \vartheta \in E^*\} \\ &= \sup\{\ell^*(\ell) : \ell^* \leq \widetilde{\phi}, \ell^* \in (E^*)^*\} \end{aligned}$$

since $(E^*)^* = E$. That is, the epigraph of s is the convex hull of the epigraph of $\widetilde{\phi}$ (cf. [Roc70, Corollary 12.1.1]). Now we observe by the Caratheodory theorem [Roc70, Corollary 17.1.3] that

$$s(\ell) = \inf\left\{\sum_{i=1}^{n+1} \lambda_i \widetilde{\phi}(\ell_i) : \ \ell_i \in \widetilde{B}, \ \lambda_i \geq 0, \ \sum_{i=1}^{n+1} \lambda_i \ell_i = \ell\right\}.$$

The infimum is attained since $\widetilde{B}$ is closed. The result follows since each $\ell_i \in \widetilde{B}$ is a non-negative multiple of some element $\bar{\ell}_i$ of B with $\phi(\bar{\ell}_i) = s(\bar{\ell}_i)$. □

From this theorem we obtain a useful result for the normal cone:

Theorem B.6. *Let E be of dimension n, and suppose A is defined by* (B.1). *Then for any $x \in \partial A$, $\mathcal{N}_x A$ is the convex cone generated by $B \cap \mathcal{N}_x A$. That is, for any $\ell \in \mathcal{N}_x A$ there exist $k \le n+1$ and $\ell_1, \ldots, \ell_k \in B \cap \mathcal{N}_x A$ and $\lambda_1, \ldots, \lambda_k \ge 0$ such that $\ell = \sum_{i=1}^k \lambda_i \ell_i$.*

Proof. Let $\ell \in \mathcal{N}_x A$. By Theorem B.5 there exist $\ell_1, \ldots, \ell_{n+1}$ and $\lambda_i \ge 0$ such that $s(\ell) = \sum_{i=1}^{n+1} \lambda_i \phi(\ell_i)$ and $\ell = \sum_{i=1}^{n+1} \lambda_i \ell_i$. Since $\ell \in \mathcal{N}_x A$ we have

$$\ell(x) = s(\ell) = \sum_{i=1}^{n+1} \lambda_i s(\ell_i) \ge \sum_{i=1}^{n+1} \lambda_i \ell_i(x) = \ell(x),$$

so that equality holds throughout, and $s(\ell_i) = \ell_i(x)$ (hence $\ell_i \in \mathcal{N}_x A$) for each i with $\lambda_i > 0$. □

This in turn gives a useful characterisation of the tangent cone:

Theorem B.7. *Let E be of dimension n, and suppose A is defined by* (B.1). *Then for any $x \in \partial A$,*

$$\mathcal{T}_x A = \bigcap_{\ell \in B:\, \ell(x) = \phi(\ell)} \{z \in E : \ \ell(z) \le 0\}$$

and the interior of $\mathcal{T}_x A$ is given by the intersection of the corresponding open half-spaces.

Proof. Any point z in $\mathcal{T}_x A$ satisfies $\ell(z) \le 0$ for every $\ell \in E \setminus \{0\}$ with $\ell(x) = s(\ell)$. In particular, if $\ell \in B$ and $\ell(x) = \phi(x)$, then $\phi(x) = s(\ell)$ and $\ell(z) \le 0$. Conversely, if $\ell(z) \le 0$ for all $\ell \in B$ with $\ell(x) = \phi(\ell)$ (equivalently, for all $\ell \in B \cap \mathcal{N}_x A$) and ϑ is any element of $\mathcal{N}_x A$, then by Theorem B.6 there exist $\ell_i \in B \cap \mathcal{N}_x A$ and $\lambda_i > 0$ for $i = 1, \ldots, k$ such that $\vartheta = \sum_{i=1}^k \lambda_i \ell_i$, and so $\vartheta(z) = \sum_{i=1}^k \lambda_i \ell_i(z) \le 0$. Since this is true for all $\vartheta \in \mathcal{N}_x A$, z is in $\mathcal{T}_x A$. □

Appendix C
Canonically Identifying Tensor Spaces with Lie Algebras

In studying the algebraic decomposition of the curvature tensor, one needs to make several natural identification between tensor spaces and Lie algebras. By doing so, one is able to use the Lie algebra structure in conjunction with the tensor space construction to elucidate the structure of the quadratic terms in the curvature evolution equation.

C.1 Lie Algebras

A *Lie algebra* consists of a finite-dimensional vector space V over a field $\mathbb{F}$ with a bilinear *Lie bracket* $[,] : (X, Y) \mapsto [X, Y]$ that satisfies the properties:

1. $[X, X] = 0$
2. $[X, [Y, Z]] + [Y, [Z, X]] + [Z, [X, Y]] = 0$

for all vectors X, Y and Z.

Any *algebra* $\mathscr{A}$ over a field $\mathbb{F}$ can be made into a Lie algebra by defining the bracket

$$[X, Y] := X \cdot Y - Y \cdot X.$$

A special case of this arises when $\mathscr{A} = \mathrm{End}(V)$ is the algebra of operator endomorphisms of a vector space V. In which case the corresponding Lie algebra is called the *general Lie algebra* $\mathfrak{gl}(V)$. Concretely, setting $V = \mathbb{R}^n$ gives the *general linear Lie algebra* $\mathfrak{gl}(n, \mathbb{R})$ of all $n \times n$ real matrices with bracket $[X, Y] := XY - YX$. Furthermore, the *special linear Lie algebra* $\mathfrak{sl}(n, \mathbb{R})$ is the set of real matrices of trace 0; it is a subalgebra of $\mathfrak{gl}(n, \mathbb{R})$. The *special orthogonal Lie algebra* $\mathfrak{so}(n, \mathbb{R}) = \{X \in \mathfrak{sl}(n, \mathbb{R}) : X^T = -X\}$ is the set of skew-symmetric matrices.

C.2 Tensor Spaces as Lie Algebras

Suppose $U = (U, \langle \cdot, \cdot \rangle)$ is a real N-dimensional inner product space with orthonormal basis $(e_\alpha)_{\alpha=1}^N$. Let $E_{\alpha\beta}$ be the matrix of zero's with a 1 in the (α, β)-th entry. The matrix product then satisfies $E_{\alpha\beta}E_{\lambda\eta} = \delta_{\beta\lambda}E_{\alpha\eta}$.

The tensor space $U \otimes U$ is equipped with an inner product

$$\langle x \otimes y, u \otimes v \rangle = \langle x, u \rangle \langle y, v \rangle .$$

The set $(e_\alpha \otimes e_\beta)_{\alpha,\beta=1}^N$ forms an orthonormal basis. We identify $U \otimes U \simeq \mathfrak{gl}(N, \mathbb{R})$ by defining the linear transformation

$$x \otimes y : z \mapsto \langle y, z \rangle x \tag{C.1}$$

for any $x \otimes y \in U \otimes U$. The map simply identifies y with its dual. Under this identification, the inner product on $\mathfrak{gl}(N, \mathbb{R})$ is given by the trace norm:

$$\langle A, B \rangle = \operatorname{tr} A^T B,$$

for any $A, B \in \mathfrak{gl}(N, \mathbb{R})$. To see why, observe that $e_\alpha \otimes e_\beta \simeq E_{\alpha\beta}$ and so

$$\operatorname{tr} E_{\alpha\beta}^T E_{\lambda\eta} = \operatorname{tr} E_{\beta\alpha}E_{\lambda\eta} = \operatorname{tr} \delta_{\alpha\lambda}E_{\beta\eta} = \delta_{\alpha\lambda}\delta_{\beta\eta} = \langle e_\alpha \otimes e_\beta, e_\lambda \otimes e_\eta \rangle .$$

C.3 The Space of Second Exterior Powers as a Lie Algebra

Consider the n-dimensional real inner product space $V = (V, \langle \cdot, \cdot \rangle)$ with orthonormal basis $(e_i)_{i=1}^n$. As usual, let (e^i) be the corresponding dual basis for V^*. Define $\bigwedge^2 V = V \otimes V / \mathcal{I}$ to be the quotient algebra of the tensor space $V \otimes V$ by the ideal $\mathcal{I}$ generated from $x \otimes x$ for $x \in V$. In which case

$$x \wedge y = x \otimes y \ (\operatorname{mod} \mathcal{I}),$$

for any $x, y \in V$. The space $\bigwedge^2 V$ is called the *second exterior power* of V and elements $x \wedge y$ are referred to as *bivectors*.[1] The canonical inner product

[1] The geometric interpretation of $x \wedge y$ is that of an oriented area element in the plane spanned by x and y. The object $x \wedge y$ is referred to as a bivector as it is a two-dimensional analog to a one-dimensional vector. Whereas a vector is often utilised to represent a one-dimensional directed quantity (often visualised geometrically as a directed line-segment), a bivector is used to represent a two-dimensional directed quantity (often visualised as an oriented plane-segment).

on $\bigwedge^2 V$ is given by

$$\langle x \wedge y, u \wedge v \rangle = \langle x, u \rangle \langle y, v \rangle - \langle x, v \rangle \langle y, u \rangle . \tag{C.2}$$

With respect to this, the set $(e_i \wedge e_j)_{i<j}$ forms an orthonormal basis for the $n(n-1)/2$-dimensional vector space $\bigwedge^2 V$. We identify $\bigwedge^2 V \simeq \mathfrak{so}(n)$ by mapping $e_i \wedge e_j$ to the linear map $L(e_i \wedge e_j)$ of rank 2 which is a rotation with angle $\pi/2$ in the (i,j)-th plane. This is equivalent defining the linear transformation

$$x \wedge y : z \mapsto \langle y, z \rangle x - \langle x, z \rangle y. \tag{C.3}$$

Under this identification, the inner product on $\mathfrak{so}(n)$ is given by the trace norm

$$\langle A, B \rangle = \frac{1}{2} \mathrm{tr}\, A^T B = -\frac{1}{2} \mathrm{tr}\, AB$$

where $A, B \in \mathfrak{so}(n)$. To see this, note that

$$\begin{aligned}(e_i \wedge e_j)^T \cdot (e_k \wedge e_\ell) &= (E_{ji} - E_{ij})(E_{k\ell} - E_{\ell k}) \\ &= \delta_{ik} E_{j\ell} - \delta_{i\ell} E_{jk} + \delta_{j\ell} E_{ik} - \delta_{jk} E_{i\ell}\end{aligned}$$

and so $\mathrm{tr}\,(e_i \wedge e_j)^T \cdot (e_k \wedge e_\ell) = 2(\delta_{ik}\delta_{j\ell} - \delta_{i\ell}\delta_{jk}) = 2 \langle e_i \wedge e_j, e_k \wedge e_\ell \rangle$.

Example C.1. When $n = 3$ and $V = \mathbb{R}^3$ we observe that

$$e_2 \wedge e_3 \longmapsto R_x = E_{23} - E_{32} = \begin{pmatrix} 0 & & \\ & 0 & 1 \\ & -1 & 0 \end{pmatrix}$$

$$e_1 \wedge e_3 \longmapsto R_y = E_{13} - E_{31} = \begin{pmatrix} 0 & 0 & 1 \\ 0 & 0 & 0 \\ -1 & 0 & 0 \end{pmatrix}$$

$$e_1 \wedge e_2 \longmapsto R_z = E_{12} - E_{21} = \begin{pmatrix} 0 & 1 & \\ -1 & 0 & \\ & & 0 \end{pmatrix}$$

where R_x, R_y, R_z are $\pi/2$-rotations about the x, y and z axis. Whence any $X \in \mathfrak{so}(3)$ can be written as

$$X = \begin{pmatrix} 0 & c & b \\ -c & 0 & a \\ -b & -a & 0 \end{pmatrix} = aR_x + bR_y + cR_z,$$

since $X^T = -X$ and $\mathrm{tr} X = 0$ by definition. Furthermore, if $Y = uR_x + vR_y + wR_z$ then the inner product $\langle X, Y \rangle = au + bv + cw = (a,b,c) \cdot (u,v,w)$ is the usual Euclidean inner product.

C.3.1 The space $\bigwedge^2 V^*$ as a Lie Algebra

As done in the above passage, $\bigwedge^2 V^* = V^* \otimes V^*/\mathcal{I}$ is the quotient algebra of $V^* \otimes V^*$ by the ideal $\mathcal{I} = \langle x \otimes x \,|\, x \in V^* \rangle$. The canonical inner product given by (C.2), except now applied to dual vectors. The wedge $\wedge$ is an anti-symmetric bilinear product with the additional property that

$$(e^i \wedge e^j)(e_k, e_\ell) = \det \begin{pmatrix} e^i(e_k) & e^i(e_\ell) \\ e^j(e_k) & e^j(e_\ell) \end{pmatrix} = \delta_{ik}\delta_{j\ell} - \delta_{i\ell}\delta_{jk}.$$

Any $\varphi \in \bigwedge^2 V^*$ may be written as

$$\varphi = \frac{1}{2}\sum_{i,j} \varphi_{ij} e^i \wedge e^j = \sum_{i<j} \varphi_{ij} e^i \wedge e^j \tag{C.4}$$

where $\varphi_{ij} := \varphi(e_i, e_j)$. Moreover, the pairing of bivectors with its dual is given by $(e^i \wedge e^j)(e_k \wedge e_\ell) = (e^i \wedge e^j)(e_k, e_\ell)$ in order to preserve orthonormality.

Remark C.2. A quick consistency check confirms the summation convention used in (C.4) allows the coefficients φ_{ij} that appear in the sum to agree with the component $\varphi(e_i, e_j)$. Indeed, we observe that

$$\left(\frac{1}{2}\sum_{i,j} \varphi_{ij} e^i \wedge e^j\right)(e_k, e_\ell) = \frac{1}{2}\varphi_{ij}(\delta_{ik}\delta_{j\ell} - \delta_{i\ell}\delta_{jk}) = \frac{1}{2}(\varphi_{k\ell} - \varphi_{\ell k}) = \varphi_{k\ell}$$

which is equal to $\varphi(e_k, e_\ell)$ by definition. Furthermore one also find that $\langle \varphi, e^k \wedge e^\ell \rangle = \frac{1}{2}\sum_{i,j} \varphi_{ij} \langle e^i \wedge e^j, e^k \wedge e^\ell \rangle = \varphi_{k\ell}$. Thus the convention is consistent.

We identify $\bigwedge^2 V^*$ with the Lie algebra $\mathfrak{so}(n)$ by sending $e^i \wedge e^j \mapsto E_{ij} - E_{ji}$ as before. This equips $\bigwedge^2 V^*$ with a Lie algebra structure. In particular the bracket

$$\begin{aligned}
[e^i \wedge e^j, e^k \wedge e^\ell] &= (e^i \wedge e^j) \cdot (e^k \wedge e^\ell) - (e^k \wedge e^\ell) \cdot (e^i \wedge e^j) \\
&= (E_{ij} - E_{ji})(E_{k\ell} - E_{\ell k}) - (E_{k\ell} - E_{\ell k})(E_{ij} - E_{ji}) \\
&= E_{ij}E_{k\ell} - E_{ij}E_{\ell k} - E_{ji}E_{k\ell} + E_{ji}E_{\ell k} \\
&\quad - E_{k\ell}E_{ij} + E_{k\ell}E_{ji} + E_{\ell k}E_{ij} - E_{\ell k}E_{ji} \\
&= \delta_{i\ell} e^j \wedge e^k + \delta_{jk} e^i \wedge e^\ell - \delta_{ik} e^j \wedge e^\ell - \delta_{j\ell} e^i \wedge e^k
\end{aligned}$$

In which case, given any $\phi, \psi \in \bigwedge^2 V^*$ one computes

$$\begin{aligned}
[\phi, \psi] &= \frac{1}{4}\phi_{ij}\psi_{k\ell}[e^i \wedge e^j, e^k \wedge e^\ell] \\
&= \frac{1}{4}\phi_{ij}\psi_{k\ell}(\delta_{i\ell}e^j \wedge e^k + \delta_{jk}e^i \wedge e^\ell - \delta_{ik}e^j \wedge e^\ell - \delta_{j\ell}e^i \wedge e^k) \\
&= \frac{1}{4}\left(\phi_{pj}\psi_{kp}e^j \wedge e^k + \phi_{ip}\psi_{p\ell}e^i \wedge e^\ell - \phi_{pj}\psi_{p\ell}e^j \wedge e^\ell - \phi_{ip}\psi_{kp}e^i \wedge e^k\right) \\
&= \frac{1}{2}\sum_{i,j}(\phi_{ip}\psi_{pj} - \psi_{ip}\phi_{pj})e^i \wedge e^j
\end{aligned}$$

Therefore we (naturally) define the components of the bracket, with respect to the basis $(e_i \wedge e_j)_{i<j}$, by

$$[\phi, \psi]_{ij} := \phi_{ip}\psi_{pj} - \psi_{ip}\phi_{pj} \tag{C.5}$$

for any $\phi, \psi \in \bigwedge^2 V^*$.

C.3.1.1 Structure Constants

Now suppose (φ^α) is an orthonormal basis for $\bigwedge^2 V^*$. The *structure constants* $c_\gamma^{\alpha\beta}$ for the bracket (C.5), with respect to the basis (φ^α), are defined by

$$[\varphi^\alpha, \varphi^\beta] = c_\gamma^{\alpha\beta}\varphi^\gamma.$$

As (φ^α) are orthonormal, the structure constants can be directly computed from

$$c_\gamma^{\alpha\beta} = \left\langle [\varphi^\alpha, \varphi^\beta], \varphi^\gamma \right\rangle.$$

It is easy to check that the tri-linear form $\left\langle [\varphi^\alpha, \varphi^\beta], \varphi^\gamma \right\rangle$ is fully antisymmetric, thus the structure constants $c_\gamma^{\alpha\beta}$ are anti-symmetric in all three components. Moreover, if (σ_α) orthonormal basis for $\bigwedge^2 V$ dual to (φ^α), then the corresponding structure constants $c_{\alpha\beta}^\gamma$ are given by

$$[\sigma_\alpha, \sigma_\beta] = c_{\alpha\beta}^\gamma \sigma_\gamma.$$

From the identification of $\bigwedge^2 V$ with $\bigwedge^2 V^*$ we also have $c_{\alpha\beta}^\gamma = c_\gamma^{\alpha\beta}$.

References

[AM94] Uwe Abresch and Wolfgang T. Meyer, *Pinching below 1/4, injectivity radius, and conjugate radius*, Journal of Differential Geometry **40** (1994), no. 3, 643–691.

[AM97] ———, *Injectivity radius estimates and sphere theorems*, Comparison geometry, 1997, pp. 1–47.

[AN07] Ben Andrews and Huy T. Nguyen, *Four-manifolds with 1/4-pinched flag curvature*, 2007. preprint.

[AS53] W. Ambrose and I. M. Singer, *A theorem on holonomy*, Transactions of the American Mathematical Society **75** (1953), no. 3, 428–443.

[Ban87] Shigetoshi Bando, *Real analyticity of solutions of Hamilton's equation*, Mathematishe Zeitschrift **195** (1987), no. 1, 93–97.

[Ber00] Marcel Berger, *Riemannian geometry during the second half of the twentieth century*, Reprint of the 1998 original, University Lecture Series, vol. 17, AMS, 2000.

[Ber03] ———, *A Panoramic View of Riemannian Geometry*, Springer-Verlag Berlin Heidelberg, 2003.

[Ber60a] ———, *Les variétés Riemanniennes 1/4-pincées*, Annali della Scuola Normale Superiore di Pisa, Classe di Scienze 3e série **14** (1960), no. 2, 161–170.

[Ber60b] ———, *Sur quelques variétés riemanniennes suffisamment pincées*, Bulletin de la Société Mathématique de France **88** (1960), 57–71.

[Bes08] Arthur L. Besse, *Einstein manifolds*, Classics in Mathematics, Springer-Verlag, 2008. Reprint of the 1987 edition.

[BL00] Yoav Benyamini and Joram Lindenstrauss, *Geometric nonlinear functional analysis, Volume 1*, American Mathematical Society Colloquium Publications, vol. 48, AMS, 2000.

[Bre08] Simon Brendle, *A general convergence result for the Ricci flow in higher dimensions*, Duke Mathematical Journal **145** (2008), no. 3, 585–601. arXiv:0706.1218v4 [math.DG].

[BS09a] Simon Brendle and Richard Schoen, *Manifolds with 1/4-pinched curvature are space forms*, Journal of the American Mathematical Society **22** (2009), no. 1, 287–307. arXiv:0705.0766v3 [math.DG].

[BS09b] ———, *Sphere theorems in geometry*, to appear in Surveys in Differential Geometry, 2009. arXiv:0904.2604v1 [math.DG].

[BW08] Christoph Böhm and Burkhard Wilking, *Manifolds with positive curvature operators are space forms*, Annals of Mathematics **167** (2008), no. 3, 1079–1097. arXiv:math/0606187v1 [math.DG].

[Car10] Mauro Carfora, *Renormalization Group and the Ricci flow*, Milan Journal of Mathematics **78** (2010), 319–353. arXiv:1001.3595v1 [hep-th].

[CCCY03] H. D. Cao, B. Chow, S. C. Chu, and S. T. Yau (eds.), *Collected papers on Ricci flow*, Series in Geometry and Topology, vol. 37, International Press, 2003.

[CCG+07] Bennett Chow, Sun-Chin Chu, David Glickenstein, Christine Guenther, James Isenberg, Tom Ivey, Dan Knopf, Peng Lu, Feng Luo, and Lei Ni, *The Ricci flow: Techniques and applications — Part I: Geometric aspects*, Mathematical Surveys and Monographs, vol. 135, AMS, 2007.

[CCG+08] ———, *The Ricci flow: Techniques and applications — Part II: Analytic aspects*, Mathematical Surveys and Monographs, vol. 144, AMS, 2008.

[CE08] Jeff Cheeger and David G. Ebin, *Comparison theorems in Riemannian geometry,* Revised reprint of the 1975 original, AMS Chelsea Publishing, 2008.

[CGT82] Jeff Cheeger, Mikhail Gromov, and Michael Taylor, *Finite propagation speed, kernel estimates for functions of the Laplace operator, and the geometry of complete Riemannian manifolds*, Journal of Differential Geometry **17** (1982), no. 1, 15–53.

[Che70] Jeff Cheeger, *Finiteness theorems for Riemannian manifolds*, American Journal of Mathematics **92** (1970), 61–74.

[Che91] Haiwen Chen, *Pointwise* 1/4*-pinched* 4*-manifolds*, Annals of Global Analysis and Geometry **9** (1991), no. 2, 161–176.

[CK04] Bennett Chow and Dan Knopf, *The Ricci flow: An Introduction*, Mathematical Surveys and Monographs, vol. 110, AMS, 2004.

[CLN06] Bennett Chow, Peng Lu, and Lei Ni, *Hamilton's Ricci flow*, Graduate Studies in Mathematics, vol. 77, AMS, 2006.

[Con08] Lawrence Conlon, *Differentiable manifolds*, 2nd ed., Modern Birkhäuser Classics, Birkhäuser Boston, 2008.

[CTZ08] Bing-Long Chen, Siu-Hung Tang, and Xi-Ping Zhu, *Complete classification of compact four-manifolds with positive isotropic curvature*, 2008. arXiv:0810.1999v1 [math.DG].

[CZ06] Bing-Long Chen and Xi-Ping Zhu, *Ricci flow with surgery on four-manifolds with positive isotropic curvature*, Journal of Differential Geometry **74** (2006), no. 2, 177–264. arXiv:math/0504478v3 [math.DG].

[Dac04] Bernard Dacorogna, *Introduction to the calculus of variations,* Translated from the 1992 French original, Imperial College Press, London, 2004.

[dC92] Manfredo Perdigão do Carmo, *Riemannian geometry,* Translated from the second Portuguese edition by Francis Flaherty, Mathematics: Theory and Applications, Birkhäuser, 1992.

[DeT83] Dennis M. DeTurck, *Deforming metrics in the direction of their Ricci tensors*, Journal of Differential Geometry **18** (1983), no. 1, 157–162.

[ES64] James Eells Jr and J. H. Sampson, *Harmonic mappings of Riemannian manifolds*, American Journal of Mathematics **86** (1964), no. 1, 109–160.

[Esc86] Jost-Hinrich Eschenburg, *Local convexity and nonnegative curvature—Gromov's proof of the sphere theorem*, Inventiones Mathematicae **84** (1986), no. 3, 507–522.

[Eva98] Lawrence C. Evans, *Partial differential equations*, Graduate Studies in Mathematics, vol. 19, AMS, 1998.

[EW76] J. Eells and J. C. Wood, *Restrictions on harmonic maps of surfaces*, Topology **15** (1976), no. 3, 263–266.

[Fri85] Daniel Harry Friedan, *Nonlinear models in* $2+\varepsilon$ *dimensions*, Annals of Physics **163** (1985), no. 2, 318–419.

[GKR74] Karsten Grove, Hermann Karcher, and Ernst A. Ruh, *Jacobi fields and Finsler metrics on compact Lie groups with an application to differentiable pinching problems*, Mathematische Annalen **211** (1974), 7–21.

[Gro66] Detlef Gromoll, *Differenzierbare Strukturen und Metriken positiver Krümmung auf Sphären*, Mathematische Annalen **164** (1966), 353–371.

[Gro99] David Gross, *Renormalization groups*, Quantum fields and strings: A course for mathematicians, 1999.

[GSW88] Michael B. Green, John H. Schwarz, and Edward Witten, *Superstring Theory 1: Introduction*, Cambridge Monographs on Mathematical Physics, Cambridge University Press, 1988.

[GT83] David Gilbarg and Neil S. Trudinger, *Elliptic partial differential equations of second order*, 2nd ed., Grundlehren der mathematischen Wissenschaften, vol. 224, Springer-Verlag Berlin Heidelberg, 1983.

[GVZ08] Karsten Grove, Luigi Verdiani, and Wolfgang Ziller, *A positively curved manifold homeomorphic to T_1S^4*, 2008. arXiv:0809.2304 [math.DG].

[GW09] Roe Goodman and Nolan R. Wallach, *Symmetry, Representations, and Invariants*, Graduate Texts in Mathematics, vol. 255, Springer, 2009.

[GW88] R. E. Greene and H. Wu, *Lipschitz convergence of Riemannian manifolds*, Pacific Journal of Mathematics **131** (1988), no. 1, 119–141.

[Ham74] Richard S. Hamilton, *Harmonic maps of manifolds with boundary*, Lecture Notes in Mathematics, vol. 471, Springer-Verlag Berlin Heidelberg, 1974.

[Ham82a] ———, *The inverse function theorem of Nash and Moser*, Bulletin (new series) of the American Mathematical Society **7** (1982), no. 1, 65–222.

[Ham82b] ———, *Three-manifolds with positive Ricci curvature*, Journal of Differential Geometry **17** (1982), no. 2, 255–306.

[Ham86] ———, *Four-manifolds with positive curvature operator*, Journal of Differential Geometry **24** (1986), no. 2, 153–179.

[Ham93] ———, *The Harnack estimate for the Ricci flow*, Journal of Differential Geometry **37** (1993), no. 1, 225–243.

[Ham95a] ———, *A compactness property for solutions of the Ricci flow*, American Journal of Mathematics **117** (1995), no. 3, 545–572.

[Ham95b] ———, *The formation of singularities in the Ricci flow*, Surveys in differential geometry, 1995, pp. 7–136. Proceedings of the Conference on Geometry and Topology held at Harvard University, Cambridge, Massachusetts, April 23–25, 1993.

[Ham97] ———, *Four-manifolds with positive isotropic curvature*, Communications in Analysis and Geometry **5** (1997), no. 1, 1–92.

[Har67] Philip Hartman, *On homotopic harmonic maps*, Canadian Journal of Mathematics **19** (1967), 673–687.

[Hel62] Sigurdur Helgason, *Differential geometry and symmetric spaces*, Pure and Applied Mathematics, vol. 12, Academic Press, 1962.

[HK78] Ernst Heintze and Hermann Karcher, *A general comparison theorem with applications to volume estimates for submanifolds*, Annales Scientifiques de l'École Normale Supérieure. Quatrième Série **11** (1978), no. 4, 451–470.

[Hop01] Heinz Hopf, *Collected Papers: Gesammelte Abhandlungen* (Beno Eckmann, ed.), Springer-Verlag Berlin Heidelberg, 2001.

[Hop25] ———, *Ueber Zusammenhänge zwischen Topologie und Metrik von Mannigfaltigkeiten*, PhD thesis, Friedrich-Wilhelms-Universität zu Berlin (1925).

[Hop26] ———, *Zum Clifford-Kleinschen Raumproblem*, Mathematische Annalen **95** (1926), no. 1, 313–339.

[Hui85] Gerhard Huisken, *Ricci deformation of the metric on a Riemannian manifold*, Journal of Differential Geometry **21** (1985), no. 1, 47–62.

[Jos08] Jürgen Jost, *Riemannian geometry and geometric analysis*, 5th ed., Universitext, Springer-Verlag Berlin Heidelberg, 2008.

[Jos83] ———, *Harmonic mappings between Riemannian manifolds*, Proceedings of the Centre for Mathematics and its Applications, vol. 4, Australian National University, Canberra, 1983.

[Kar70] Hermann Karcher, *A short proof of Berger's curvature tensor estimates*, Proceedings of the American Mathematical Society **26** (1970), no. 4, 642–644.

[Kaz81] Jerry L. Kazdan, *Another proof of Bianchi's identity in Riemannian geometry*, Proceedings of the American Mathematical Society **81** (1981), no. 2, 341–342.

[KL08] Bruce Kleiner and John Lott, *Notes on Perelman's papers*, Geometry & Topology **12** (2008), no. 5, 2587–2855. arXiv:math/0605667 [math.DG].

[Kli59] Wilhelm Klingenberg, *Contributions to Riemannian geometry in the large*, The Annals of Mathematics **69** (1959), no. 3, 654–666.

[Kli61] ———, *Über Riemannsche Mannigfaltigkeiten mit positiver Krümmung*, Commentarii Mathematici Helvetici **35** (1961), no. 1, 47–54.

[KM63] Michel A. Kervaire and John W. Milnor, *Groups of homotopy spheres I*, The Annals of Mathematics **34** (1963), no. 3, 504–537.

[KN96] Shoshichi Kobayashi and Katsumi Nomizu, *Foundations of differential geometry. Vol. I*, Wiley Classics Library, John Wiley & Sons Inc., 1996. Reprint of the 1963 original, A Wiley-Interscience Publication.

[Lee02] John M. Lee, *Introduction to smooth manifolds*, Graduate Texts in Mathematics, vol. 218, Springer, 2002.

[Lee97] ———, *Riemannian manifolds: An introduction to curvature*, Graduate Texts in Mathematics, vol. 176, Springer, 1997.

[LP03] Joram Lindenstrauss and David Preiss, *On Fréchet differentiability of Lipschitz maps between Banach spaces*, Annals of Mathematics **157** (2003), no. 1, 257–288. arXiv:math/0402160v1 [math.FA].

[LP87] John M. Lee and Thomas H. Parker, *The Yamabe problem*, Bulletin (New Series) of the American Matmatical Society **17** (1987), no. 1, 37–91.

[Mar86] Christophe Margerin, *Pointwise pinched manifolds are space forms*, Geometric measure theory and the calculus of variations, 1986, pp. 307–328.

[Mar94a] ———, *Une caractérisation optimale de la structure différentielle standard de la sphére en terme de courbure pour (presque) toutes les dimensions. I. Les énoncés*, Comptes Rendus de l'Académie des Sciences. Série I. Mathématique **319** (1994), no. 6, 605–607.

[Mar94b] ———, *Une caractérisation optimale de la structure différentielle standard de la sphére en terme de courbure pour (presque) toutes les dimensions. II. Le schéma de la preuve*, Comptes Rendus de l'Académie des Sciences. Série I. Mathématique **319** (1994), no. 7, 713–716.

[Mil56] John W. Milnor, *On manifolds homeomorphic to the* 7*-sphere*, The Annals of Mathematics **64** (1956), no. 2, 399–405.

[MM88] Mario J. Micallef and John Douglas Moore, *Minimal two-spheres and the topology of manifolds with positive curvature on totally isotropic two-planes*, Annals of Mathematics **127** (1988), no. 1, 199–227.

[Mok88] Ngaiming Mok, *The uniformization theorem for compact Kähler manifolds of nonnegative holomorphic bisectional curvature*, Journal of Differential Geometry **27** (1988), no. 2, 179–214.

[Mos60] Jürgen Moser, *A new proof of De Giorgi's theorem concerning the regularity problem for elliptic differential equations*, Communications on Pure and Applied Mathematics **13** (1960), 457–468.

[MT07] John W. Morgan and Gang Tian, *Ricci flow and the Poincaré conjecture*, Clay Mathematics Monographs, vol. 3, American Mathematical Society, 2007. arXiv:math/0607607v2 [math.DG].

[Ngu08] Huy T. Nguyen, *Invariant curvature cones and the Ricci flow*, PhD thesis, The Australian National University (2008).

[Ngu10] ———, *Isotropic Curvature and the Ricci Flow*, International Mathematics Research Notices **2010** (2010), no. 3, 536–558. doi:10.1093/imrn/rnp147.

[Ni04] Lei Ni, *The entropy formula for linear heat equation*, Journal of Geometric Analysis **14** (2004), no. 1, 87–100.

[Nis86] Seiki Nishikawa, *On deformation of Riemannian metrics and manifolds with positive curvature operator*, Curvature and Topology of Riemannian Manifolds, 1986, pp. 202–211.

[Nom56] Nomizu, Katsumi, *Lie groups and differential geometry*, Publications of the Mathematical Society of Japan, vol. 2, The Mathematical Society of Japan, 1956.

[NW07] Lei Ni and Baoqiang Wu, *Complete manifolds with nonnegative curvature operator*, Proceedings of the American Mathematical Society **135** (2007), no. 9, 3021–3028.

[Oku66] Okubo, Tanjiro, *On the differential geometry of frame bundles*, Annali di Matematica Pura ed Applicata, Serie Quarta **72** (1966), no. 1, 29–44.

[OSW06] T. Oliynyk, V. Suneeta, and E. Woolgar, *A gradient flow for worldsheet nonlinear sigma models*, Nuclear Physics B **739** (2006), 441–458. arXiv:hep-th/0510239v2.

[Per02] Grisha Perel'man, *The entropy formula for the Ricci flow and its geometric applications*, 2002. arXiv:math/0211159v1 [math.DG].

[Pet06] Peter Petersen, *Riemannian geometry*, 2nd ed., Graduate Texts in Mathematics, vol. 171, Springer, 2006.

[Pet87] Stefan Peters, *Convergence of Riemannian manifolds*, Compositio Mathematica **62** (1987), no. 1, 3–16.

[Pol98] Joseph Polchinski, *String theory 1: An introduction to the bosonic string*, Cambridge Monographs on Mathematical Physics, Cambridge University Press, 1998.

[PRS08] Stefano Pigola, Marco Rigoli, and Alberto G. Setti, *Vanishing and finiteness results in geometric analysis, A generalization of the Bochner technique*, Progress in Mathematics, vol. 266, Birkhäuser, 2008.

[PW08] Peter Petersen and Frederick Wilhelm, *An exotic sphere with positive sectional curvature*, 2008. arXiv:0805.0812 [math.DG].

[Rau51] Harry Ernest Rauch, *A contribution to differential geometry in the large*, The Annals of Mathematics **54** (1951), no. 1, 38–55.

[Roc70] Ralph Tyrell Rockafellar, *Convex analysis*, Princeton Mathematical Series, vol. 28, Princeton University Press, 1970.

[Ruh71] Ernst A. Ruh, *Curvature and differentiable structure on spheres*, Commentarii Mathematici Helvetici **46** (1971), 127–136.

[Ruh73] ———, *Krümmung und differenzierbare Struktur auf Sphären, II*, Mathematische Annalen **205** (1973), 113–129.

[Ruh82] ———, *Riemannian manifolds with bounded curvature ratios*, Journal of Differential Geometry **17** (1982), no. 4, 643–653.

[Sch07] Richard Schoen, *Manifolds with pointwise 1/4-pinched curvature,* Pacific Northwest Geometry Seminar, 2007 Fall Meeting, University of Oregon, 2007.

[Sch86] Friedrich Schur, *Ueber den Zusammenhang der Räume constanten Riemann'schen Krümmungsmaasses mit den projectiven Räumen*, Mathematische Annalen **27** (1886), no. 4, 537–567.

[Shi89] Wan-Xiong Shi, *Deforming the metric on complete Riemannian manifolds*, Journal of Differential Geometry **30** (1989), 223–301.

[SSK71] K. Shiohama, M. Sugimoto, and H. Karcher, *On the differentiable pinching problem*, Mathematische Annalen **195** (1971), 1–16.

[Str08] Michael Struwe, *Variational methods*, 4th ed., Ergebnisse der Mathematik und ihrer Grenzgebiete. 3. Folge, vol. 34, Springer-Verlag Berlin Heidelberg, 2008.

[Tao06] Terence Tao, *Perel'man's proof of the Poincaré conjecture: A nonlinear PDE perspective*, 2006. arXiv:math/0610903v1 [math.DG].

[Tao08] ———, *Research Blog on MATH 285G: Perelmans proof of the Poincaré conjecture,* Lecture Notes, University of California, Los Angeles, 2008. http://wordpress.com/tag/285g-poincare-conjecture/.

[Top05] Peter Topping, *Diameter control under Ricci flow*, Communications in Analysis and Geometry **13** (2005), no. 5, 1039–1055.

[Top06] ———, *Lectures on the Ricci flow*, London Mathematical Society Lecture Note Series, vol. 325, Cambridge University Press, 2006.

[Top58] V. A. Toponogov, *Riemannian spaces having their curvature bounded below by a positive number*, Doklady Akademii Nauk SSSR **120** (1958), no. 4, 719–721.

[Váz84] J. L. Vázquez, *A strong maximum principle for some quasilinear elliptic equations*, Applied Mathematics and Optimization. An International Journal with Applications to Stochastics **12** (1984), no. 3, 191–202.

[Wal07] Nolan Wallach, *The decomposition of the space of curvature operators*, 2007. http://math.ucsd.edu/~nwallach/curvature.pdf.

[Won61] Wong, Yung-Chow, *Recurrent tensors on a linearly connected differentiable manifold*, Transactions of the American Mathematical Society **99** (1961), no. 2, 325–341.

Index

Lecture Notes in Mathematics

For information about earlier volumes
please contact your bookseller or Springer
LNM Online archive: springerlink.com

Vol. 1824: J. A. Navarro González, J. B. Sancho de Salas, $\mathcal{C}^{\infty}$ – Differentiable Spaces (2003)

Vol. 1825: J. H. Bramble, A. Cohen, W. Dahmen, Multiscale Problems and Methods in Numerical Simulations, Martina Franca, Italy 2001. Editor: C. Canuto (2003)

Vol. 1826: K. Dohmen, Improved Bonferroni Inequalities via Abstract Tubes. Inequalities and Identities of Inclusion-Exclusion Type. VIII, 113 p, 2003.

Vol. 1827: K. M. Pilgrim, Combinations of Complex Dynamical Systems. IX, 118 p, 2003.

Vol. 1828: D. J. Green, Grbner Bases and the Computation of Group Cohomology. XII, 138 p, 2003.

Vol. 1829: E. Altman, B. Gaujal, A. Hordijk, Discrete-Event Control of Stochastic Networks: Multimodularity and Regularity. XIV, 313 p, 2003.

Vol. 1830: M. I. Gil', Operator Functions and Localization of Spectra. XIV, 256 p, 2003.

Vol. 1831: A. Connes, J. Cuntz, E. Guentner, N. Higson, J. E. Kaminker, Noncommutative Geometry, Martina Franca, Italy 2002. Editors: S. Doplicher, L. Longo (2004)

Vol. 1832: J. Azéma, M. Émery, M. Ledoux, M. Yor (Eds.), Séminaire de Probabilités XXXVII (2003)

Vol. 1833: D.-Q. Jiang, M. Qian, M.-P. Qian, Mathematical Theory of Nonequilibrium Steady States. On the Frontier of Probability and Dynamical Systems. IX, 280 p, 2004.

Vol. 1834: Yo. Yomdin, G. Comte, Tame Geometry with Application in Smooth Analysis. VIII, 186 p, 2004.

Vol. 1835: O.T. Izhboldin, B. Kahn, N.A. Karpenko, A. Vishik, Geometric Methods in the Algebraic Theory of Quadratic Forms. Summer School, Lens, 2000. Editor: J.-P. Tignol (2004)

Vol. 1836: C. Năstăsescu, F. Van Oystaeyen, Methods of Graded Rings. XIII, 304 p, 2004.

Vol. 1837: S. Tavaré, O. Zeitouni, Lectures on Probability Theory and Statistics. Ecole d'Eté de Probabilités de Saint-Flour XXXI-2001. Editor: J. Picard (2004)

Vol. 1838: A.J. Ganesh, N.W. O'Connell, D.J. Wischik, Big Queues. XII, 254 p, 2004.

Vol. 1839: R. Gohm, Noncommutative Stationary Processes. VIII, 170 p, 2004.

Vol. 1840: B. Tsirelson, W. Werner, Lectures on Probability Theory and Statistics. Ecole d'Eté de Probabilités de Saint-Flour XXXII-2002. Editor: J. Picard (2004)

Vol. 1841: W. Reichel, Uniqueness Theorems for Variational Problems by the Method of Transformation Groups (2004)

Vol. 1842: T. Johnsen, A. L. Knutsen, K_3 Projective Models in Scrolls (2004)

Vol. 1843: B. Jefferies, Spectral Properties of Noncommuting Operators (2004)

Vol. 1844: K.F. Siburg, The Principle of Least Action in Geometry and Dynamics (2004)

Vol. 1845: Min Ho Lee, Mixed Automorphic Forms, Torus Bundles, and Jacobi Forms (2004)

Vol. 1846: H. Ammari, H. Kang, Reconstruction of Small Inhomogeneities from Boundary Measurements (2004)

Vol. 1847: T.R. Bielecki, T. Bjrk, M. Jeanblanc, M. Rutkowski, J.A. Scheinkman, W. Xiong, Paris-Princeton Lectures on Mathematical Finance 2003 (2004)

Vol. 1848: M. Abate, J. E. Fornaess, X. Huang, J. P. Rosay, A. Tumanov, Real Methods in Complex and CR Geometry, Martina Franca, Italy 2002. Editors: D. Zaitsev, G. Zampieri (2004)

Vol. 1849: Martin L. Brown, Heegner Modules and Elliptic Curves (2004)

Vol. 1850: V. D. Milman, G. Schechtman (Eds.), Geometric Aspects of Functional Analysis. Israel Seminar 2002-2003 (2004)

Vol. 1851: O. Catoni, Statistical Learning Theory and Stochastic Optimization (2004)

Vol. 1852: A.S. Kechris, B.D. Miller, Topics in Orbit Equivalence (2004)

Vol. 1853: Ch. Favre, M. Jonsson, The Valuative Tree (2004)

Vol. 1854: O. Saeki, Topology of Singular Fibers of Differential Maps (2004)

Vol. 1855: G. Da Prato, P.C. Kunstmann, I. Lasiecka, A. Lunardi, R. Schnaubelt, L. Weis, Functional Analytic Methods for Evolution Equations. Editors: M. Iannelli, R. Nagel, S. Piazzera (2004)

Vol. 1856: K. Back, T.R. Bielecki, C. Hipp, S. Peng, W. Schachermayer, Stochastic Methods in Finance, Bressanone/Brixen, Italy, 2003. Editors: M. Fritelli, W. Runggaldier (2004)

Vol. 1857: M. Émery, M. Ledoux, M. Yor (Eds.), Séminaire de Probabilités XXXVIII (2005)

Vol. 1858: A.S. Cherny, H.-J. Engelbert, Singular Stochastic Differential Equations (2005)

Vol. 1859: E. Letellier, Fourier Transforms of Invariant Functions on Finite Reductive Lie Algebras (2005)

Vol. 1860: A. Borisyuk, G.B. Ermentrout, A. Friedman, D. Terman, Tutorials in Mathematical Biosciences I. Mathematical Neurosciences (2005)

Vol. 1861: G. Benettin, J. Henrard, S. Kuksin, Hamiltonian Dynamics – Theory and Applications, Cetraro, Italy, 1999. Editor: A. Giorgilli (2005)

Vol. 1862: B. Helffer, F. Nier, Hypoelliptic Estimates and Spectral Theory for Fokker-Planck Operators and Witten Laplacians (2005)

Vol. 1863: H. Führ, Abstract Harmonic Analysis of Continuous Wavelet Transforms (2005)

Vol. 1864: K. Efstathiou, Metamorphoses of Hamiltonian Systems with Symmetries (2005)

Vol. 1865: D. Applebaum, B.V. R. Bhat, J. Kustermans, J. M. Lindsay, Quantum Independent Increment Processes I. From Classical Probability to Quantum Stochastic Calculus. Editors: M. Schürmann, U. Franz (2005)

Vol. 1866: O.E. Barndorff-Nielsen, U. Franz, R. Gohm, B. Kümmerer, S. Thorbjønsen, Quantum Independent Increment Processes II. Structure of Quantum Lévy

Processes, Classical Probability, and Physics. Editors: M. Schürmann, U. Franz, (2005)
Vol. 1867: J. Sneyd (Ed.), Tutorials in Mathematical Biosciences II. Mathematical Modeling of Calcium Dynamics and Signal Transduction. (2005)
Vol. 1868: J. Jorgenson, S. Lang, $Pos_n(R)$ and Eisenstein Series. (2005)
Vol. 1869: A. Dembo, T. Funaki, Lectures on Probability Theory and Statistics. Ecole d'Eté de Probabilités de Saint-Flour XXXIII-2003. Editor: J. Picard (2005)
Vol. 1870: V.I. Gurariy, W. Lusky, Geometry of Mntz Spaces and Related Questions. (2005)
Vol. 1871: P. Constantin, G. Gallavotti, A.V. Kazhikhov, Y. Meyer, S. Ukai, Mathematical Foundation of Turbulent Viscous Flows, Martina Franca, Italy, 2003. Editors: M. Cannone, T. Miyakawa (2006)
Vol. 1872: A. Friedman (Ed.), Tutorials in Mathematical Biosciences III. Cell Cycle, Proliferation, and Cancer (2006)
Vol. 1873: R. Mansuy, M. Yor, Random Times and Enlargements of Filtrations in a Brownian Setting (2006)
Vol. 1874: M. Yor, M. Émery (Eds.), In Memoriam Paul-Andr Meyer - Sminaire de Probabilits XXXIX (2006)
Vol. 1875: J. Pitman, Combinatorial Stochastic Processes. Ecole d'Et de Probabilits de Saint-Flour XXXII-2002. Editor: J. Picard (2006)
Vol. 1876: H. Herrlich, Axiom of Choice (2006)
Vol. 1877: J. Steuding, Value Distributions of L-Functions (2007)
Vol. 1878: R. Cerf, The Wulff Crystal in Ising and Percolation Models, Ecole d'Et de Probabilités de Saint-Flour XXXIV-2004. Editor: Jean Picard (2006)
Vol. 1879: G. Slade, The Lace Expansion and its Applications, Ecole d'Et de Probabilits de Saint-Flour XXXIV-2004. Editor: Jean Picard (2006)
Vol. 1880: S. Attal, A. Joye, C.-A. Pillet, Open Quantum Systems I, The Hamiltonian Approach (2006)
Vol. 1881: S. Attal, A. Joye, C.-A. Pillet, Open Quantum Systems II, The Markovian Approach (2006)
Vol. 1882: S. Attal, A. Joye, C.-A. Pillet, Open Quantum Systems III, Recent Developments (2006)
Vol. 1883: W. Van Assche, F. Marcellàn (Eds.), Orthogonal Polynomials and Special Functions, Computation and Application (2006)
Vol. 1884: N. Hayashi, E.I. Kaikina, P.I. Naumkin, I.A. Shishmarev, Asymptotics for Dissipative Nonlinear Equations (2006)
Vol. 1885: A. Telcs, The Art of Random Walks (2006)
Vol. 1886: S. Takamura, Splitting Deformations of Degenerations of Complex Curves (2006)
Vol. 1887: K. Habermann, L. Habermann, Introduction to Symplectic Dirac Operators (2006)
Vol. 1888: J. van der Hoeven, Transseries and Real Differential Algebra (2006)
Vol. 1889: G. Osipenko, Dynamical Systems, Graphs, and Algorithms (2006)
Vol. 1890: M. Bunge, J. Funk, Singular Coverings of Toposes (2006)
Vol. 1891: J.B. Friedlander, D.R. Heath-Brown, H. Iwaniec, J. Kaczorowski, Analytic Number Theory, Cetraro, Italy, 2002. Editors: A. Perelli, C. Viola (2006)
Vol. 1892: A. Baddeley, I. Bárány, R. Schneider, W. Weil, Stochastic Geometry, Martina Franca, Italy, 2004. Editor: W. Weil (2007)
Vol. 1893: H. Hanßmann, Local and Semi-Local Bifurcations in Hamiltonian Dynamical Systems, Results and Examples (2007)
Vol. 1894: C.W. Groetsch, Stable Approximate Evaluation of Unbounded Operators (2007)
Vol. 1895: L. Molnár, Selected Preserver Problems on Algebraic Structures of Linear Operators and on Function Spaces (2007)
Vol. 1896: P. Massart, Concentration Inequalities and Model Selection, Ecole d'Été de Probabilités de Saint-Flour XXXIII-2003. Editor: J. Picard (2007)
Vol. 1897: R. Doney, Fluctuation Theory for Lévy Processes, Ecole d'Été de Probabilités de Saint-Flour XXXV-2005. Editor: J. Picard (2007)
Vol. 1898: H.R. Beyer, Beyond Partial Differential Equations, On linear and Quasi-Linear Abstract Hyperbolic Evolution Equations (2007)
Vol. 1899: Séminaire de Probabilités XL. Editors: C. Donati-Martin, M. Émery, A. Rouault, C. Stricker (2007)
Vol. 1900: E. Bolthausen, A. Bovier (Eds.), Spin Glasses (2007)
Vol. 1901: O. Wittenberg, Intersections de deux quadriques et pinceaux de courbes de genre 1, Intersections of Two Quadrics and Pencils of Curves of Genus 1 (2007)
Vol. 1902: A. Isaev, Lectures on the Automorphism Groups of Kobayashi-Hyperbolic Manifolds (2007)
Vol. 1903: G. Kresin, V. Maz'ya, Sharp Real-Part Theorems (2007)
Vol. 1904: P. Giesl, Construction of Global Lyapunov Functions Using Radial Basis Functions (2007)
Vol. 1905: C. Prévôt, M. Röckner, A Concise Course on Stochastic Partial Differential Equations (2007)
Vol. 1906: T. Schuster, The Method of Approximate Inverse: Theory and Applications (2007)
Vol. 1907: M. Rasmussen, Attractivity and Bifurcation for Nonautonomous Dynamical Systems (2007)
Vol. 1908: T.J. Lyons, M. Caruana, T. Lévy, Differential Equations Driven by Rough Paths, Ecole d'Été de Probabilités de Saint-Flour XXXIV-2004 (2007)
Vol. 1909: H. Akiyoshi, M. Sakuma, M. Wada, Y. Yamashita, Punctured Torus Groups and 2-Bridge Knot Groups (I) (2007)
Vol. 1910: V.D. Milman, G. Schechtman (Eds.), Geometric Aspects of Functional Analysis. Israel Seminar 2004-2005 (2007)
Vol. 1911: A. Bressan, D. Serre, M. Williams, K. Zumbrun, Hyperbolic Systems of Balance Laws. Cetraro, Italy 2003. Editor: P. Marcati (2007)
Vol. 1912: V. Berinde, Iterative Approximation of Fixed Points (2007)
Vol. 1913: J.E. Marsden, G. Misiołek, J.-P. Ortega, M. Perlmutter, T.S. Ratiu, Hamiltonian Reduction by Stages (2007)
Vol. 1914: G. Kutyniok, Affine Density in Wavelet Analysis (2007)
Vol. 1915: T. Bıyıkoğlu, J. Leydold, P.F. Stadler, Laplacian Eigenvectors of Graphs. Perron-Frobenius and Faber-Krahn Type Theorems (2007)
Vol. 1916: C. Villani, F. Rezakhanlou, Entropy Methods for the Boltzmann Equation. Editors: F. Golse, S. Olla (2008)
Vol. 1917: I. Veselić, Existence and Regularity Properties of the Integrated Density of States of Random Schrdinger (2008)
Vol. 1918: B. Roberts, R. Schmidt, Local Newforms for GSp(4) (2007)

Vol. 1919: R.A. Carmona, I. Ekeland, A. Kohatsu-Higa, J.-M. Lasry, P.-L. Lions, H. Pham, E. Taflin, Paris-Princeton Lectures on Mathematical Finance 2004. Editors: R.A. Carmona, E. inlar, I. Ekeland, E. Jouini, J.A. Scheinkman, N. Touzi (2007)
Vol. 1920: S.N. Evans, Probability and Real Trees. Ecole d'Été de Probabilités de Saint-Flour XXXV-2005 (2008)
Vol. 1921: J.P. Tian, Evolution Algebras and their Applications (2008)
Vol. 1922: A. Friedman (Ed.), Tutorials in Mathematical BioSciences IV. Evolution and Ecology (2008)
Vol. 1923: J.P.N. Bishwal, Parameter Estimation in Stochastic Differential Equations (2008)
Vol. 1924: M. Wilson, Littlewood-Paley Theory and Exponential-Square Integrability (2008)
Vol. 1925: M. du Sautoy, L. Woodward, Zeta Functions of Groups and Rings (2008)
Vol. 1926: L. Barreira, V. Claudia, Stability of Nonautonomous Differential Equations (2008)
Vol. 1927: L. Ambrosio, L. Caffarelli, M.G. Crandall, L.C. Evans, N. Fusco, Calculus of Variations and Non-Linear Partial Differential Equations. Cetraro, Italy 2005. Editors: B. Dacorogna, P. Marcellini (2008)
Vol. 1928: J. Jonsson, Simplicial Complexes of Graphs (2008)
Vol. 1929: Y. Mishura, Stochastic Calculus for Fractional Brownian Motion and Related Processes (2008)
Vol. 1930: J.M. Urbano, The Method of Intrinsic Scaling. A Systematic Approach to Regularity for Degenerate and Singular PDEs (2008)
Vol. 1931: M. Cowling, E. Frenkel, M. Kashiwara, A. Valette, D.A. Vogan, Jr., N.R. Wallach, Representation Theory and Complex Analysis. Venice, Italy 2004. Editors: E.C. Tarabusi, A. D'Agnolo, M. Picardello (2008)
Vol. 1932: A.A. Agrachev, A.S. Morse, E.D. Sontag, H.J. Sussmann, V.I. Utkin, Nonlinear and Optimal Control Theory. Cetraro, Italy 2004. Editors: P. Nistri, G. Stefani (2008)
Vol. 1933: M. Petkovic, Point Estimation of Root Finding Methods (2008)
Vol. 1934: C. Donati-Martin, M. Émery, A. Rouault, C. Stricker (Eds.), Séminaire de Probabilités XLI (2008)
Vol. 1935: A. Unterberger, Alternative Pseudodifferential Analysis (2008)
Vol. 1936: P. Magal, S. Ruan (Eds.), Structured Population Models in Biology and Epidemiology (2008)
Vol. 1937: G. Capriz, P. Giovine, P.M. Mariano (Eds.), Mathematical Models of Granular Matter (2008)
Vol. 1938: D. Auroux, F. Catanese, M. Manetti, P. Seidel, B. Siebert, I. Smith, G. Tian, Symplectic 4-Manifolds and Algebraic Surfaces. Cetraro, Italy 2003. Editors: F. Catanese, G. Tian (2008)
Vol. 1939: D. Boffi, F. Brezzi, L. Demkowicz, R.G. Durán, R.S. Falk, M. Fortin, Mixed Finite Elements, Compatibility Conditions, and Applications. Cetraro, Italy 2006. Editors: D. Boffi, L. Gastaldi (2008)
Vol. 1940: J. Banasiak, V. Capasso, M.A.J. Chaplain, M. Lachowicz, J. Miękisz, Multiscale Problems in the Life Sciences. From Microscopic to Macroscopic. Będlewo, Poland 2006. Editors: V. Capasso, M. Lachowicz (2008)
Vol. 1941: S.M.J. Haran, Arithmetical Investigations. Representation Theory, Orthogonal Polynomials, and Quantum Interpolations (2008)
Vol. 1942: S. Albeverio, F. Flandoli, Y.G. Sinai, SPDE in Hydrodynamic. Recent Progress and Prospects. Cetraro, Italy 2005. Editors: G. Da Prato, M. Rckner (2008)
Vol. 1943: L.L. Bonilla (Ed.), Inverse Problems and Imaging. Martina Franca, Italy 2002 (2008)
Vol. 1944: A. Di Bartolo, G. Falcone, P. Plaumann, K. Strambach, Algebraic Groups and Lie Groups with Few Factors (2008)
Vol. 1945: F. Brauer, P. van den Driessche, J. Wu (Eds.), Mathematical Epidemiology (2008)
Vol. 1946: G. Allaire, A. Arnold, P. Degond, T.Y. Hou, Quantum Transport. Modelling, Analysis and Asymptotics. Cetraro, Italy 2006. Editors: N.B. Abdallah, G. Frosali (2008)
Vol. 1947: D. Abramovich, M. Mariño, M. Thaddeus, R. Vakil, Enumerative Invariants in Algebraic Geometry and String Theory. Cetraro, Italy 2005. Editors: K. Behrend, M. Manetti (2008)
Vol. 1948: F. Cao, J-L. Lisani, J-M. Morel, P. Mus, F. Sur, A Theory of Shape Identification (2008)
Vol. 1949: H.G. Feichtinger, B. Helffer, M.P. Lamoureux, N. Lerner, J. Toft, Pseudo-Differential Operators. Quantization and Signals. Cetraro, Italy 2006. Editors: L. Rodino, M.W. Wong (2008)
Vol. 1950: M. Bramson, Stability of Queueing Networks, Ecole d'Eté de Probabilits de Saint-Flour XXXVI-2006 (2008)
Vol. 1951: A. Moltó, J. Orihuela, S. Troyanski, M. Valdivia, A Non Linear Transfer Technique for Renorming (2009)
Vol. 1952: R. Mikhailov, I.B.S. Passi, Lower Central and Dimension Series of Groups (2009)
Vol. 1953: K. Arwini, C.T.J. Dodson, Information Geometry (2008)
Vol. 1954: P. Biane, L. Bouten, F. Cipriani, N. Konno, N. Privault, Q. Xu, Quantum Potential Theory. Editors: U. Franz, M. Schuermann (2008)
Vol. 1955: M. Bernot, V. Caselles, J.-M. Morel, Optimal Transportation Networks (2008)
Vol. 1956: C.H. Chu, Matrix Convolution Operators on Groups (2008)
Vol. 1957: A. Guionnet, On Random Matrices: Macroscopic Asymptotics, Ecole d'Eté de Probabilits de Saint-Flour XXXVI-2006 (2009)
Vol. 1958: M.C. Olsson, Compactifying Moduli Spaces for Abelian Varieties (2008)
Vol. 1959: Y. Nakkajima, A. Shiho, Weight Filtrations on Log Crystalline Cohomologies of Families of Open Smooth Varieties (2008)
Vol. 1960: J. Lipman, M. Hashimoto, Foundations of Grothendieck Duality for Diagrams of Schemes (2009)
Vol. 1961: G. Buttazzo, A. Pratelli, S. Solimini, E. Stepanov, Optimal Urban Networks via Mass Transportation (2009)
Vol. 1962: R. Dalang, D. Khoshnevisan, C. Mueller, D. Nualart, Y. Xiao, A Minicourse on Stochastic Partial Differential Equations (2009)
Vol. 1963: W. Siegert, Local Lyapunov Exponents (2009)
Vol. 1964: W. Roth, Operator-valued Measures and Integrals for Cone-valued Functions and Integrals for Cone-valued Functions (2009)
Vol. 1965: C. Chidume, Geometric Properties of Banach Spaces and Nonlinear Iterations (2009)
Vol. 1966: D. Deng, Y. Han, Harmonic Analysis on Spaces of Homogeneous Type (2009)
Vol. 1967: B. Fresse, Modules over Operads and Functors (2009)
Vol. 1968: R. Weissauer, Endoscopy for GSP(4) and the Cohomology of Siegel Modular Threefolds (2009)

Vol. 1969: B. Roynette, M. Yor, Penalising Brownian Paths (2009)
Vol. 1970: M. Biskup, A. Bovier, F. den Hollander, D. Ioffe, F. Martinelli, K. Netočný, F. Toninelli, Methods of Contemporary Mathematical Statistical Physics. Editor: R. Kotecký (2009)
Vol. 1971: L. Saint-Raymond, Hydrodynamic Limits of the Boltzmann Equation (2009)
Vol. 1972: T. Mochizuki, Donaldson Type Invariants for Algebraic Surfaces (2009)
Vol. 1973: M.A. Berger, L.H. Kauffmann, B. Khesin, H.K. Moffatt, R.L. Ricca, De W. Sumners, Lectures on Topological Fluid Mechanics. Cetraro, Italy 2001. Editor: R.L. Ricca (2009)
Vol. 1974: F. den Hollander, Random Polymers: École d'Été de Probabilités de Saint-Flour XXXVII – 2007 (2009)
Vol. 1975: J.C. Rohde, Cyclic Coverings, Calabi-Yau Manifolds and Complex Multiplication (2009)
Vol. 1976: N. Ginoux, The Dirac Spectrum (2009)
Vol. 1977: M.J. Gursky, E. Lanconelli, A. Malchiodi, G. Tarantello, X.-J. Wang, P.C. Yang, Geometric Analysis and PDEs. Cetraro, Italy 2001. Editors: A. Ambrosetti, S.-Y.A. Chang, A. Malchiodi (2009)
Vol. 1978: M. Qian, J.-S. Xie, S. Zhu, Smooth Ergodic Theory for Endomorphisms (2009)
Vol. 1979: C. Donati-Martin, M. Émery, A. Rouault, C. Stricker (Eds.), Śeminaire de Probablitiés XLII (2009)
Vol. 1980: P. Graczyk, A. Stos (Eds.), Potential Analysis of Stable Processes and its Extensions (2009)
Vol. 1981: M. Chlouveraki, Blocks and Families for Cyclotomic Hecke Algebras (2009)
Vol. 1982: N. Privault, Stochastic Analysis in Discrete and Continuous Settings. With Normal Martingales (2009)
Vol. 1983: H. Ammari (Ed.), Mathematical Modeling in Biomedical Imaging I. Electrical and Ultrasound Tomographies, Anomaly Detection, and Brain Imaging (2009)
Vol. 1984: V. Caselles, P. Monasse, Geometric Description of Images as Topographic Maps (2010)
Vol. 1985: T. Linß, Layer-Adapted Meshes for Reaction-Convection-Diffusion Problems (2010)
Vol. 1986: J.-P. Antoine, C. Trapani, Partial Inner Product Spaces. Theory and Applications (2009)
Vol. 1987: J.-P. Brasselet, J. Seade, T. Suwa, Vector Fields on Singular Varieties (2010)
Vol. 1988: M. Broué, Introduction to Complex Reflection Groups and Their Braid Groups (2010)
Vol. 1989: I.M. Bomze, V. Demyanov, Nonlinear Optimization. Cetraro, Italy 2007. Editors: G. di Pillo, F. Schoen (2010)
Vol. 1990: S. Bouc, Biset Functors for Finite Groups (2010)
Vol. 1991: F. Gazzola, H.-C. Grunau, G. Sweers, Polyharmonic Boundary Value Problems (2010)
Vol. 1992: A. Parmeggiani, Spectral Theory of Non-Commutative Harmonic Oscillators: An Introduction (2010)
Vol. 1993: P. Dodos, Banach Spaces and Descriptive Set Theory: Selected Topics (2010)
Vol. 1994: A. Baricz, Generalized Bessel Functions of the First Kind (2010)
Vol. 1995: A.Y. Khapalov, Controllability of Partial Differential Equations Governed by Multiplicative Controls (2010)
Vol. 1996: T. Lorenz, Mutational Analysis. A Joint Framework for Cauchy Problems *In* and *Beyond* Vector Spaces (2010)
Vol. 1997: M. Banagl, Intersection Spaces, Spatial Homology Truncation, and String Theory (2010)
Vol. 1998: M. Abate, E. Bedford, M. Brunella, T.-C. Dinh, D. Schleicher, N. Sibony, Holomorphic Dynamical Systems. Editors: G. Gentili, J. Guenot, G. Patrizio (2010)
Vol. 1999: H. Schoutens, The Use of Ultraproducts in Commutative Algebra (2010)
Vol. 2000: H. Yserentant, Regularity and Approximability of Electronic Wave Functions (2010)
Vol. 2001: T. Duquesne, O. Reichmann, K.-i. Sato, C. Schwab, Lévy Matters I. Editors: O.E. Barndorff-Nielson, J. Bertoin, J. Jacod, C. Klüppelberg (2010)
Vol. 2002: C. Pötzsche, Geometric Theory of Discrete Nonautonomous Dynamical Systems (2010)
Vol. 2003: A. Cousin, S. Crépey, O. Guéant, D. Hobson, M. Jeanblanc, J.-M. Lasry, J.-P. Laurent, P.-L. Lions, P. Tankov, Paris-Princeton Lectures on Mathematical Finance 2010. Editors: R.A. Carmona, E. Cinlar, I. Ekeland, E. Jouini, J.A. Scheinkman, N. Touzi (2010)
Vol. 2004: K. Diethelm, The Analysis of Fractional Differential Equations (2010)
Vol. 2005: W. Yuan, W. Sickel, D. Yang, Morrey and Campanato Meet Besov, Lizorkin and Triebel (2011)
Vol. 2006: C. Donati-Martin, A. Lejay, W. Rouault (Eds.), Séminaire de Probabilités XLIII (2011)
Vol. 2007: G. Gromadzki, F.J. Cirre, J.M. Gamboa, E. Bujalance, Symmetries of Compact Riemann Surfaces (2010)
Vol. 2008: P.F. Baum,G. Cortiñas, R. Meyer, R. Sánchez-García, M. Schlichting, B. Toën, Topics in Algebraic and Topological K-Theory (2011)
Vol. 2009: J.-L. Colliot-Thélène, P.S. Dyer, P. Vojta, Arithmetic Geometry. Cetraro, Italy 2007. Editors: P. Corvaja, C. Gasbarri (2011)
Vol. 2010: A. Farina, A. Klar, R.M.M. Mattheij, A. Mikelić, Mathematical Models in the Manufacturing of Glass. Montecatini Terme, Italy 2008. Editors: A. Fasano, J.R. Ockendon (2011)
Vol. 2011: B. Andrews, C. Hopper, The Ricci Flow in Riemannian Geometry. A Complete Proof of the Differentiable 1/4-Pinching Sphere Theorem (2011)

Recent Reprints and New Editions

Vol. 1702: J. Ma, J. Yong, Forward-Backward Stochastic Differential Equations and their Applications. 1999 – Corr. 3rd printing (2007)
Vol. 830: J.A. Green, Polynomial Representations of GL_n, with an Appendix on Schensted Correspondence and Littelmann Paths by K. Erdmann, J.A. Green and M. Schoker 1980 – 2nd corr. and augmented edition (2007)
Vol. 1693: S. Simons, From Hahn-Banach to Monotonicity (Minimax and Monotonicity 1998) – 2nd exp. edition (2008)
Vol. 470: R.E. Bowen, Equilibrium States and the Ergodic Theory of Anosov Diffeomorphisms. With a preface by D. Ruelle. Edited by J.-R. Chazottes. 1975 – 2nd rev. edition (2008)
Vol. 523: S.A. Albeverio, R.J. Høegh-Krohn, S. Mazzucchi, Mathematical Theory of Feynman Path Integral. 1976 – 2nd corr. and enlarged edition (2008)
Vol. 1764: A. Cannas da Silva, Lectures on Symplectic Geometry 2001 – Corr. 2nd printing (2008)

LECTURE NOTES IN MATHEMATICS

Edited by J.-M. Morel, F. Takens, B. Teissier, P.K. Maini

Editorial Policy (for Multi-Author Publications: Summer Schools/Intensive Courses)

1. Lecture Notes aim to report new developments in all areas of mathematics and their applications - quickly, informally and at a high level. Mathematical texts analysing new developments in modelling and numerical simulation are welcome. Manuscripts should be reasonably self-contained and rounded off. Thus they may, and often will, present not only results of the author but also related work by other people. They should provide sufficient motivation, examples and applications. There should also be an introduction making the text comprehensible to a wider audience. This clearly distinguishes Lecture Notes from journal articles or technical reports which normally are very concise. Articles intended for a journal but too long to be accepted by most journals, usually do not have this "lecture notes" character.
2. In general SUMMER SCHOOLS and other similar INTENSIVE COURSES are held to present mathematical topics that are close to the frontiers of recent research to an audience at the beginning or intermediate graduate level, who may want to continue with this area of work, for a thesis or later. This makes demands on the didactic aspects of the presentation. Because the subjects of such schools are advanced, there often exists no textbook, and so ideally, the publication resulting from such a school could be a first approximation to such a textbook. Usually several authors are involved in the writing, so it is not always simple to obtain a unified approach to the presentation.

   For prospective publication in LNM, the resulting manuscript should not be just a collection of course notes, each of which has been developed by an individual author with little or no co-ordination with the others, and with little or no common concept. The subject matter should dictate the structure of the book, and the authorship of each part or chapter should take secondary importance. Of course the choice of authors is crucial to the quality of the material at the school and in the book, and the intention here is not to belittle their impact, but simply to say that the book should be planned to be written by these authors jointly, and not just assembled as a result of what these authors happen to submit.

   This represents considerable preparatory work (as it is imperative to ensure that the authors know these criteria before they invest work on a manuscript), and also considerable editing work afterwards, to get the book into final shape. Still it is the form that holds the most promise of a successful book that will be used by its intended audience, rather than yet another volume of proceedings for the library shelf.
3. Manuscripts should be submitted either online at www.editorialmanager.com/lnm/ to Springer's mathematics editorial, or to one of the series editors. Volume editors are expected to arrange for the refereeing, to the usual scientific standards, of the individual contributions. If the resulting reports can be forwarded to us (series editors or Springer) this is very helpful. If no reports are forwarded or if other questions remain unclear in respect of homogeneity etc, the series editors may wish to consult external referees for an overall evaluation of the volume. A final decision to publish can be made only on the basis of the complete manuscript; however a preliminary decision can be based on a pre-final or incomplete manuscript. The strict minimum amount of material that will be considered should include a detailed outline describing the planned contents of each chapter.

   Volume editors and authors should be aware that incomplete or insufficiently close to final manuscripts almost always result in longer evaluation times. They should also be aware that parallel submission of their manuscript to another publisher while under consideration for LNM will in general lead to immediate rejection.

4. Manuscripts should in general be submitted in English. Final manuscripts should contain at least 100 pages of mathematical text and should always include
   - a general table of contents;
   - an informative introduction, with adequate motivation and perhaps some historical remarks: it should be accessible to a reader not intimately familiar with the topic treated;
   - a global subject index: as a rule this is genuinely helpful for the reader.

   Lecture Notes volumes are, as a rule, printed digitally from the authors' files. We strongly recommend that all contributions in a volume be written in the same LaTeX version, preferably LaTeX2e. To ensure best results, authors are asked to use the LaTeX2e style files available from Springer's web-server at

   ftp://ftp.springer.de/pub/tex/latex/svmonot1/ (for monographs) and
   ftp://ftp.springer.de/pub/tex/latex/svmultt1/ (for summer schools/tutorials).

   Additional technical instructions are available on request from: lnm@springer.com.
5. Careful preparation of the manuscripts will help keep production time short besides ensuring satisfactory appearance of the finished book in print and online. After acceptance of the manuscript authors will be asked to prepare the final LaTeX source files and also the corresponding dvi-, pdf- or zipped ps-file. The LaTeX source files are essential for producing the full-text online version of the book. For the existing online volumes of LNM see: http://www.springerlink.com/openurl.asp?genre=journal&issn=0075-8434.

   The actual production of a Lecture Notes volume takes approximately 12 weeks.
6. Volume editors receive a total of 50 free copies of their volume to be shared with the authors, but no royalties. They and the authors are entitled to a discount of 33.3% on the price of Springer books purchased for their personal use, if ordering directly from Springer.
7. Commitment to publish is made by letter of intent rather than by signing a formal contract. Springer-Verlag secures the copyright for each volume. Authors are free to reuse material contained in their LNM volumes in later publications: a brief written (or e-mail) request for formal permission is sufficient.

Addresses:

Professor J.-M. Morel, CMLA,
École Normale Supérieure de Cachan,
61 Avenue du Président Wilson,
94235 Cachan Cedex, France
E-mail: Jean-Michel.Morel@cmla.ens-cachan.fr

Professor F. Takens, Mathematisch Instituut,
Rijksuniversiteit Groningen, Postbus 800,
9700 AV Groningen, The Netherlands
E-mail: F.Takens@rug.nl

Professor B. Teissier,
Institut Mathématique de Jussieu,
UMR 7586 du CNRS,
Équipe "Géométrie et Dynamique",
175 rue du Chevaleret,
75013 Paris, France
E-mail: teissier@math.jussieu.fr

For the "Mathematical Biosciences Subseries" of LNM:

Professor P.K. Maini, Center for Mathematical Biology,
Mathematical Institute, 24-29 St Giles,
Oxford OX1 3LP, UK
E-mail: maini@maths.ox.ac.uk

Springer, Mathematics Editorial I, Tiergartenstr. 17,
69121 Heidelberg, Germany,
Tel.: +49 (6221) 487-8259
Fax: +49 (6221) 4876-8259
E-mail: lnm@springer.com